Statics and Strength of Materials

Statics and Strength of Materials

Foundations for Structural Design

Barry Onouye
Department of Architecture
University of Washington

PEARSON

Prentice
Hall

Upper Saddle River, New Jersey
Columbus, Ohio

Library of Congress Cataloging-in-Publication Data

Onouye, Barry.

   Statics and strength of materials: foundations for structural design / Barry Onouye.—
1st ed.

      p. cm.

   Includes index.

   ISBN 0-13-111837-4

  1.   Structural design. 2.   Statics. 3.   Strength of materials.   I.   Title.

   TA658.O67 2005

   624.1'771--dc22

                                                   2004015580

Executive Editor: Ed Francis
Editorial Assistant: Jennifer Day
Project Coordination: Carlisle Publishers Services
Production Editor: Holly Shufeldt
Design Coordinator: Diane Ernsberger
Cover Designer: Thomas Mack
Production Manager: Deidra Schwartz
Marketing Manager: Mark Marsden

This book was set in Palatino by Carlisle Communications, Ltd. It was printed and bound by Courier Kendallville, Inc. The cover was printed by Phoenix Color Corp.

Pearson Education Ltd.
Pearson Education Singapore Pte. Ltd.
Pearson Education Canada, Ltd.
Pearson Education—Japan

Pearson Education Australia Pty. Limited
Pearson Education North Asia Ltd.
Pearson Educación de Mexico, S.A. de C.V.
Pearson Education Malaysia Pte. Ltd.

10 9 8 7 6 5 4 3

0-13-111837-4

Preface

The writing of this text began as a workbook to complement my first book, *Statics and Strength of Materials for Architecture and Building Construction,* by Barry Onouye and Kevin Kane, published by Prentice Hall. However, as the writing project progressed, editor Ed Francis and other reviewers suggested that it be converted into a text. This text provides an extensive number of worked examples and numerous practice problems for the student to enhance his/her understanding of the concepts and principles presented.

A primary aim of this book has been to develop and present basic structural concepts in an easily understood manner using "building" examples and numerous illustrations (over 900) to supplement the written text. Much of this material has been "field tested," revised, and modified over a course of 30 years of teaching. The body of this work would not have been possible without the valuable insights given to me over the years by several thousand students.

Introducing structural theory without relying on a predominantly mathematical treatment has been challenging, to say the least, and a non-calculus engineering alternative to the topic seemed essential if the target audience (students of architecture, building construction, and some engineering technology programs) were to remain interested. Early on it was decided that a heavily illustrated, visual approach was essential in connecting and linking structural theory to real buildings and components. Using examples and problems that are commonly found in buildings and structures around us seemed to be a logical way of introducing mathematically based material in a nonthreatening way.

This text is organized along the lines of traditional textbooks on statics and strength of materials because it seems to be the most logical approach. A sound understanding of statics and strength of materials establishes the *foundation* for the study of more advanced topics in structural analysis, timber, steel, and reinforced concrete design. Numerical calculations are included as a way of explaining and testing one's understanding of the principles involved. Many fully worked example problems are included, with additional problems for student practice.

Organizationally, the book consists of two parts: statics in Chapters 2 through 5, and strength of materials in Chapters 6 through 10. *Retaining walls* in Chapter 4 and *load tracing* in Chapter 5 are not customarily covered in statics, but were intentionally included to illustrate the power of the basic principle of mechanics and the use of free-body diagrams. *Gravity load tracing* is often covered in subsequent structures courses, but the fundamentals can be introduced at this stage without much difficulty.

A heavy emphasis is placed on the use of free-body diagrams in understanding the forces acting on a structural member. All problems begin with a pictorial representation of a structural component or assembly and are accompanied by a free-body diagram. Illustrations are used extensively to ensure that the student sees the connection between the real building object and its abstraction. Chapter 4 uses the principles discussed in the previous chapter to solve an array of determinate structural frameworks. The discussion of load tracing in Chapter 5 examines the overall structural condition with regard to gravity loads. This chapter illustrates the interaction of one member with other members and introduces the concept of load paths that develop within a building.

Chapter 6 introduces the concepts of stress, strain, and material properties as they relate to materials commonly used in the building industry. Cross-sectional properties are covered in Chapter 7, again with an emphasis on commonly used beam and column shapes. Chapters 8, 9, and 10 develop the basis for beam and column analysis and design. Elastic theory has been utilized throughout, and the allowable stress method has been employed for the design of beams and columns.

Some simplifications have been introduced to beam and column design equations to eliminate the complexity unwarranted for preliminary design purposes. Sizing of beams and columns is well within the range of a final, closely engineered element sized by the more complex formulas. It is assumed that students will take subsequent courses in timber, steel, and concrete; therefore, building code equations and criteria have not been incorporated in these chapters.

No attempt was made to include the study of indeterminate beams and frames since it would require substantial development beyond the purview of statics and strength of materials. Indeterminate structures are probably among the more important structural topics for building designers since most of the commercial and institutional buildings of moderate size are of this type. Indeterminate structural behavior using one of the many available structural analysis software packages is emerging as a critical area of study for all future building designers.

This text is intended to be used for two 15-week semester classes in architecture, building construction, building inspection, and engineering technology programs. Chapter 5 might be of interest and use to the civil engineering student who wants to better understand building components in a larger context. Also, Chapters 9 and 10 might be useful to builders in need of quick preliminary methods of sizing beams and columns. Although this text might be used for self-study, it is this author's belief that the principles and concepts should be presented and elaborated upon by the instructor in class.

Many of the topics covered in the text can be demonstrated in model form in class. Slides of actual buildings representing the subject being covered help to reinforce the idea through visual images. Previous teaching experience has been convincing about the need to use a variety of media and techniques to illustrate a concept. Structures should by no means be a "dry" subject.

As part of an ongoing effort by the United States to convert from the U.S. customary system of units to the international system of units (SI metric units), a substantial number of example and practice problems in this text use the SI units. Design tables in Appendix A include section properties for beam and column design in both U.S. customary and SI metric units. Appendix B contains a basic math review for students who have not been actively using algebra, geometry, or trigonometry in recent years. There is no attempt made to teach the basic topics in math but only to serve as a refresher. The derivation of both the bending stress and general shear stress equations are found in Appendix C.

ACKNOWLEDGMENTS

Particular thanks are due to Edward L. Bernstein, Alabama A & M University; Hollis G. Bray, Jr., University of Louisiana, Monroe; Elliott Colchamiro, NYC College of Technology (CUNY); Majed Dabdoub, Northern Kentucky University; Dr. Madan Mehta, University of Texas, Arlington; Ronald Nichols, Alfred State College; Dr. Thomas Rogers, Northern Arizona University; and Dr. Jason C. Shih, Louisiana State University, Baton Rouge, for their assistance with the text review.

Barry Onouye
July, 2004

Contents

1 Introduction

1.1 INTRODUCTION

A building structure is an assemblage of elements used to channel and direct loads present in a building to the foundation (see Figure 1.1). Primarily, structures must resist against the pull of gravity, but must also be able to withstand the pressures of wind and the inertial forces resulting from earthquakes. The study of the foundations of building structures may be divided into three distinct but related areas: statics, the strength of materials, and structural design.

Statics is the analysis of forces on an individual structural element or an entire structural assembly that is in a state of balance or *equilibrium*. The study of statics enables us to determine all forces and reactions to the applied on one or multiple members of a framework in equilibrium. Sir Isaac Newton's (1642–1727) three laws are the fundamental principles of mechanics that make up the foundation of what we refer to as statics.

Strength of materials constitutes the basis for structural design. Formulas and procedures used in the design of beams, trusses, columns, walls, and foundations incorporate the material attributes together with the cross-sectional properties of the components being designed. Strength of materials, also called mechanics of materials, relates the external forces applied to the structural assembly or element with the internal forces (*stresses*) within the component(s).

Structural design involves the evaluation and calculation of the strength of structural elements used to support and resist loads in the framework of a building. Material strength characteristics are evaluated to determine their suitability in supporting loads placed upon it. The determination of the size and cross-sectional shape of beams and columns to safely carry internal forces (*stresses*) are also part of structural design. Appropriate loads are assigned to the building framework and a variety of structural analyses are performed to assess the critical magnitude and intensity of the forces for each structural component. Member design begins with the tracing of loads from the roof elements and progresses down to the column and/or wall supports below. In multistory structures, the process continues progressively downward through each floor level and finally to the basement and foundations. After all of the separate phases have been examined and modified in an iterative manner, the structural elements within the system are then checked

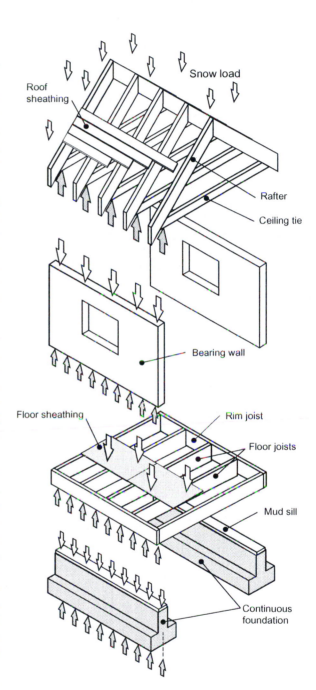

Figure 1.1   Building structure is an assemblage of parts used to channel and direct loads to the foundation.

Figure 1.2   Complex framing requires a team of consultants to work closely together from the beginning of the project. (Experience Music Project, Seattle Center.)

mathematically by the structural consultant to ensure the safety and economy of the structure. The process of conceiving and visualizing a structure is truly an art. Good structural design is more related to correct intuitive sense than to sets of complex mathematical equations. Mathematics is merely a convenient and validating tool by which the designer determines the physical sizes and proportions of the elements to be used in the intended structure.

Nowadays, with the design of any large structure involving a team of designers working jointly with specialists and consultants, the architect is required to function as a coordinator and still maintain a leadership role even in the initial structural scheme (Figure 1.2). The architect needs to have a broad, general understanding of the structure with its various problems and must also sufficiently understand the fundamental principles of structural behavior to provide useful approximations of member sizes. The structural principles influence the form of the building, and a logical solution, often an economical one as well, is always based on a correct interpretation of these principles. A responsibility of the builder (constructor) is to have the knowledge, experience, and inventiveness to resolve complex structural and constructional issues without losing sight of the spirit of the design.

This book, *Statics and Strength of Materials: Foundations for Structural Design,* covers the analysis of statically determinate systems using the fundamental principles of free-body diagrams and equations of equilibrium. Although during recent years there has been an incredible emphasis on the use of computers to analyze structures using matrix analysis, it is the author's opinion that a classical approach for a beginning course is necessary. An understanding of physical phenomena before embarking on the application of sophisticated mathematical analysis is the aim of this book, as reliance on the computer for answers that one does not fully understand is a risky proposition at best. Application of the basic principles of statics and strength of materials will enable the student to gain a clearer and, it is hoped, more intuitive sense about structure.

Criteria for the Selection of Structural Systems

Most building projects begin with a client program outlining the functional and spatial requirements to be accommodated. Architects typically interpret and prioritize this information, coordinating architectural design work with the work of other consultants on the project. The architect and structural engineer must satisfy a wide range of factors in determining the most appropriate structural system. Several of these factors are discussed here.

Nature and magnitude of loads

The weight of most building materials and the self-weight of structural elements (dead loads) can be calculated from reference tables listing the densities of various materials. Building codes establish design values for the weight of the occupants and furnishings (live loads) and other temporary loads like snow, wind, and earthquake. (See Chapter 5.1 and Figure 1.3.)

Building use/function

Sports facilities require long, clear span areas free of columns. Light wood framing is well suited to the relatively small rooms and spans found in residential construction. See Figure 1.4.

Site conditions

Topography and soil conditions often determine the design of the foundation system, which in turn influences the way loads are transmitted though walls and columns. Climatic variables such as wind speed and snowfall also affect design loads. Significant movement (thermal expansion and contraction) can result from extreme temperature fluctuations. Seismic forces, used to calculate building-code design loads, vary in different parts of the country.

Building system integration

All building systems (lighting, heating/cooling, ventilation, plumbing, fire sprinklers, electrical) have a rational basis that governs their arrangement. It is generally more elegant and cost-effective to coordinate these systems with the structure to avoid conflict and compromise in their performance. This is especially the case where the structure is exposed and dropped-ceiling spaces are not available for duct and pipe runs (Figure 1.5).

Fire resistance

Building codes require that building components and structural systems meet minimum fire-resistance standards. The combustibility of materials and their ability to carry design loads when subjected to intense heat are tested to ensure that buildings involved in fires can be safely evacuated in a given period of time. Wood is naturally combustible, but heavy timber construction maintains much of its strength for an extended period of time in a fire. Steel can be weakened to the point of failure unless protected by fireproof coverings. Concrete and masonry are considered noncombustible and are not significantly weakened in fires. The levels of fire resistance vary from unrated construction to four hours and are based on the type of occupancy and size of a building.

Construction variables

Cost and construction time are almost always relevant issues. Several structural systems will often accommodate the load, span, and fire-resistance requirements for a building. Local availability of materials and skilled construction trades typically affect cost and schedule. The

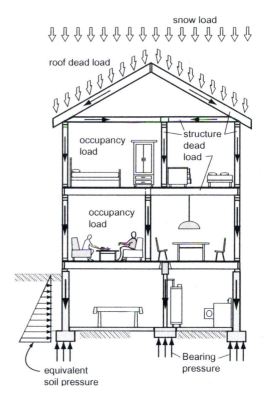

Figure 1.3   Dead and live loads on a simple structure.

Figure 1.4   Long-span structure—skating rink.

Figure 1.5   Integration of mechanical and electrical systems.

selected system is often refined to achieve the most economical framing arrangement or construction method, and the use of heavy equipment such as cranes or concrete trucks and pumps may be restricted by availability or site access.

Architectural form and space

Social and cultural factors that influence the architect's conception of form and space extend to the selection and use of appropriate materials. Where structure is exposed, the location, scale, hierarchy, and direction of framing members contribute significantly to the expression of the building.

2 Fundamentals of Statics

2.1   NEWTON'S LAWS

A force can be simply defined as a push or a pull, indicated by an arrow, exerted by one body or source on another, tending to produce a change in the motion of the body being acted upon (Figure 2.1). Forces in building structures are a result of gravity-induced loads (structure/material weights, occupancy loads), wind pressures (Figure 2.2), and inertial loads from earthquakes. Sir Isaac Newton summarized the effects of force on a body in three basic laws:

> First Law: Any body at rest will remain at rest and any body in motion will move uniformly in a straight line unless acted upon by a force.
>
> (*This is equilibrium—the state of a body in which it is at rest and there is no net force acting on the body. The study of bodies in equilibrium is* statics.)
>
> Second Law: The time rate of change of momentum is equal to the force producing it, and the change takes place in the direction in which the force is acting. ($F = ma$)
>
> Third Law: For every force of action, there is a reaction that is equal in magnitude, opposite in direction, and has the same line of action. (*Basic concept of force*)

Forces acting on a body generally cause two effects:

1. It will cause the body to move if it is at rest or change the motion of the body if it is already in motion.
2. It will cause a deformation of the body.

The study of *statics* in Chapters two to five will be concerned with bodies at rest. Bodies that deform under the application of forces, called *deformable bodies* (Figure 2.3a), are the purview of strength of materials (Chapters six to ten). In statics, structural elements (bodies) are theoretically assumed as *rigid bodies* (Figure 2.3b), displaying negligible or no noticeable deformation under load.

Forces shown in the statics problems found in the next four chapters are considered externally applied forces. External forces represent the action of other bodies on the rigid body (truss, beam, column, or framework). In Figure 2.4, forces F_1, F_2, and F_3 are externally applied forces on the truss (rigid

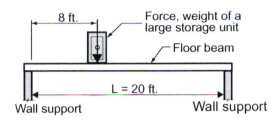

Figure 2.1   Downward-acting force exerted on a floor beam.

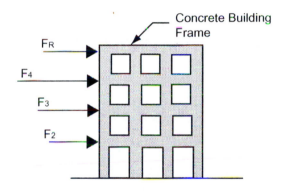

Figure 2.2   Wind forces horizontally applied to a building frame.

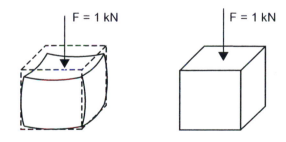

Figure 2.3   (a) Foam block (Deformable body subject to large deformations). (b) Stone (Rigid body with negligible deformation).

5

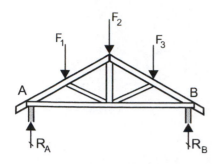

Figure 2.4   Roof truss with external loads and reactions.

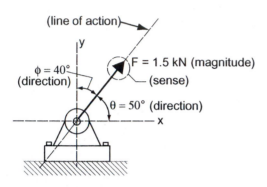

Figure 2.5   An anchor bracket with a pulling force.

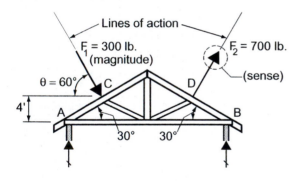

Figure 2.6   A roof truss with loads on the windward and leeward slopes.

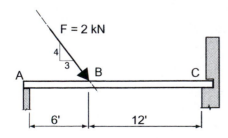

Figure 2.7   A force direction indicated by a slope.

body) and forces R_A and R_B represent external reaction forces developed at the truss supports. Internal forces are developed in each respective truss member in response to the external forces and reactions. Internal forces occur between bodies within a system and may also occur within the members themselves, holding together the particles forming the rigid body. A detailed study of internal forces will be considered in the subsequent chapters dealing with the strength of materials.

2.2   CHARACTERISTICS OF A FORCE

Forces are *vectors*, having the additional property of direction. Vectors are quantities that have both magnitude and direction and must be combined according to certain rules of addition. Other examples of vectors are acceleration and velocity. Quantities that can be completely specified by a number and dimensional unit (magnitude) are called *scalars*. Length, time, density, and temperature are examples of scalars. Scalars may be combined algebraically.

A force is characterized by its (a) *point of application*, (b) *magnitude*, and (c) *direction*. In diagrammatic form a force is represented by an arrow with the arrowhead indicating the sense of the force (Figures 2.5 and 2.6).

a. The *point of application* defines the point where the force is applied. In statics, the point of application does not imply the exact molecule on which a force is applied but a location that, in general, describes the origin of a force. The location of a force is often shown by dimension lines to a reference point.

b. *Magnitude* refers to the quantity of force; a numerical measure of its intensity. Basic units of force that will be used throughout are the pound (lb. or #) and the kilo pound (kilopound or kip; k = 1000 lb.). In metric (S.I.) units, force is expressed as Newton (N) or kilonewton (kN = 1,000 N). A 32,500-lb. force may be expressed as 32.5 k and a 49,200-N force is the same as 49.2 kN.

c. The *direction of a force* is defined by its line of action and sense. The line of action represents an infinite straight line along which the force is acting. If a force is applied such that the line of action is neither vertical nor horizontal, some reference system must be established. Most commonly accepted is the angular symbol of θ (theta) or ϕ (phi) to denote the number of degrees the line of action of the force is in relation to the horizontal or vertical axis, respectively. Only one, θ or ϕ, need be indicated. An alternative to angular designations is a slope relationship (Figure 2.7).

Principle of Transmissibility

An important principle of statics that applies to rigid bodies is the *principle of transmissibility*. The principle states that the external effects on a body remain unchanged when a force F, acting at point A, is replaced by a force F of equal magnitude at point B, provided that both forces have the same sense and line of action.

In Figure 2.8a, the external forces R_1 and R_2 represent the reactions of the wall in supporting the roof truss, opposing the applied load F applied at the top peak of the roof. Moving the applied force to the center of the lower chord (Figure 2.8b), with the magnitude, sense, and line of action remaining constant, the reactions R_1 and R_2 are unchanged. Individual truss members, however, will be subject to different internal forces due to the change in the point of application.

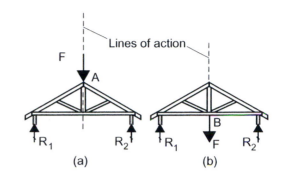

Figure 2.8 *Load at A and B on a roof truss.*

Types of Force Systems

Structural elements and frameworks are generally subjected to various combinations of forces. Forces, viewed collectively, are generally referred to as a force system. Force systems are often identified by the type or types of systems on which they act: (a) *collinear*, (b) *coplanar parallel*, (c) *coplanar concurrent*, (d) *noncoplanar parallel*, (e) *noncoplanar concurrent*, or (f) *noncoplanar, nonconcurrent* (Figure 2.9a-f). When forces act along a straight line they are called collinear; and when they are randomly distributed in space they are *noncoplanar, nonconcurrent,* and *nonparallel*. Forces that intersect at a common point are called *concurrent*, while forces with parallel lines of action are called *parallel*. If the forces are neither concurrent nor parallel, they fall under the classification of general force systems.

Concurrent force systems can act on a *particle* (point) or a rigid body, whereas parallel and general force systems can act only on a rigid body or a system of rigid bodies.

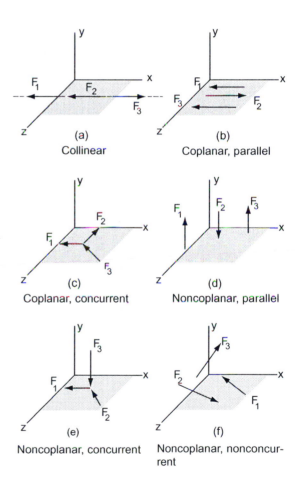

Figure 2.9 *Types of force systems.*

2.3 VECTOR ADDITION

In the analysis of structural elements or systems, it is often necessary to add the effects of several forces in determining the net resultant on the member (or system). The process of adding forces is called *vector addition* and the net force is called the *resultant*. In the case of scalar quantities where only magnitudes are considered, the process of addition involves a simple arithmetical summation. Vectors, however, have magnitude and direction; thus, they require a special procedure for combining them.

The vector sum of a series of collinear forces requires the algebraic sum of the forces to account for the directions of each respective force.

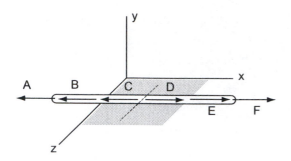

Figure 2.10   Forces on a rope during a tug-of-war contest.

Example Problem 2.1 (Figure 2.10)

In a tug-of-war contest six contestants (three on each team) are pulling along the axis of a rope. Since all forces are applied along a single line or axis, this constitutes a collinear system. Determine the *resultant* of all six forces.

$A = 35$ lb., $B = 32$ lb., $C = 29$ lb., $D = 34$ lb., $E = 37$ lb., and $F = 27$ lb.

Solution:

$$\sum F_x = -A - B - C + D + E + F$$

$$\sum F_x = -35\ \text{lb.} - 32\ \text{lb.} - 29\ \text{lb.} + 34\ \text{lb.}$$

$$+ 37\ \text{lb.} + 27\ \text{lb.} = +2\ \text{lb.}$$

The resultant force is equal to 2 lb. to the right.

Vector Addition—Graphical

An important characteristic of coplanar, concurrent vectors is that they must be added according to the *parallelogram law*. The parallelogram law states that:

> *If two forces act concurrently at a point, the resultant force can be represented by the diagonal of the parallelogram formed by the sides, parallel and proportional to the two forces.*

The graphical method of the parallelogram law involves the construction, to scale, of a parallelogram using forces (vectors) *A* and *B* as the legs (Figure 2.11). Complete the parallelogram and draw in the diagonal connecting the origin and the opposite corner. The diagonal represents the vector addition of *A* and *B*. A convenient scale is used in drawing *A* and *B*, whereby the magnitude of *R* is scaled off using the same scale. To complete the representation, the angle θ must be designated from some reference axis; in this case, the horizontal axis *x*.

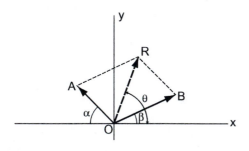

Figure 2.11   Graphical vector addition using the parallelogram law.

Another graphical vector addition approach, which preceded the parallelogram law by 100 years or so, is the *triangle rule* or *tip-to-tail method*. In Figure 2.11, the side of the parallelogram opposite force *A* is equal to *A* in both magnitude and direction, with the same being true for force *B*. It is possible to construct, to scale, only half of the parallelogram by arranging the forces in a tip-to-tail fashion (Figure 2.12) with the net result being a triangle. The sum of the two vectors, *A* and *B*, may be found by arranging them in a tip-to-tail sequence with the tip of *A* to the tail of *B* or vice versa. Drawing the other half of the parallelogram will yield the complementary triangle and the same resultant force *R*. The resultant *R* can be obtained by drawing a line beginning at the tail of the first vector and ending at the tip of the last vector. Both magnitude and direction are scaled directly from the drawing and the sense of the resultant moves from the tail of the first force to the tip of the last force. The order in which the vectors are drawn is unimportant, where

$$A + B = B + A.$$

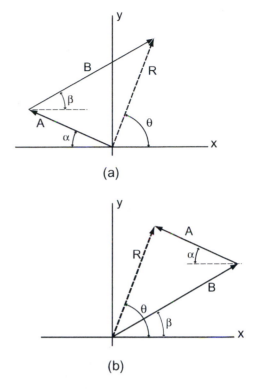

(a)

(b)

Figure 2.12   *Graphical vector addition using the tip-to-tail method.*

Example Problem 2.2

A utility pole supports two tension forces *A* and *B* with the directions shown. Using the parallelogram law and the tip-to-tail methods, determine the resultant force for *A* and *B* (magnitude and direction).

Scale: 1" = 200 lb.

Solution:

Employing the parallelogram method (Figure 2.13b) to determine the resultant for the two forces *A* and *B*, begin by drawing the two forces, using the suggested scale, with a reference *x* and *y* axis passing through the origin at point *O*. Accuracy in the drawings will have a direct impact on the accuracy of the results. Construct a line parallel to *A*, passing through the tip of force *B*; similarly, draw a line parallel to *B*, completing the parallelogram. Carefully connect the origin at *O* to the diagonally opposite corner of the parallelogram. Using the same scale, measure the length of the resultant line and, with a protractor, determine the angle θ between the reference *x* axis and the resultant force. The resultant *R* should yield a magnitude of 630 lbs. and a direction of θ = 83° relative to the *x* axis.

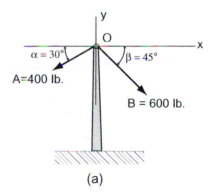

(a)

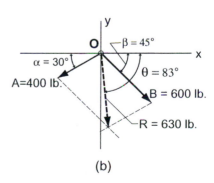

(b)

Figure 2.13   *(a) Pole with two concurrent forces. (b) Parallelogram method.*

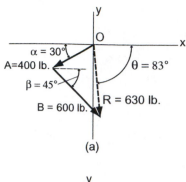

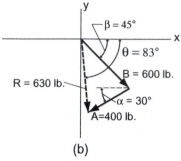

Figure 2.14 *Tip-to-tail method.*

In the tip-to-tail method (Figures 2.14a and 2.14b) involving two concurrent forces, half of the parallelogram is constructed, resulting in a triangle. Either triangle can be constructed with the resultant magnitude and direction being the same.

Begin the tip-to-tail process by constructing a reference *x* and *y* axis with the origin at **O**. In Example (a), draw force **A** to scale. At the tip of force **A** construct, to scale, force **B**. Draw a line from the origin at **O** to the tip of the last force **B**. This line represents the resultant **R** of forces **A** + **B** and can be scaled to obtain the magnitude and direction.

As in the parallelogram method, the resultant is, **R** = 630 lbs. and θ = 83°.

By changing the order of forces being drawn, the triangle that results will be the other half of the parallelogram. The process begins by drawing force **B** to scale with its tail at the origin point **O**. Force **A** is then drawn to scale with its tail at the tip of force **B**.

The resultant is the line that is generated from the reference origin at **O** to the tip of force **A**. Again, the resultant is **R** = 630 lbs. and θ = 83°.

Example Problem 2.3

Two workers are pulling a large crate as shown in Figure 2.15a. If the resultant force required to move the crate along its axis line is 120 lb., determine the tension each worker must exert. Solve graphically using the parallelogram and tip-to-tail-tail methods. Scale: 1" = 40 lb.

Solution:

Using the parallelogram law (Figure 2.15b), begin by constructing the resultant force of 120 lb. (horizontally to the right) to scale. The sides of the parallelogram have unknown magnitudes, but the directions are known. Close the parallelogram at the tip of the resultant by drawing line **A**' parallel to **A**, and extending it to intersect with **B**. The magnitude of **B** may now be determined. Similarly, line **B**' may be constructed and the magnitude of force **A** determined. From scaling, **A** = 79 lb. and **B** = 53 lb.

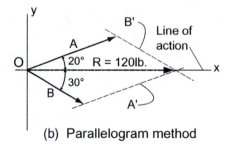

(a) Plan View

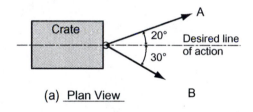

(b) Parallelogram method

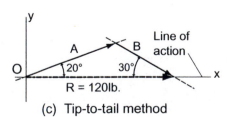

(c) Tip-to-tail method

Figure 2.15 *(a) Plan view. (b) Parallelogram method. (c) Tip-to-tail method.*

Fundamentals of Statics

11

In the tip-to-tail method (Figure 2.15c), begin by drawing the desired horizontal line of action. Then, at the origin point **O**, draw a scaled length of line along the line of action to represent the resultant force **R** = 120 lbs. At reference point **O**, also draw the line of action of force **A** as a temporary construction line. The length of line **A** is unknown at this point but the direction is given as 20°. Draw the line of action of force **B**, with a 30° inclination from the horizontal, passing through the tip of the resultant **R**. The intersection of the lines of action of forces **A** and **B** defines the location of the tip of force **A** and the tail of force **B**. Scaling the lengths of each force results in **A** = 79 lb. and **B** = 53 lb.

Graphical Addition of Three or More Vectors

The sum of any number of vectors may be obtained by repeatedly applying the parallelogram law (Figure 2.16) to successive pairs of vectors until all of the given vectors are replaced by a single resultant vector.

Note: The graphical method of vector addition requires that all vectors be coplanar.

Assume that three coplanar forces, **A**, **B**, and **C**, are acting at point **O** as shown in Figure 2.16 a and the resultant of all three is desired. In Figure 2.16 b and Figure 2.16 c, the parallelogram law is applied successively until the final resultant force **R'** is obtained. The addition of vectors **A** and **B** yields the intermediate resultant **R**, where **R** is then added vectorially to vector **C**, resulting in **R'**.

A simpler solution may be obtained by using the tip-to-tail method as shown in Figure 2.17. Again, the vectors are drawn to scale but not necessarily in any particular sequence. Example problems in this chapter which involve three or more vectors will utilize the simpler tip-to-tail method to obtain resultant forces.

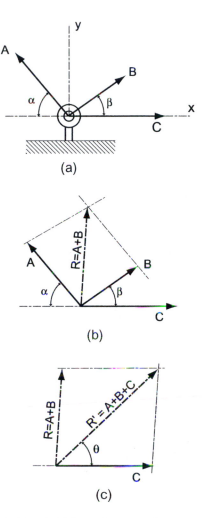

(a)

(b)

(c)

Figure 2.16 (a) Three forces acting on an eyebolt. (b) Vector sum of forces A and B (c) Vector sum of forces A, B, and C.

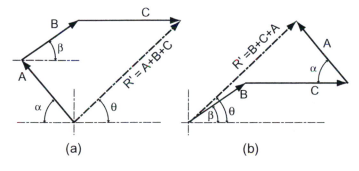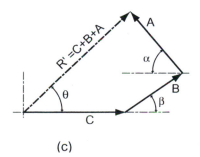

(a)   (b)   (c)

Figure 2.17   Vector sum using the tip-to-tail method.

Example Problem 2.4

A tent stake is subjected to three pulling forces, as shown in Figure 2.18. Using the graphical tip-to-tail method, determine the resultant of forces A, B, and C (magnitude and direction).

Suggested scale: $\frac{1}{8}$" = 1 lb. or 1" = 8 lb.

Solution:

In the tip-to-tail method for this problem, where all three forces have known directions and magnitudes, the sequence of drawing the forces is unimportant. The solution that results by using the sequence $A + B + C$ is $R = 50$ lbs. The resultant direction is given by the angle $\theta = 53.1°$ from the horizontal reference axis.

The same resultant is obtained if the sequence used is $B + C + A$ or $C + A + B$, as shown in Figures 2.18c, and 2.18d. These two additional solutions can be used to compare the solution for $A + B + C$, illustrating that no matter what the path sequence, the beginning and end points are the same.

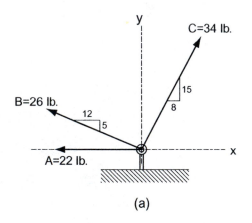

(a)

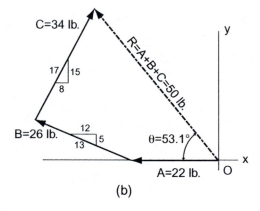

(b)

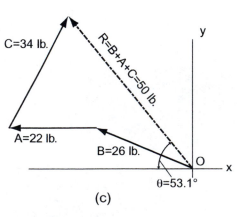

(c)

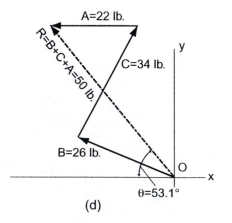

(d)

Figure 2.18

Example Problem 2.5

Three stay cables are used in supporting a bridge deck, as shown in Figure 2.19. The three cables are coplanar and concurrent at the top of the tower at D and generate a large downward compressive force. To minimize the effect of bending in the tower, it is necessary to tension the cable DC such that the resultant of cable forces DA, DB, and DC remain vertically downward. Using the tip-to-tail graphical method and a scale of 1 mm = 10 kN, solve for the cable force in DC and the resultant of all three cables.

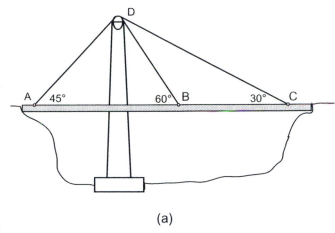

(a)

Solution:

Since cable forces DA and DB have known magnitudes and directions they will be drawn first. Cable force DC has a known direction but its magnitude is unknown. As part of the requirement of the problem, the resultant needs to remain vertical, with a known direction. The procedure will be to draw, to scale, forces DA and DB. The tail of force DA will be placed at the origin D with a reference x and y axis. At the tip of force DB, construct the line of action of DC. The intersection of the line of action of DC with the vertical y axis (which is the line of action of the resultant) determines the extent of force DC.

Scale force DC and the resultant R.

$DC = 384$ kN and $R = 852$ kN.

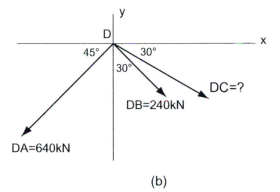

(b)

Vector Addition—Analytical Method

In the previous section, resultants of a coplanar, concurrent force system were determined by using graphical techniques (parallelogram and tip-to-tail methods). This section will examine an analytical method to determine resultants for two or more concurrent forces. Analytical vector addition usually consists of two methods: (1) use of the cosine and sine laws and (2) the component method. This book will fully explore the component method and skip entirely the use of trigonometric laws.

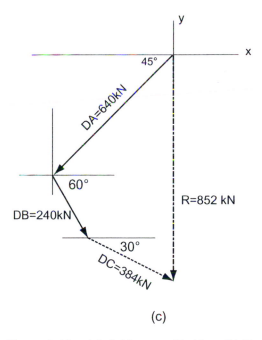

(c)

Figure 2.19   (a) Cable stayed bridge. (b) Force diagram. (c) Force polygon—tip-to-tail.

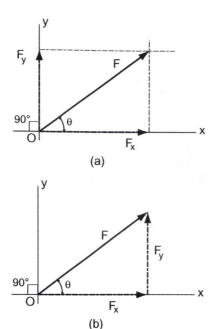

Figure 2.20   Rectangular components of a force.

Components of a Force

Before introducing the analytical method of vector addition, a procedure needs to be established for dealing with the varied directions of the forces in a concurrent force system.

Combining force directions in a systematic way can be accomplished by *resolving* each force into its respective *components* along the x and y axes. Resolution of a force into its components has the reverse effect of vector addition. The original force represents the diagonal and the components are the sides of the parallelogram (Figure 2.20). Components of a force are generally perpendicular to each other and are called *rectangular components*. The x and y axes of a rectangular coordinate system are most often assumed to be horizontal and vertical, respectively. However, they may be chosen in any two mutually perpendicular directions for convenience.

A force *F* with a direction θ from the horizontal x axis can be resolved into its rectangular components F_x and F_y, as shown in Figure 2.20. Both F_x and F_y are trigonometric functions of *F* and θ, where:

$$\cos \theta = \frac{F_x}{F}; \quad \therefore F_x = F \cos \theta;$$

$$\sin \theta = \frac{F_y}{F}; \quad \therefore F_y = F \sin \theta$$

In effect, the force components F_x and F_y form the legs of a parallelogram with the diagonal representing the original force *F*. Therefore, applying the Pythagorean theorem for right triangles:

$$F = \sqrt{F_x^2 + F_y^2} \quad \text{and} \quad \tan \theta = \frac{F_y}{F_x} \quad \text{or,}$$

$$\theta = \tan^{-1}\left(\frac{F_y}{F_x}\right)$$

Force directions indicated by a slope relationship (Figure 2.21) result in components expressed as ratios of the original force. Using a similar triangle relationship:

$$\frac{F_x}{4} = \frac{F_y}{3} = \frac{F}{5}; \quad \text{and} \quad F_x = \frac{4F}{5} \quad \text{and} \quad F_y = \frac{3F}{5}$$

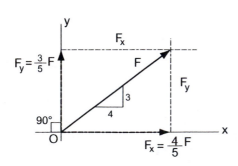

Figure 2.21   Rectangular components for a force with a slope relationship.

Example Problem 2.6

A cantilever beam, shown Figure 2.22, is subjected to a force **F** at its free end **B**.

Resolve the force into its rectangular components along the *x* and *y* axes.

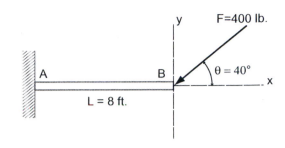

Figure 2.22   *Inclined load on a cantilever beam.*

Solution: (Figure 2.23a-e)

Draw, to a reasonably accurate scale, the force **F** at an angle of $\theta = 40°$ from the horizontal *x* axis. From the tail end of the force arrow, construct perpendicular lines to the respective *x* and *y* axes. The parallelogram that results has the original force **F** as the diagonal (resultant).

Component F_x represents the side of the parallelogram along the *x* axis and F_y is the vertical side along the *y* axis. Indicate the rectangular component forces F_x and F_y on the parallelogram, as shown in the diagram. Note that the sense (arrowhead) of each component force is shown at the same location as the original force **F**. This may not seem like a particularly important point at this time but in future work with force components, the direction of the components are vitally important.

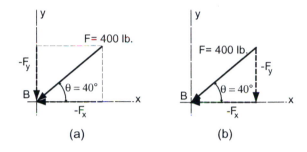

(a)   (b)

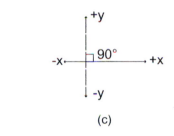

(c)

Examining half of the parallelogram in Figure 2.23a or the right triangle formed in the tip-to-tail analogy in Figure 2.23b, the equations for the component forces can be expressed as:

$$F_x = F \cos 40° = (400 \text{ lb.})(.766) = 306.4 \text{ lb. } (\leftarrow)$$

$$F_y = F \sin 40° = (400 \text{ lb.})(.643) = 257.1 \text{ lb. } (\downarrow)$$

The directions of the respective components are generally assigned a plus (+) or minus (−) based on the conventions of the Cartesian coordinate system. Forces directed to the right, along the *x* axis, are assumed as positive (+) forces and forces going to the left are considered negative (−). Similarly, forces directed upward along the *y* axis are assigned a positive (+) value while downward-acting forces are negative(−). See Figure 2.23c.

This same problem will be examined using the angle $\phi = 50°$ referenced from the vertical *y* axis. See Figures 2.23d, and 2.23e. Note that since the included angle in the right triangle changes, the side adjacent and side opposite also changes. Therefore, the rectangular force components are expressed as:

$$F_x = F \sin \phi = (400 \text{ lb.})(.766) = 306.4 \text{ lb. } (\leftarrow)$$

$$F_y = F \cos \phi = (400 \text{ lb.})(.643) = 257.1 \text{ lb. } (\downarrow)$$

Notice that the magnitudes and directions in both examples remain the same.

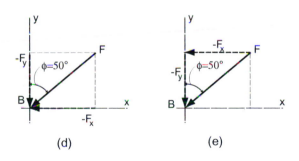

(d)   (e)

Figure 2.23   *(a) Parallelogram. (b) Tip-to-tail. (c) Sign convention. (d) Parallelogram. (e) Tip-to-tail.*

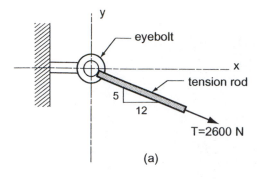

(a)

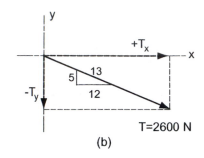

(b)

Figure 2.24 Inclined load on an eyebolt.

Example Problem 2.7

A large eyebolt (Figure 2.24) is used in supporting a canopy over the entry to an office building. The tension developed in the support rod is equal to 2600 newtons. Determine the rectangular components of the force if the rod is at a 5 in 12 slope.

Solution:

The slope of the tension rod is a valid way of defining the direction of a force.

Component forces T_x and T_y can be expressed as ratios of the original force $T = 2600$ N. A slope of 5 in 12 (rise to run) has a hypotenuse of 13. Using similar triangle relationships, the components are equal to:

$$\frac{T_x(\text{horiz.})}{12(\text{horiz.})} = \frac{T_y(\text{vert.})}{5(\text{vert.})} = \frac{T(\text{hypot.})}{13(\text{hypot.})}$$

$$\therefore T_x = \frac{12}{13}T = \frac{12}{13}(2600 \text{ N}) = +2400 \text{ N} (\rightarrow)$$

and

$$T_y = \frac{5}{13}T = \frac{5}{13}(2600 \text{ N} = -1000 \text{ N} (\downarrow)$$

Based on the sign convention discussed earlier, the horizontal component T_x is positive (+) along the x axis and T_y is negative, going in a downward direction along the y axis.

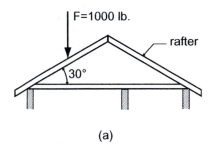

(a)

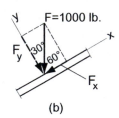

(b)

Figure 2.25 Vertical load on a roof rafter.

Example Problem 2.8

A vertical force F is applied to a roof rafter, as shown in Figure 2.25a. Resolve the force F into two components, one perpendicular to the rafter and the other along its length. The rafter forms a slope of $\theta = 30°$ from the horizontal ceiling tie.

Solution:

In this example, since the desired components are perpendicular and parallel to the rafter slope, it is convenient to rotate the coordinate axis such that the x axis lies along the rafter length.

Based on this rotated axis, the force F forms an angle of 60° from the x axis (along the rafter slope). Force F_x is the side adjacent to the 60° angle and F_y is opposite the 60° angle. Therefore:

$$F_x = F \cos 60° = (1000 \text{ lb.})(0.50)$$

$$= 500 \text{ lb.} (\rightarrow) \text{ along the } x \text{ axis.}$$

$$F_y = F \sin 60° = (1000 \text{ lb.})(0.866) = 866 \text{ lb.} (\downarrow)$$

The force F_x is trying to slide the rafter off the wall while the force F_y is responsible for causing the bending in the rafter.

Component Method of Vector Addition

Analytical vector addition using the component method is probably the most general and useful method for determining the resultant of two or more concurrent, coplanar forces as in Figure 2.26. The basic idea in this methodology is to resolve all concurrent forces in the system into x and y components and combine algebraically all components in the respective x and y directions (Figure 2.27). Summing all of the forces along the x axis yields the resultant R_x and similarly, summing y component forces results in R_y (Figure 2.28). Component resultants, R_x and R_y, form the two sides of a rectangle in which the diagonal R represents the final resultant of the entire concurrent force system.

A general procedure for determining the resultant of a coplanar, concurrent force system using the component method is as follows:

- Resolve each force into its respective x and y components. Forces with their tail at the origin will result in components with their tails at the origin. If a force is pushing toward the concurrent point (arrowhead at the origin), the components of the force will also be pushing at the origin.
- Note the direction of each component force, plus or minus, based on a sign convention.
  The most commonly used sign convention for a rectangular coordinate system defines a direction to the right (horizontally) as the positive x direction and upward (vertically) as denoting a positive y direction.
- Sum the horizontal component forces algebraically, noting the plus or minus direction for each force. The general expression for the resultant R_x may be written as:

$$R_x = \sum F_x$$

- Sum the vertical component forces algebraically, such that:

$$R_y = \sum F_y$$

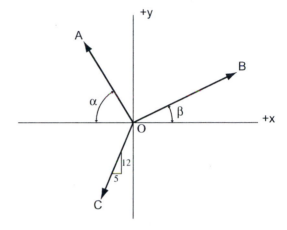

Figure 2.26   Three coplanar, concurrent forces at O.

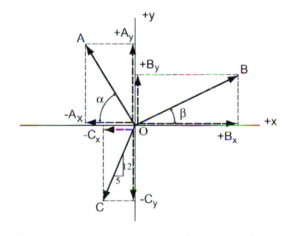

Figure 2.27   Forces A, B, and C resolved onto their respective x and y components.

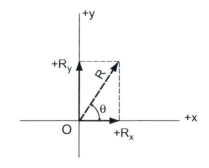

Figure 2.28   Final resultant R from R_x and R_y.

- Note the sign for each component resultant. A positive R_x means that the horizontal resultant is directed to the right and a negative R_x would be shown going to the left.
  Similarly, a positive R_y is directed upward and a negative R_y is acting downward.
- Sketch, to a relative scale, the resultants R_x and R_y on an x-y coordinate axis. Component resultants R_x and R_y form the two sides of a rectangle. Construct the other two sides of the rectangle and draw the diagonal that passes through the origin and the diagonally opposite corner. This diagonal represents the final resultant of all of the concurrent forces in the system.
- The final resultant R can be computed as:

$$R = \sqrt{(R_x)^2 + (R_y)^2}$$

The magnitude of the resultant can be obtained by using the Pythagorean theorem, and the direction of the resultant force is obtained by using the trigonometric function:

$$\theta = \tan^{-1}\left(\frac{R_y}{R_x}\right)$$

Example Problem 2.9 (Figure 2.29)

This is the same problem as Example Problem 2.2, which was solved earlier using the graphical methods.

Solution:

Construct an x-y axis system and draw to relative scale forces A and B. Show the x and y components of each force, paying close attention to the directions (arrowheads) of each component.

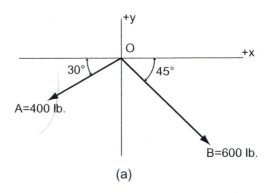

(a)

$$-A_x = -A \cos 30° = -(400 \text{ lb.})(0.866) = -346.4 \text{ lb.}$$

$$-A_y = -A \sin 30° = -(400 \text{ lb.})(0.50) = -200 \text{ lb.}$$

$$+B_x = +B \cos 45° = +(600 \text{ lb.})(0.707) = +424.2 \text{ lb.}$$

$$-B_y = -B \sin 45° = -(600 \text{ lb.})(0.707) = -424.2 \text{ lb.}$$

Component resultants can now be determined by algebraically summing forces in the respective x and y directions.

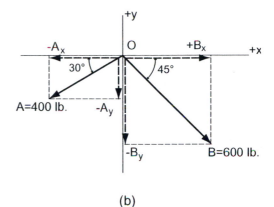

(b)

$$R_x = \sum F_x = -A_x + B_x$$

$$= -346.4 \text{ lb.} + 424.2 \text{ lb.} = +77.8 \text{ lb.}$$

$$R_y = \sum F_y = -A_y - B_y$$

$$= -200 \text{ lb.} - 424.2 \text{ lb.} = -624.2 \text{ lb.}$$

The magnitude of the final resultant R can be determined by using the Pythagorean theorem, where:

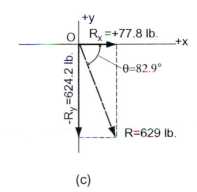

(c)

$$R = \sqrt{(R_x)^2 + (R_y)^2}$$

$$= \sqrt{(+77.8)^2 + (-624.2)^2} = 629 \text{ lb.}$$

Figure 2.29 (a) *Forces A and B concurrent at point O.* (b) *Forces resolved into components along the x and y axes.* (c) *Components R_x and R_y combined to produce the final resultant R.*

Direction for the final resultant can be determined by:

$$\tan \theta = \left(\frac{R_y}{R_x}\right) \qquad \therefore \theta = \tan^{-1}\left(\frac{R_y}{R_x}\right)$$

$$\theta = \tan^{-1}\left(\frac{624.2}{77.8}\right) = 82.9°$$

Note that the sign for R_y was ignored since the diagram of the resultant in Figure 2.29c is very descriptive of the direction.

The answers obtained by the component method are identical with the graphical results obtained earlier. In subsequent problems the graphical tip-to-tail method will be used to check the analytical results of the component method.

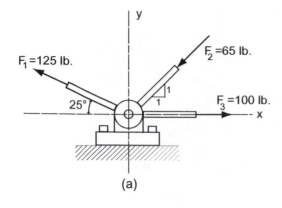

(a)

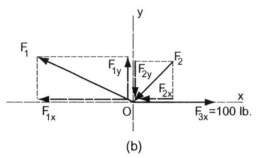

(b)

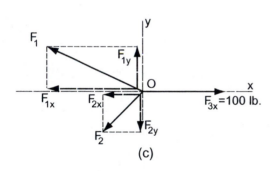

(c)

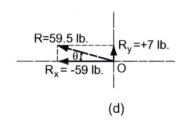

(d)

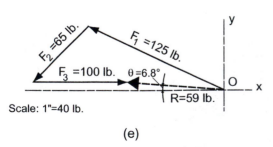

Scale: 1"=40 lb.

(e)

Figure 2.30 (a) Three concurrent forces on an anchor device. (b) Resolution of forces into components. (c) Alternate resolution of forces into components. (d) Resultant force R. (e) Graphical tip-to-tail solution.

Example Problem 2.10

An anchoring device, shown in Figure 2.30a, is subjected to the three forces as shown. Determine analytically, using the component method, the resultant force being resisted by the anchor.

Solution:

First, resolve each force into its respective component parts. Note that in this example, force F_2 is pushing at the origin point O. A drawing (Figure 2.30b) to show the forces and components at the point of concurrency will be useful in visualizing how the forces are acting.

$$-F_{1x} = -F_1 \cos 25° = -(125 \text{ lb.})(0.906) = -113 \text{ lb.}$$

$$+F_{1y} = F_1 \sin 25° = (125 \text{ lb.})(0.423) = +53 \text{ lb.}$$

$$-F_{2x} = -F_2\left(\frac{1}{\sqrt{2}}\right) = -(65 \text{ lb.})\left(\frac{1}{\sqrt{2}}\right) = -46 \text{ lb.}$$

$$-F_{2y} = -F_2\left(\frac{1}{\sqrt{2}}\right) = -(65 \text{ lb.})\left(\frac{1}{\sqrt{2}}\right) = -46 \text{ lb.}$$

$$+F_{3x} = +F_3 = +100 \text{ lb.}$$

An alternate force diagram can be used by applying the *principle of transmissibility*. Force F_2 can be slid along its line of action through the origin point O and shown as a pulling force. Figure 2.30c shows all three forces as tension, or pulling, forces at O. Component forces for force F_2 remain the same in both magnitude and direction. The external effects on the anchor device remain unchanged.

Resultants along the x and y directions are obtained by the equations:

$$R_x = \sum F_x = -F_{1x} - F_{2x} + F_{3x}$$

$$R_x = -113 \text{ lb.} - 46 \text{ lb.} + 100 \text{ lb.} = -59 \text{ lb. } (\leftarrow)$$

$$R_y = \sum F_y = +F_{1y} - F_{2y}$$

$$R_y = +53 \text{ lb.} - 46 \text{ lb.} = +7 \text{ lb. } (\uparrow)$$

Combining the component resultants into the final resultant R using the Pythagorean theorem yields:

$$R = \sqrt{(R_x)^2 + (R_y)^2}$$
$$= \sqrt{(-59 \text{ lb.})^2 + (+7 \text{ lb.})^2} = 59.5 \text{ lb.}$$

The direction of the resultant, relative to the x axis, is:

$$\tan \theta = \frac{R_y}{R_x} = \frac{7}{59} = 0.119$$

$$\theta = \tan^{-1}(0.119) = 6.8°$$

Example Problem 2.11

Example Problem 2.11 shown in Figure 2.31a will be solved using the component method and verify the results obtained earlier from the graphical approach.

Remember that the problem requires all three forces having a vertical resultant to minimize the bending in the tower.

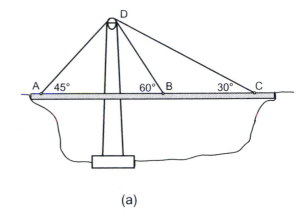

(a)

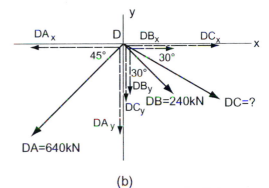

(b)

Solution:

A good first step in setting up a problem involving a concurrent force system is to draw a coordinate axis with the forces and respective components (see Figure 2.31b). Indicate clearly the direction of each force and the directions of the components. When three or more concurrent forces are involved, it is sometimes more convenient to set up a table which lists the forces and respective components. This tabular approach will be illustrated in this example.

Figure 2.31   (a) Cable-stayed bridge. (b) Force diagram.

Force = F	F_x	F_y
DA	$-DA \sin 45° = -(640 \text{ kN})(.707) = -452.5 \text{ kN}$	$-DA \cos 45° = -(640 \text{ kN})(.707) = -452.5 \text{ kN}$
DB	$+DB \sin 30° = +(240 \text{ kN})(0.5) = +120 \text{ kN}$	$-DB \cos 30° = -(240 \text{ kN})(.866) = -207.8 \text{ kN}$
DC	$+DC \cos 30° = +0.866DC$	$-DC \sin 30° = -0.5DC$
	$R_x = \sum F_x = -452.5 \text{ kN} + 120 \text{ kN} + .866DC = 0$	$R_y = \sum F_y = -452.5 \text{ kN} - 207.8 \text{ kN} - 0.5DC$

The resultant R_x is found by summing the component forces listed under the F_x column and similarly, R_y is the sum of the vertical components in the F_y column.

Solve for the unknown DC using the R_x equation. Rewriting the equation for R_x:

$$DC = \frac{+452.5 \text{ kN} - 120 \text{ kN}}{0.866} = +384 \text{ kN}$$

Substituting for DC in the R_y equation:

$$R_y = -852 \text{ kN}$$

These results check with the answers obtained in the graphical solution for Example Problem 2.5.

2.4   MOMENT OF A FORCE

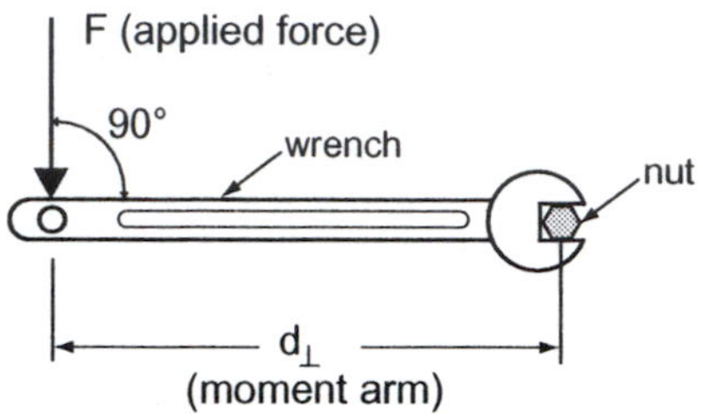

When a force is applied to a rigid body the tendency will be for the body to move along a line of action (*translation*) or to *rotate*. The tendency of the force to produce a *rotation* or *twist* of the body about an axis or point is referred to as the *moment* of a force. Examples that involve moments are tightening or loosening a nut with a wrench, turning the steering wheel in a car, pulling a bent nail out of a piece of lumber with a crowbar, or opening or closing a door.

Moments are present whether there is an actual rotation of the rigid body or only a tendency of the body to rotate. The unsuccessful attempt of pushing on the end of a wrench to loosen a "frozen" nut is an example of the nut having a tendency to rotate but with no actual rotation occurring (Figure 2.32).

Quantitatively, the *moment of a force* with respect to a *reference point* is equal to the product of the force and the perpendicular distance of the force from the point (Figure 2.33). The perpendicular distance from the line of action of the force to the reference point is often referred to as the *moment arm*. Moments may be evaluated at any reference point, depending on the information desired by the designer.

The moment equation can be written as:

$$M_i = (F) \times (d_\perp)$$

Where:

M_i =Moment about any reference point '*i*'. Units for moment are expressed as pound-feet (lb.-ft) or kip-feet (k-ft). The S.I. units for moment are Newton-meters (N-m) or kilonewton-meters (kN-m).

F =Applied force in pounds (lbs.), kilopounds (k), Newtons (N), or kilonewtons (kN).

$d_\perp$ =Perpendicular distance from the reference point '*i*' to the line of action of the applied force. Units are in inches (in.), feet (ft.), or meters (m).

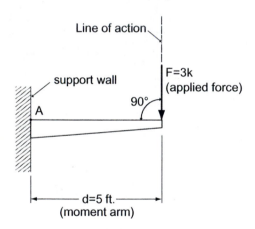

Figure 2.32   Force F *applied at a perpendicular distance 'd' from the center of the nut.*

Figure 2.33   A cantilever beam with a load applied at the free end.

Moments are a product of a force (a vector) and thus are vector quantities as well. Directions for moments about a reference point are either *clockwise* or *counterclockwise* rotations. A commonly used sign convention is to assume counterclockwise rotation as positive (+) and clockwise rotation as negative (−).

In the example of the cantilever beam supporting a 3-k load (Figure 2.33), the moment evaluated at the supporting wall at A is:

$$M_A = -(F) \times (d_\perp) = -(3k)(5 \text{ ft.}) = -15 \text{ k-ft. (clockwise)}$$

The clockwise direction is determined through observation.

Example Problem 2.12 (Figure 2.34)

A 1-meter-long lever is inclined at a 5:12 slope with a vertically applied force of 200 N. Determine the moment about pivot point A caused by the applied force.

Solution:

The moment of the force about point A is found by multiplying the magnitude of the force times the perpendicular distance from A to the line of action of the force.

$$d_\perp = \frac{12}{13}(L) = \frac{12}{13}(1 \text{ m}) = 0.923 \text{ m}$$

$$M_A = -F \times d = -(200 \text{ N})(0.923 \text{ m}) = -184.6 \text{ N-m}$$

Since the rotation is in the clockwise direction, the answer is expressed as a negative value based on the adopted sign convention.

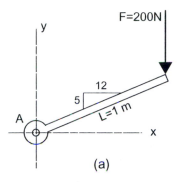

(a)

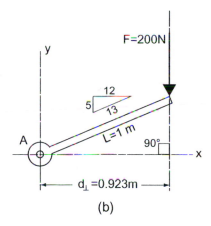

(b)

Figure 2.34   Vertical force on a lever.

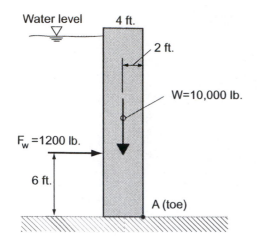

Figure 2.35   Forces on a concrete gravity dam.

Example Problem 2.13 (Figure 2.35)

A 1-foot-wide slice of a 4-foot-thick concrete gravity dam weighs 10,000 pounds and the equivalent force due to water pressure behind the dam is equal to 1200 pounds. The stability of the dam against overturning is evaluated about the "toe" at *A*.

Determine the resultant moment at *A* due to the two forces shown. Is the dam stable?

Solution:

The resultant moment about *A* is obtained by algebraically summing the moments due to each force and noting the positive and negative directions of each moment.

$$M_A = -(F_w) \times (6 \text{ ft.}) + (W) \times (2 \text{ ft.})$$
$$M_A = -(1200 \text{ lb.})(6 \text{ ft.}) + (10,000 \text{ lb.})(2 \text{ ft.})$$
$$= +12,800 \text{ lb.-ft.}$$

The resultant moment yielded a positive sign; thus, the moment tendency is in the counterclockwise direction and the dam is quite stable.

Example Problem 2.14 (Figure 2.36)

A utility pole is supporting a cable with a tension force of 250 lb., directed at a 15° angle from a horizontal reference. Determine the moment developed at the base of the pole at *A* due to the cable force.

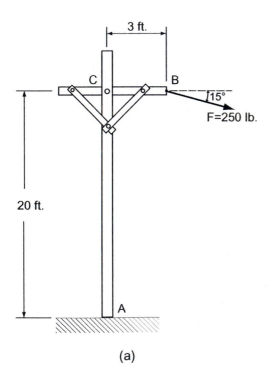

(a)

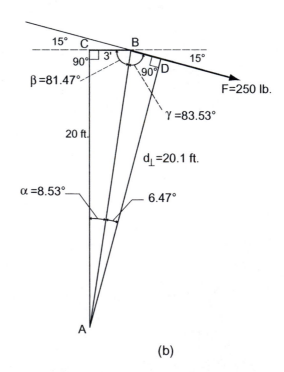

(b)

Figure 2.36   (a) Tension force on a utility. (b) Determining the perpendicular moment arm of force F.

Solution:

The key to solving this problem is the determination of the perpendicular moment arm for the cable force $F = 250$ lb. A recommendation, which might help you in visualizing the trigonometry involved in this problem, is to draw to a reasonably accurate scale the dimensions of the utility pole and the force with the proper inclination shown. This is by no means intended to be a graphical solution but the correctly drawn diagram will assist you in seeing the trigonometric relationships.

A right triangle is formed between the points A, B, and C. Since the dimensions AB and BC are known, the angles α and β can be found.

$$\alpha = \tan^{-1}\left(\frac{BC}{AC}\right) = \tan^{-1}\left(\frac{3 \text{ ft.}}{20 \text{ ft.}}\right) = 8.53°$$

Then, $\beta = 90° - \alpha = 90° - 8.53° = 81.47°$

The length of the line from A to B, the hypotenuse of the right triangle, is equal to:

$$\cos \alpha = \left(\frac{AC}{AB}\right) = \left(\frac{20 \text{ ft.}}{AB}\right);$$

$$AB = \left(\frac{20 \text{ ft.}}{\cos 8.53°}\right) = \left(\frac{20 \text{ ft.}}{0.989}\right) = 20.2 \text{ ft.}$$

Construct the line AD, which represents the $d_\perp$ distance from the reference point at A to the line of action of the applied force F. A new right triangle is formed with the dimension of the hypotenuse AB already determined. Angle γ can be solved by the following relationship:

$$\gamma = 180° - \beta - 15° = 180° - 81.47° - 15° = 83.53°$$

The perpendicular moment arm can now be determined by using the trigonometric relationship:

$$\sin \gamma = \frac{d_\perp}{AB};$$

$$\therefore d_\perp = AB \sin \gamma = (20.2 \text{ ft.})(\sin 83.53°) = 20.1 \text{ ft.}$$

Finally, the moment about the base at A can be calculated as:

$$M_A = -(F) \times (d_\perp) = -(250 \text{ lb.}) \times (20.1 \text{ ft.})$$
$$= -5025 \text{ lb.-ft.}$$

Principle of Moments—Varignon's Theorem

In the previous Example Problem 2.14, the major difficulty in determining the moment about A was the determination of the perpendicular moment arm. Instead of finding the perpendicular distance, this same problem can be solved much more simply by using what is known as the *principle of moments* or *Varignon's theorem (after French mathematician Pierre Varignon; 1654–1722).* Varignon's theorem states that:

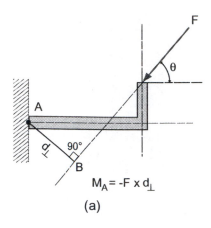

$$M_A = -F \times d_\perp$$

(a)

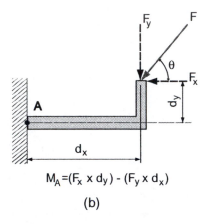

$$M_A = (F_x \times d_y) - (F_y \times d_x)$$

(b)

*Figure 2.37 (a) Load on an upturned beam.
(b) Resolution of forces into F_x and F_y.*

*The moment of a force about a point is equal to the
algebraic sum of the moments of the components of the
force with respect to the same point.*

In the upturned cantilever beam shown in Figure 2.37, the
moment of the applied force **F** about reference point **A** is:
$M_A = -F \times d_\perp$.

Finding the distance $d_\perp$ may be quite involved; therefore,
employing Varignon's theorem might be very convenient
as an alternative. The force **F** is resolved into its horizontal and vertical components and the moment arm distances d_x and d_y (usually readily found) of the respective
components are obtained. The resulting moment M_A is
computed by algebraically summing the moments about
point **A** generated by each of the component forces. In
both cases, the moments are identical in magnitude and
direction.

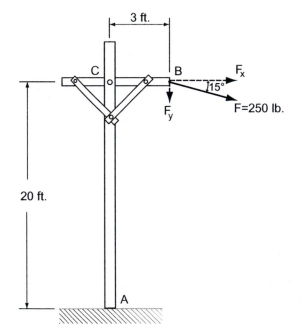

*Figure 2.38 Moments using Varignon's
theorem.*

Example Problem 2.15 (Figure 2.38)

The same problem from Example Problem 2.14 will be
solved using the principle of moments. (Figure 2.36)

Solution:

Begin the solution by resolving the force into its x and y
components.

$$F_x = F \cos 15° = (250 \text{ lb.})(0.966) = 241.5 \text{ lb.}$$

$$F_y = F \sin 15° = (250 \text{ lb.})(0.259) = 64.7 \text{ lb.}$$

Next, algebraically sum the moments about A:

$$M_A = -(F_x \times d_{\perp y}) - (F_y \times d_{\perp x})$$

The negative sign preceding each of the component moments is due to the tendency of each respective force component to produce a clockwise tendency of rotation about
point A, not because of the force direction.
Perpendicular moment arms are given as dimensions of
the pole, where:

$$d_{\perp y} = 20 \text{ ft. and } d_{\perp x} = 3 \text{ ft.}$$

Computing for M_A:

$$M_A = -(241.5 \text{ lb.})(20 \text{ ft.}) - (64.7 \text{ lb.})(3 \text{ ft.})$$
$$M_A = -(4830 \text{ lb.-ft.}) - (194 \text{ lb.-ft.}) = -5024 \text{ lb.-ft.}$$

This solution results in the same magnitude and direction of rotation for M_A found in Example Problem 2.14.

Example Problem 2.16

A wind force of 2000 pounds, at an angle of 50° from the horizontal, is applied at joint A on a roof truss as shown in Figure 2.39a. Determine the moment due to the force F about the supports at B and C and the crown at D. Use the principle of moments in determining M_B, M_C, and M_D.

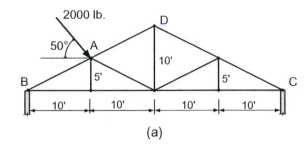

(a)

Solution:

Begin by resolving the force F into its respective x and y components as shown in Figure 2.39b. The perpendicular moment arms for the components are easily determined from the dimensions already given for the truss.

$$F_x = F \cos 50° = (2000 \text{ lb.})(0.643) = 1286 \text{ lb.}$$
$$F_y = F \sin 50° = (2000 \text{ lb.})(0.766) = 1532 \text{ lb.}$$

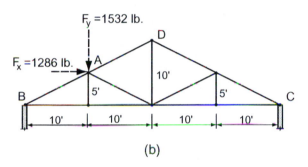

(b)

Again, the plus or minus direction for the respective component forces is unimportant but the direction of rotation caused by the force (clockwise or counterclockwise) relative to the reference point must be indicated.

Figure 2.39 (a) Symmetrical roof truss. (b) Force components at joint A.

The moment M_B, relative to support point B, is computed as:

$$M_B = -(F_x) \times (5 \text{ ft.}) - (F_y) \times (10 \text{ ft.})$$
$$M_B = -(1286 \text{ lb.})(5 \text{ ft.}) - (1532 \text{ lb.})(10 \text{ ft.})$$
$$M_B = -(6430 \text{ lb.-ft.}) - (15,320 \text{ lb.-ft.})$$
$$M_B = -21,750 \text{ lb.-ft.}$$

Using the reference point at support C, the moment M_C is calculated to be:

$$M_C = -(F_x) \times (5 \text{ ft.}) + (F_y) \times (30 \text{ ft.})$$
$$M_C = -(1286 \text{ lb.})(5 \text{ ft.}) + (1532 \text{ lb.})(30 \text{ ft.})$$
$$M_C = -(6430 \text{ lb.-ft.}) + (45,960 \text{ lb.-ft.})$$
$$M_C = +39,530 \text{ lb.-ft.}$$

The moment at the crown of the roof truss is equal to:

$$M_D = +(F_x) \times (5 \text{ ft.}) + (F_y) \times (10 \text{ ft.})$$
$$M_D = +(1286 \text{ lb.})(5 \text{ ft.}) + (1532 \text{ lb.})(10 \text{ ft.})$$
$$M_D = +(6430 \text{ lb.-ft.}) + (15{,}320 \text{ lb.-ft.})$$
$$M_D = +21{,}750 \text{ lb.-ft.}$$

Example Problem 2.17

A 8-meter vertical pole is used to support three cable forces as shown in Figure 2.40a. Determine the moment at the base of the pole at A.

Solution (Figure 2.40b):

Resolve forces F_1 and F_2 into their respective x and y components.

$$F_{1x} = F_1 \sin 30° = (1800 \text{ N})(0.5) = 900 \text{ N}$$
$$F_{1y} = F_1 \cos 30° = (1800 \text{ N})(0.866) = 1560 \text{ N}$$
$$F_{2x} = \frac{3}{5}F_2 = \frac{3}{5}(900 \text{ N}) = 540 \text{ N}$$
$$F_{2y} = \frac{4}{5}F_2 = \frac{4}{5}(900 \text{ N}) = 720 \text{ N}$$

The moment at the base of the pole at A is the algebraic sum of the moments due to force F_3 and the component forces of F_1 and F_2.

$$M_A = +(F_{1x})(8 \text{ m}) - (F_{2x})(8 \text{ m}) - (F_3)(8 \text{ m})$$
$$M_A = +(900 \text{ N})(8 \text{ m}) - (540 \text{ N})(8 \text{ m})$$
$$- (360 \text{ N})(8 \text{ m})$$
$$M_A = +(7200 \text{ N-m}) - (4320 \text{ N-m})$$
$$- (2880 \text{ N-m}) = 0$$

A zero resultant moment at A means that there is no tendency for the pole to rotate about the base for this particular combination of forces. Also, note that the vertical components of forces F_1 and F_2 did not appear in the moment equation because neither had a moment arm.

> *Forces that intersect the reference point have no moment arms and will cause no tendency for rotation about the point.*

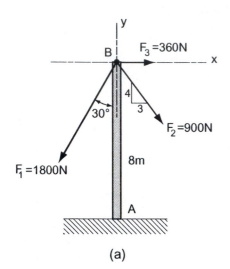

(a)

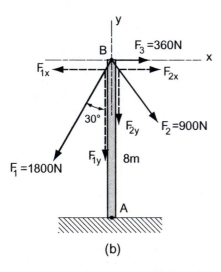

(b)

Figure 2.40 (a) Three forces on a vertical pole. (b) Forces resolved into x and y components.

Example Problem 2.18

Wind forces produce a direct pressure of 1200 lb. on the windward roof slope (at the midpoint of the rafter length) and a simultaneous suction (uplifting) force of 3000 lb. on the midpoint along the leeward rafter in Figure 2.41a. Both forces are applied perpendicular to the roof slope. Assuming the rafters on each side have a slope of 9:12, determine the overturning moment caused by the two wind forces about point A. Then, determine what the dead load weight of the building needs to be in order to counteract the overturning tendency of the find forces.

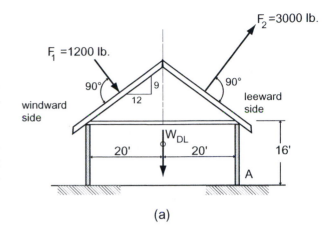

(a)

Solution:

Resolve each wind force into its respective x and y components as shown in Figure 2.41b. Note that a 9:12 slope is the same as a 3:4 slope and the slope of force F_1 is at a 4:3 relationship.

$$F_{1x} = \frac{3}{5}F_1 = \frac{3}{5}(1200 \text{ lb.}) = 720 \text{ lb.}$$

$$F_{1y} = \frac{4}{5}F_1 = \frac{4}{5}(1200 \text{ lb.}) = 960 \text{ lb.}$$

$$F_{2x} = \frac{3}{5}F_2 = \frac{3}{5}(3000 \text{ lb.}) = 1800 \text{ lb.}$$

$$F_{2y} = \frac{4}{5}F_2 = \frac{4}{5}(3000 \text{ lb.}) = 2400 \text{ lb.}$$

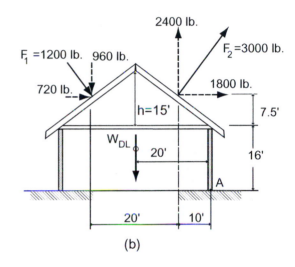

(b)

Figure 2.41   (a) Forces on a roof structure. (b) Forces resolved into x and y components.

From the geometry of the building, determine the perpendicular moment arm for each component force.

The height to the midpoint of the rafter from the top of the wall can be determined by using a similar triangular relationship in which:

$3: 4 :: h : 20$ ft.; therefore, the height of the roof from the top of the wall to the peak is equal to:

$$h = \frac{3}{4}(20 \text{ ft.}) = 15 \text{ ft.}; \qquad \frac{h}{2} = \frac{15 \text{ ft.}}{2} = 7.5 \text{ ft.}$$

Perpendicular moment arms for F_{1x} and F_{2x} are measured from A to the midheight of the rafter.

$$d_\perp = (16 \text{ ft.}) + (7.5 \text{ ft.}) = 23.5 \text{ ft.}$$

$$M_A = -(F_{1x})(23.5 \text{ ft.}) + (F_{1y})(30 \text{ ft.}) - (F_{2x})(23.5 \text{ ft.})$$
$$- (F_{2y})(10 \text{ ft.})$$

$$M_A = -(720 \text{ lb.})(23.5 \text{ ft.}) + (960 \text{ lb.})(30 \text{ ft.})$$
$$- (1800 \text{ lb.})(23.5 \text{ ft.}) - (2400 \text{ lb.})(10 \text{ ft.})$$

$$M_A = -54{,}420 \text{ lb.-ft.}$$

To stabilize the building using its dead load, equate the overturning moment due to the wind forces with the stabilizing moment.

$$M_A = 54{,}420 \text{ lb.-ft.} = (W_{DL}) \times (20 \text{ ft.})$$

The dead load weight of the building is computed as:

$$W_{DL} = \frac{(54{,}420 \text{ lb.-ft.})}{20 \text{ ft.}} = 2{,}721 \text{ lb.}$$

Couple and Moment of a Couple

Occasionally in *statics*, a unique arrangement of forces produces a tendency for a member to rotate without experiencing any translation. This force system is referred to as a *couple* (Figure 2.42). A couple is defined as two forces having the same magnitude, parallel lines of action, but opposite sense (arrowhead direction). Couples have pure rotational effects on a body with no capacity to translate (move) the body in the vertical or horizontal direction.

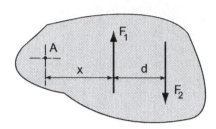

Figure 2.42   Force couple system.

The *moment of a couple, M,* is computed as the product of the force *F* times the perpendicular distance *d* between the two equal and opposite forces. The moment of a couple is a constant value and is independent of any specific reference point. A couple can be balanced only by an equal and opposite couple in the same plane.

$$M_{\text{couple}} = F \times d$$

Example Problem 2.19

The cantilevered beam shown in Figure 2.43 is subjected to two equal and opposite forces as shown. Determine the resultant moment M_A at the beam support and the moment M_B at the free end.

Solution:

Recognizing that the two applied forces on the cantilever beam constitute a couple system (equal magnitudes, parallel lines of action, and opposite sense), the moment of the couple M_A is determined as:

$$M_A = F \times d = +(2k) \times (5 \text{ ft.}) = +10 \text{ k-ft.}$$

The sign for the direction of rotation is obtained through observation and, in this case, is counterclockwise.

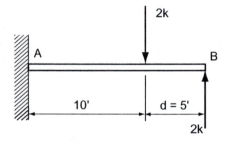

Figure 2.43   15-ft. cantilever beam.

As a check, the moment about *A* will be computed by algebraically summing the moments of the two forces.

$$M_A = -(2 \text{ k})(10 \text{ ft.}) + (2 \text{ k})(15 \text{ ft.}) = +10 \text{ k-ft.}$$

This solution verifies the moment computation for the couple system. Solving for the moment at the free end *B:*

$$M_B = F \times d = +(2 \text{ k})(5 \text{ ft.}) = +10 \text{ k-ft.}$$

Checking this answer by the more conventional moment calculation:

$$M_B = +(2 \text{ k})(5 \text{ ft.}) = +10 \text{ k-ft.}$$

The 2-kip load at **B** causes no rotation about **B** since its moment arm is equal to zero and thus does not appear in the moment equation.

Let's examine another case of this cantilever beam (Figure 2.44) that is extended another 5 feet. Again, compute the moment at the support **A** and at the free end **B**.

The moment of the couple is equal to:

$$M_A = + (2 \text{ k})(5 \text{ ft.}) = +10 \text{ k-ft.}$$

Checking this answer:

$$M_A = -(2 \text{ k})(15 \text{ ft.}) + (2 \text{ k})(20 \text{ ft.}) = +10 \text{ k-ft.}$$

The moment at the free end based on the couple is:

$$M_B = +(2 \text{ k})(5 \text{ ft.}) = +10 \text{ k-ft.}$$

Checking this answer:

$$M_B = +(2 \text{ k})(5 \text{ ft.}) = +10 \text{ k-ft.}$$

Notice that the moment of the couple is a constant, even if the reference point has moved. The sense of the moment must be obtained by visual inspection (using your intuition and judgment).

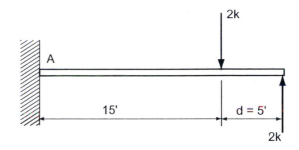

Figure 2.44   20-ft. cantilever beam.

Example Problem 2.20 (Figure 2.45)

A cantilevered truss supports three applied loads F_1, F_2, and F_3 as shown. Determine the resultant moment due to the three forces about support **A**. What is the moment due to the three forces if the moments are summed about support **B**?

Solution:

Resolve forces F_1 and F_2 into their respective x and y components. Note that the x-components of forces F_1 and F_2 form a couple system and the y-component of force F_2 and force F_3 also form a couple.

Summing moments due to the three applied forces about support point **A** result in the equation:

$$M_A = - \underset{\text{(couple)}}{(F_{1x})(6 \text{ ft.})} - \underset{\text{(couple)}}{(F_3)(16 \text{ ft.})} - (F_{1y})(16 \text{ ft.})$$

$$M_A = -(12 \text{ k})(6 \text{ ft.}) - (5 \text{ k})(16 \text{ ft.}) - (9 \text{ k})(16 \text{ ft.})$$
$$M_A = -(72 \text{ k-ft.}) - (80 \text{ k-ft.}) - (144 \text{ k-ft.})$$
$$M_A = -296 \text{ k-ft.}$$

Changing the reference point to **B:**

$$M_B = - \underset{\text{(couple)}}{(F_{1x})(6 \text{ ft.})} - \underset{\text{(couple)}}{(F_3)(16 \text{ ft.})} - (F_{1y})(16 \text{ ft.})$$

$$M_B = -(12 \text{ k})(6 \text{ ft.}) - (5 \text{ k})(16 \text{ ft.}) - (9 \text{ k})(16 \text{ ft.})$$
$$M_B = -296 \text{ k-ft.}$$

A check of the moments using each force component independently will result in the same answer for M_A and M_B.

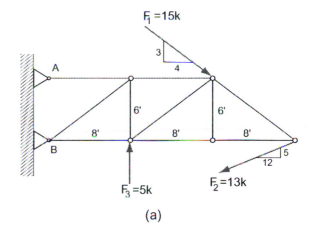

(a)

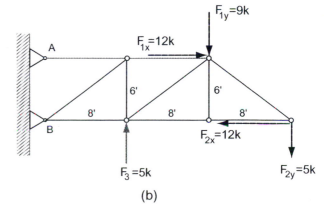

(b)

Figure 2.45   (a) Three forces on a cantilever truss. (b) Forces resolved into x and y components.

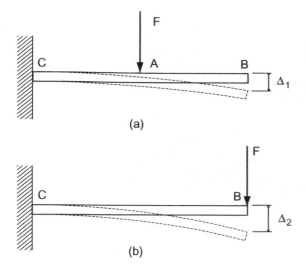

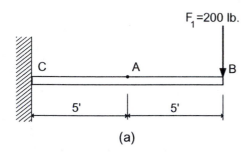

Figure 2.46 *Force moved to a new parallel line of action.*

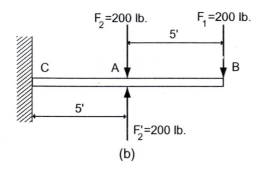

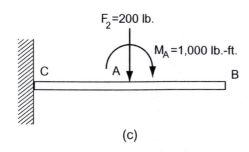

Figure 2.47 *Moving a force to another parallel line of action.*

Resolution of a Force into a Force and Couple Acting at Another Point

In the analysis of some types of problems, it may be useful to change the location of an applied force to a more convenient point on the rigid body. In a previous section we discussed the possibility of moving a force **F** along its line of action (principle of transmissibility) without changing the external effects on the body, as shown in Figure 2.8. However, we cannot move a force away from the original line of action without modifying the external effects on the rigid body, as shown in Figure 2.46.

Examination of Figure 2.46 shows that if the applied force **F** is changed from point **A** to point **B** on the cantilevered beam, differing deflections at the free end result. The deflection Δ_2 (**F** applied at point **B**) is considerably larger than Δ_1 (**F** applied at a point **A**).

Let's apply a force **F** at point **B**, as shown in Figure 2.47a. The objective is to have **F** moved to point **A** without changing the effects on the rigid body. Two forces F_2 and F_2' are applied at **A** in Figure 2.47b with a line of action parallel to that of the original force at **B**. The addition of the equal and opposite forces at **A** does not change the effect on the rigid body. We observe that the forces F_1 at **B** and F_2' at **A** are equal and opposite forces with parallel lines of action, thus forming a couple system.

By definition, the moment due to the couple is equal to $(F) \times (d)$ and is a constant value anywhere on the rigid body. The couple M_A can then be placed at any convenient location with the remaining force F_2 at **A**, as shown in Figure 2.47c.

As a verification of the procedure previously discussed, check that the moment at the support **C** remains the same between the diagram in Figures 2.47a and 2.47c.

In Figure 2.47a:

$$M_C = -(F_1) \times (10 \text{ ft.}) = -(200 \text{ lb.}) \times (10 \text{ ft.}) = -2000 \text{ lb.-ft.}$$

Checking Figure 2.47c:

$$M_C = -(F_2) \times (5 \text{ ft.}) - M_A$$

and

$$M_A = -(F) \times (d) = (200 \text{ lb.})(5 \text{ ft.}) = -1000 \text{ lb.-ft.}$$

Therefore

$$M_C = -(200 \text{ lb.})(5 \text{ ft.}) - (1000 \text{ lb.-ft.}) = -2000 \text{ lb.-ft.}$$

The preceding example may then be summarized as follows:

*Any force **F** acting on a rigid body may be moved to any given point **A** (with a parallel line of action), provided that a couple **M** is added. The moment **M** of the couple is equal to **F** times the perpendicular distance between the original line of action and the new location **A**.*

Example Problem 2.21 (Figure 2.48)

A pole *DBC* supports a light fixture that weighs 25 lbs. at the end of a 5-ft. cantilever arm and a signal light of 100 lbs., 10 ft. from **B**. In order to design the column and the supporting base connection at *C*, it is necessary to have the compressive force applied through the axis of the column. Show the equivalent force system when the forces W_1 and W_2 are moved to points *D* and *B*, respectively.

Solution:

As weight W_1 is moved to the column axis at *D*, it is accompanied by a moment couple M_1 equal to:

$$M_1 = -W_1 \times (5 \text{ ft.}) = -(25 \text{ lb.})(5 \text{ ft.}) = -125 \text{ lb.-ft.}$$

When the weight of the light fixture at *A* is moved to a parallel line of action at *B*, the moment couple M_2 that develops is equal to:

$$M_2 = -W_2 \times (10 \text{ ft.}) = -(100 \text{ lb.})(10 \text{ ft.}) = -1000 \text{ lb.-ft.}$$

The moment at the column base at *C* due to W_1 and W_2 is equal to the algebraic sum of the moment couples M_1 and M_2 and the total compressive force is equal to $W_1 + W_2$.

$$M_C = -M_1 - M_2 = -(125 \text{ lb.-ft.}) - (1000 \text{ lb.-ft.}) = -1125 \text{ lb.-ft.}$$

$$W_{Total} = W_1 + W_2 = 25 \text{ lb.} + 100 \text{ lb.} = 125 \text{ lb.}$$

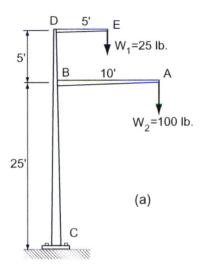

(a)

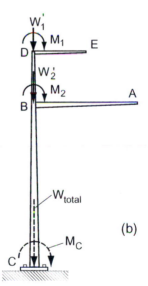

(b)

Figure 2.48   Equivalent force system on a luminaire pole.

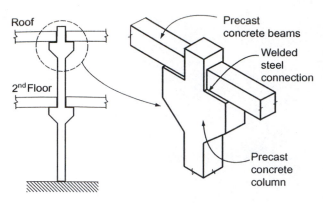

Figure 2.49   Precast concrete beam-column connection.

Example Problem 2.22 (Figures 2.49 and 2.50)

A major, precast-concrete column supports beam loads from the roof and second floor as shown. Beams are supported by seats projecting from the columns. Loads from the beams are assumed to be applied one foot from the column axis.

Determine the equivalent column load condition when all beam loads are shown acting through the column axis.

Solution:

The 12-k force produces a +12 k-ft. moment when moved to the column axis, while the 10-k force counters with a -10 k-ft. clockwise moment. The resultant moment equals +2 k-ft.

$$M_{Roof} = +(12 \text{ k-ft.}) - (10 \text{ k-ft.}) = +2 \text{ k-ft.}$$

The total compressive force at the roof level is equal to:

$$F_R = 12 \text{ k} + 10 \text{ k} = 22 \text{ k}$$

At the second-floor level:

$$M_2 +(20 \text{ k-ft.}) - (15 \text{ k-ft.}) =+ 5 \text{ k-ft.}$$
$$F_2 = 20 \text{ k} + 15 \text{ k} = 35 \text{ k}$$

The resultant effect due to the column loads at the base *A* equals:

$$M_A = + M_R + M_2 = +(2 \text{ k-ft.}) + (5 \text{ k-ft.}) = + 7 \text{ k-ft.}$$
$$F_A = -F_R - F_2 = -(22 \text{ k}) = (35 \text{ k}) = -57 \text{ k}$$

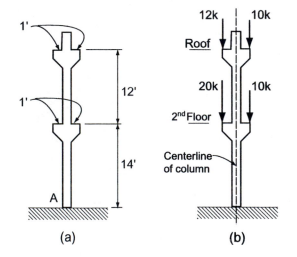

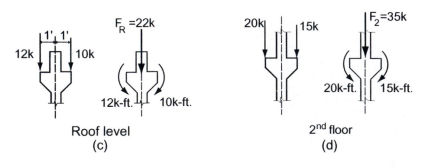

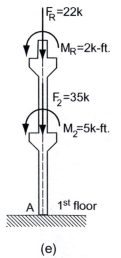

Figure 2.50   Loads and moments on a precast concrete column.

Resultant of Two Parallel Forces

Suppose we wish to represent the two forces A and B shown on the girder in Figure 2.51 with a single resultant force R, which produces an equivalent effect as the original forces. The equivalent resultant R must produce the same translational tendency as forces A and B as well as the same rotational effect, as shown in Figure 2.52.

Since forces by definition have magnitude, direction, sense, and a point of application, it is necessary to establish the exact location of the resultant R from some given reference point. Only a single location R will produce an equivalent effect as the girder with forces A and B.

The magnitude of the resultant R of the parallel forces A and B equals the algebraic summation of A and B, where $R = A + B$.

Direction of the forces must be accounted for by using a convenient sign notation, such as positive for upward-acting forces and negative for downward-acting forces.

Location of the resultant R is obtained by employing the principle of moments.

$$\sum M_C = -A \times (a) - B \times (L - b) = -R \times (x)$$

$$x = \frac{A \times (a) + B \times (L - b)}{R}$$

Example Problem 2.23 (Figure 2.53)

Determine the single resultant R (magnitude and location) that would produce an equivalent effect as the forces shown on the combined footing.

Solution:

The magnitude of the resultant is calculated as:

$$R = -20 \text{ kN} - 60 \text{ kN} = -80 \text{ kN}$$

To determine the location of R, select a convenient reference point and calculate moments about the point, where:

$$M_A = -(60 \text{ kN})(4 \text{ m}) = -240 \text{ kN-m}$$

The moment about point A due to R must be equal to the M_A of the original force system to maintain equivalence.

Therefore:

$$M_A = -(R) \times (x) = -(240 \text{ kN-m})$$

But $R = 80 \text{ k}$:

$$\therefore x = \frac{-240 \text{ kN-m}}{-80 \text{ kN}} = 3 \text{ m}$$

must be located 3 m to the right of point A.

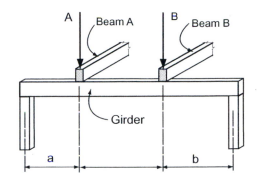

Figure 2.51   Two parallel forces acting on a girder.

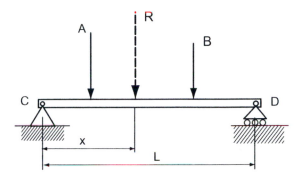

Figure 2.52   Equivalent resultant force R for forces A and B.

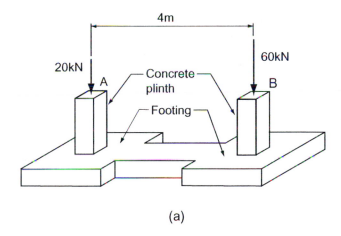

(a)

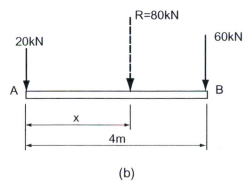

(b)

Figure 2.53   (a) Loads on a combined footing. (b) Equivalent resultant force R.

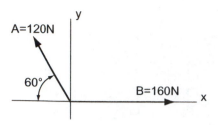

Problem 2.3.1

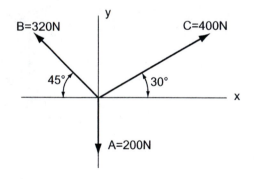

Problem 2.3.2

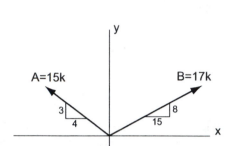

Problem 2.3.3

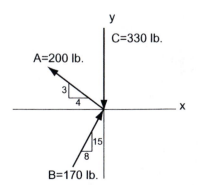

Problem 2.3.4

Supplementary Problems

Section 2.3—Vector Addition (Graphical Methods)

2.3.1   Determine the resultant (magnitude and direction) of the two forces shown using the graphical method. Scale: 1 mm = 2 N

2.3.2   Solve for the resultant (magnitude and direction) for forces *A* and *B*. Scale: $^1/_8$" = 1 k

2.3.3   Using the graphical (tip-to-tail) method, determine the resultant of the three forces shown. Use the sequence of: *A* to *B* to *C*. Scale: 1 mm = 4 N

2.3.4   Solve for the resultant using the tip-to-tail method following the sequence: *A* to *B* to *C*. Scale: $^1/_8$" = 10 lb.

2.3.5   Determine using the sequence F_1 to F_2 to F_3. Scale: 10 mm = 1 kN

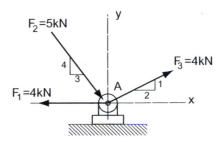

Problem 2.3.5

2.3.6   A utility pole is loaded with three forces as shown. Using the graphical tip-to-tail method, determine the force F_3 (horizontal) such that the resultant of all three forces is vertical. Also, what is the magnitude of the resultant? Suggested sequence: F_1 to F_2 to F_3. Scale: $1/4$" = 10 lb.

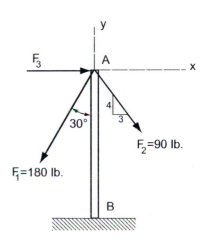

Problem 2.3.6

2.3.7   The resultant of the two forces F_1 and F_2 is vertical, down the axis of the pole AB. Determine the magnitude of F_2. Scale: 1 mm = 40 N

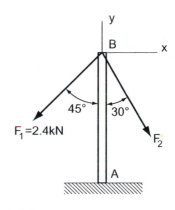

Problem 2.3.7

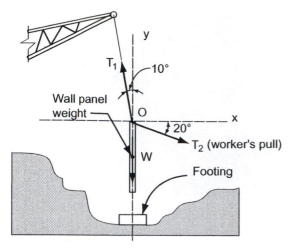

Problem 2.3.8

2.3.8 A precast concrete wall panel is being hoisted into place as shown. The wall weighs 1200 lb. with the weight passing through its center through point *O*. Determine the force T_2 necessary for the workers to guide the wall into place. Scale: $\frac{1}{4}$" = 400 lb.

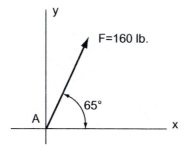

Problem 2.3.9

2.3—Resolution of forces into x and y components

2.3.9 Resolve, analytically, the force *F* into its *x* and *y* components.

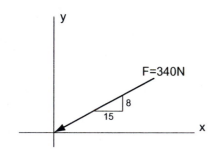

Problem 2.3.10

2.3.10 Resolve the compression force *F* into its horizontal and vertical components.

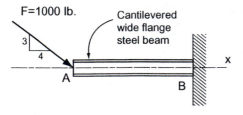

Problem 2.3.11

2.3.11 Determine the *x* and *y* components of force *F* shown.

2.3.12 If a hook can sustain a maximum withdrawal force of 1 kN in the vertical direction, determine the maximum tension T that can be exerted.

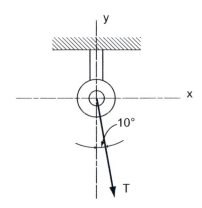

Problem 2.3.12

2.3.13 A roof purlin, supported by a rafter, must support a 300-lb. vertical snow load. Determine the components of P, perpendicular and parallel to the axis of the rafter.

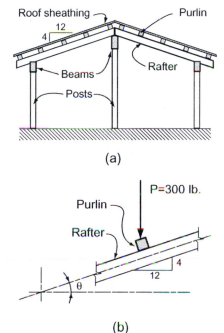

(a)

(b)

Problem 2.3.13

2.3.14 Resolve the 600-lb. forces into two components; one perpendicular to the rafter and the other parallel to the rafter.

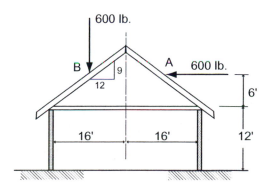

Problem 2.3.14

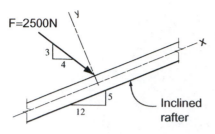

Problem 2.3.15

2.3.15 An inclined force F pushes on a rafter at a 5:12 slope. Resolve the force F into x and y components relative to the inclined rafter. Assume that the axis of the rafter is the x axis.

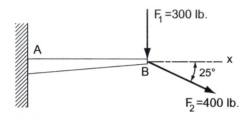

Problem 2.3.16

2.3—Vector Addition (Analytical Method)

2.3.16 Determine the resultant of forces F_1 and F_2 at the overhang end **B**. Check your results graphically. Scale: $\frac{1}{2}$" = 100 lb.

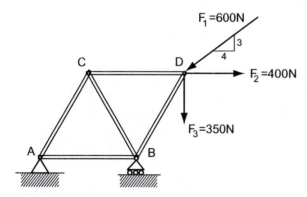

Problem 2.3.17

2.3.17 Find the resultant of the three forces applied at joint **D** of the truss shown.

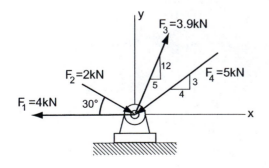

Problem 2.3.18

2.3.18 An anchoring bracket supports four forces as shown. Determine the resultant that is the equivalent force of all four forces.

2.3.19 Three members of a truss frame into a steel gusset plate as shown. All forces are concurrent at point *O*. Determine the resultant of the three forces that must be carried by the gusset plate.

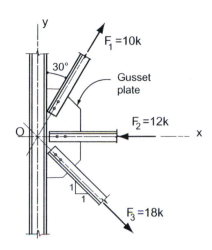

Problem 2.3.19

2.3.20 This is the same problem as in Problem 2.3.6. Using the analytical method, determine the force F_3 such that the resultant of all three forces is vertical. What is the magnitude of the resultant?

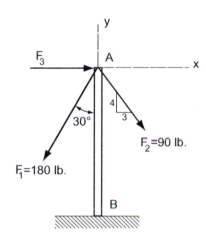

Problem 2.3.20

2.3.21 Determine the resultant at point *D* (which is supported by the crane's mast) if *AD* = 6 k, *BD* = 3 k, and *CD* = 7.5 k.

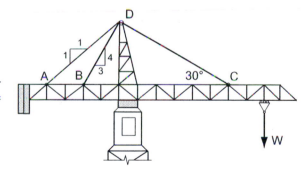

Problem 2.3.21

2.3.22 One end of a timber roof truss is supported on a brick wall but not securely fastened. The reaction of the wall is only vertical. Assuming that the maximum capacity of either the inclined or horizontal member is 600 lb., determine the maximum magnitudes of F_1 and F_2 such that their resultant is vertical through the brick wall.

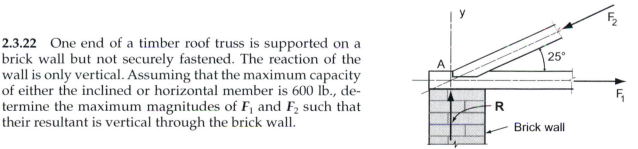

Problem 2.3.22

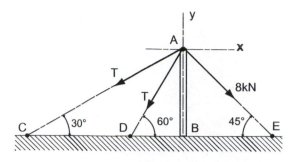

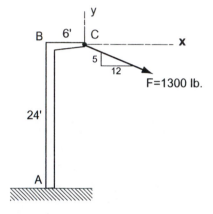

Problem 2.3.23

2.3.23 The resultant of three tensions in the guy wires anchored at the top of the tower is vertical. Find the unknown but equal tensions T in the two wires. All three wires and the tower are in the same vertical plane.

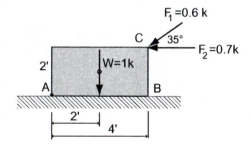

Problem 2.4.1

F=1300 lb.

Section 2.4—Moment of a Force—Principle of Moments

2.4.1 A bent strain pole supports a cable force F, inclined at a 5:12 slope. Determine the moment at points B and A due to the applied force.

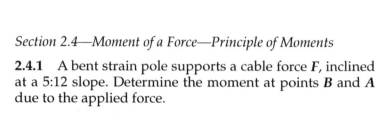

Problem 2.4.2

2.4.2 A 1000-lb. crate is subjected to two applied forces at C. Determine the moment about points A and B due to forces F_1, F_2, and the weight W.

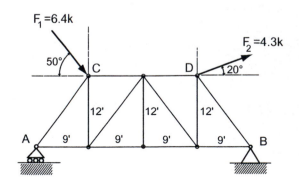

Problem 2.4.3

2.4.3 A roof truss is subjected to two forces as shown. Determine the moment due to forces F_1 and F_2 about support A and support B.

2.4.4 A painter is standing at midheight on a ladder inclined at an angle of 65° from the horizontal. Determine the horizontal force B_x (reaction from the wall surface) necessary such that the resultant moment at A is equal to zero.

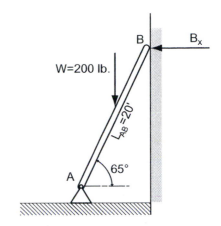

Problem 2.4.4

2.4.5 A gravity dam weighing 8 k is subjected to water pressure on both sides. Determine the resultant moment about the "toe" of the dam at B.

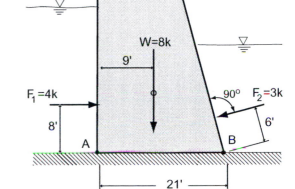

Problem 2.4.5

2.4.6 Determine the resultant moment at support points A and B due to the forces acting on the truss as shown.

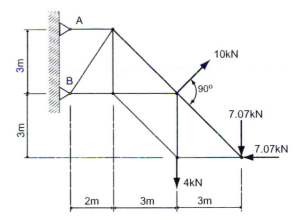

Problem 2.4.6

2.4.7 A hoisting boom is configured as shown and supports a 1000-lb. load at an angle of 20° off of vertical. Determine the moment at supports A and B.

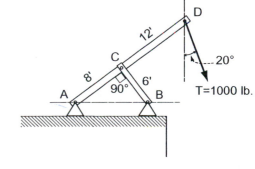

Problem 2.4.7

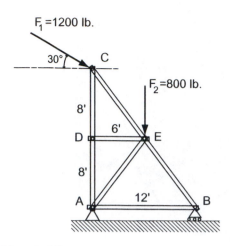

$F_1 = 1200$ lb.

30°

C

8'

6'

D   E

8'

A   12'   B

$F_2 = 800$ lb.

Problem 2.4.8

2.4.8 A vertical truss supports two applied forces F_1 and F_2. Determine the moment at supports A and B.

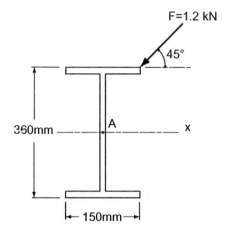

F=1.2 kN

45°

360mm   A   x

150mm

Problem 2.4.9

2.4.9 Determine the moment due to force F about the center point A of a wide-flange beam cross-section. The resulting moment is a twisting action on the beam section known as *torsion*.

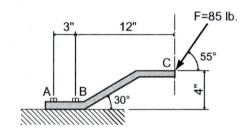

3"   12"   F=85 lb.

C   55°

A   B   4"

30°

Problem 2.4.10

2.4.10 An 85-lb. force is applied to the bent plate as shown. Determine an equivalent force-couple system at A and at B.

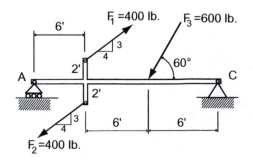

6' $F_1 = 400$ lb. $F_3 = 600$ lb.

3

4

2'   60°

A   C

2'

3   6'   6'

4

$F_2 = 400$ lb.

Problem 2.4.11

Section 2.4—Moment of a Couple(s)

2.4.11 Three forces are acting on the beam as shown. Determine the resultant moment about A from forces F_1, F_2, and F_3. What is the moment at support C?

2.4.12   A very unusual cantilever beam supports five applied forces as shown. Determine the resultant moment at support *A*.

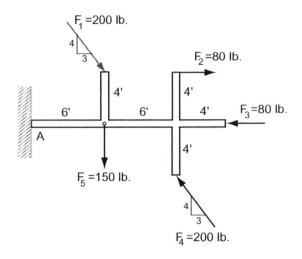

Problem 2.4.12

Section 2.4—Resultant of Parallel Forces

2.4.13   Determine the magnitude of the resultant of the parallel force system shown. Locate the resultant with respect to support point *A*.

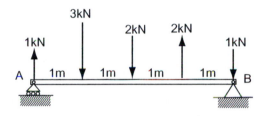

Problem 2.4.13

2.4.14   A steel column supports a load from a crane runway girder in an industrial building. The load is eccentrically placed 15 in. from the centerline of the column. Replace the force *P* with a force and couple at the column's centerline.

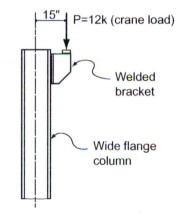

Problem 2.4.14

2.4.15   Determine the magnitude of force F_3 if the resultant of all three forces is directly above the support at *B*.

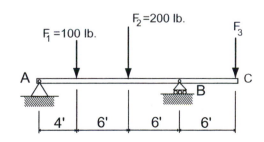

Problem 2.4.15

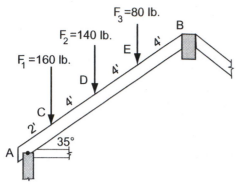

Problem 2.4.16

2.4.16 A 14-ft. rafter supports three vertical loads as shown. Determine the resultant of the parallel force system relative to the support point *A*.

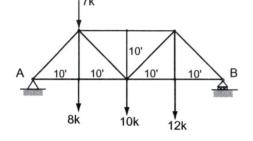

Problem 2.4.17

2.4.17 The 40-ft. span truss supports four vertical loads at the panel points. Determine the resultant magnitude and locate its position relative to the support at *A*.

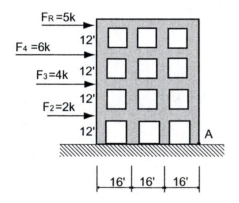

Problem 2.4.18

2.4.18 A four-story building is subjected to wind forces concentrated at each of the floor levels. Solve for the resultant wind force and its location above the ground.

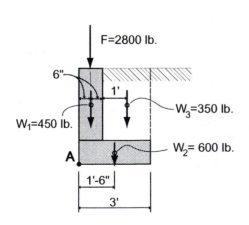

Problem 2.4.19

2.4.19 Determine the resultant of the four loads acting on a footing support as shown. Locate the resultant force relative to the toe of the footing at *A*.

3
Equilibrium of Two-Dimensional Systems

3.1 EQUILIBRIUM OF A PARTICLE— CONCURRENT FORCE SYSTEM

Equilibrium refers essentially to a state of rest or balance. Recall Newton's first law, which states:

> *Any body at rest will remain at rest and any body in motion will move uniformly in straight lines unless acted upon by a force.*

The concept of a body or particle at rest unless acted upon by some force indicates an initial state of *static equilibrium*, whereby the net effect of all forces on the body or particle is zero. Equilibrium or nonmotion (static) is simply a special case of motion (Figures 3.1 and 3.2). In the subsequent chapters all of the problems are presumed to be in a state of rest or static equilibrium.

The mathematical requirement necessary to establish a condition of two-dimensional equilibrium can be stated as:

$$R_x = \sum F_x = 0$$
$$R_y = \sum F_y = 0$$
$$M_i = \sum M_i = 0$$

Various types of problems require the selection of only one or maybe all of the equations of equilibrium. However, for any one particular type of problem, the minimum number of equations of equilibrium necessary to justify a state of balance is also the maximum number of equations of equilibrium permitted. Since various force systems require differing types and numbers of equations of equilibrium, each will be discussed separately.

Collinear Force System

A collinear force system involves the action of forces along the same line of action. There is no restriction on the direction or on the magnitude of each force as long as all forces act along the same line.

An example of a collinear force system is a construction crane supporting three weights in which no movement is taking place, as shown in Figure 3.3. If we assume that the

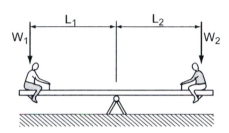

Figure 3.1 $W_1 = W_2$; $L_1 = L_2$ $W_1 x L_1 = W_2 x L_2$ *(Rotational equilibrium).*

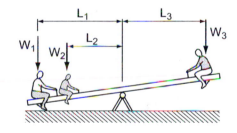

Figure 3.2 $W_1 x L_1 + W_2 x L_2 > W_3 x L_3$ *Rotational imbalance (nonequilibrium).*

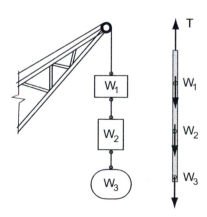

Figure 3.3 *Three loads supported by a crane.*

force exerted by each of the weights and the resisting cable force is along the axis of the vertical cable, then all forces are collinear. In equation form:

$$\sum F_y = -W_1 - W_2 - W_3 + T = 0$$

$$\therefore T = W_1 + W_2 + W_3$$

Concurrent Force System—Equilibrium of a Particle

In the preceding chapter, we discussed the graphical as well as analytical methods for determining the resultant of several forces acting on a particle. In most problems, there exists the condition where the resultant of several concurrent forces acting on a body or particle is zero. For these cases, we say that the body or particle is in equilibrium. The definition of this condition may be stated as follows:

When the resultant of all concurrent forces acting on a particle is zero, the particle is in a state of equilibrium.

An example of a coplanar, concurrent force system is a weight suspended from two cables, as shown in Figure 3.4. Cable forces *CA*, *CB*, and *CD* intersect at a common point *C*. Using the concurrent point *C* as the origin of an *x-y* co-ordinate system, a force diagram (Figure 3.5) of the forces at *C* is drawn.

We found in Section 2.3 that by resolving each force (for a series of concurrent forces) into their respective primary *x* and *y* components, we can algebraically determine the resultant R_x and R_y for the system. In order to justify a condition of equilibrium in a coplanar (two-dimensional), concurrent force system, two equations of equilibrium are written to show that the resultants in the *x* and *y* directions are equal to zero.

$$R_x = \sum F_x = 0$$
$$R_y = \sum F_y = 0$$

These two conditions must be satisfied before equilibrium is established and no translation in either the *x* or *y* direction is permitted. In a concurrent force system, since all forces intersect at a common point, the tendency for rotation does not exist.

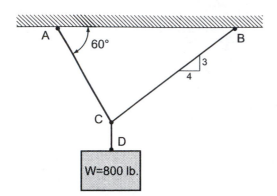

Figure 3.4   Three cables concurrent at C.

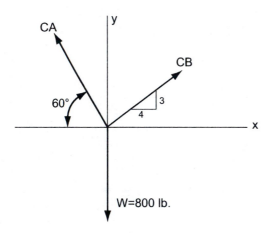

Figure 3.5   Force diagram of the cable forces.

Free-Body Diagram of Particles

An essential step in solving equilibrium problems involves the drawing of free-body diagrams. The free-body diagram, or FBD, is the essential key to modern mechanics. Everything in mechanics is reduced to forces in a free-body diagram. This method of simplification is very efficient in reducing apparently complex mechanisms into concise force systems.

What, then, is a free-body diagram of a particle? The free-body diagram of a particle simply shows all of the concurrent forces emanating from the concurrent point. An $x-y$ coordinate axis system is overlayed onto the forces, indicating the direction and magnitude of each force, if they are known. The equations of equilibrium are written based on the information provided by the free-body diagram. Examples of free-body diagrams of concurrent forces systems are shown in Figure 3.6b and Figure 3.7b.

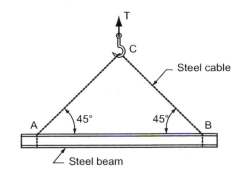

Figure 3.6   (a) Beam being hoisted by a crane.

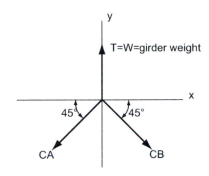

Figure 3.6   (b) FBD of the concurrent point C.

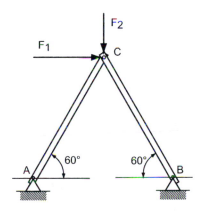

Figure 3.7   (a) A-frame supporting two forces at C.

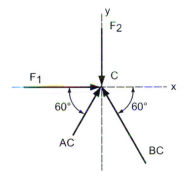

Figure 3.7   (b) FBD of the concurrent joint C.

Example Problem 3.1: Equilibrium of a Particle

Two cables, shown in Figure 3.8, are used to support a weight $W = 800$ lb., suspended at concurrent point C. Determine the tension developed in cables CA and CB for the system to be in equilibrium. Solve this problem analytically and check the answer graphically.

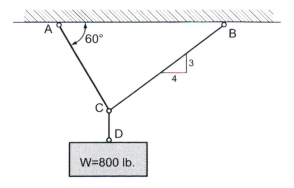

Figure 3.8   Three cables concurrent at C.

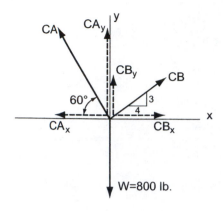

Figure 3.9 Analytical solution.

Solution: (Analytical—Figure 3.9)

a. The first step is the construction of the FBD showing all forces present at the concurrent point *C*.

b. Using the *x-y* coordinate axis system, resolve each angular force into their respective *x* and *y* components. Remember that forces directed to the right and up are positive in direction and forces to the left or down are assigned a negative sign.

$$CA_x = -CA \cos 60° = -0.5CA$$

$$CA_y = +CA \sin 60° = +0.866CA$$

$$CB_x = +\frac{4}{5}CB = +0.8CB$$

and

$$CB_y = +\frac{3}{5}CB = +0.6CB$$

c. For point (particle) *C* to be in equilibrium, the two force equations of equilibrium must be satisfied.

$$\sum F_x = -CA_x + CB_x = 0$$

$$-0.5CA + 0.80CB = 0 \ldots \text{ Eq. (1)}$$

and

$$\sum F_y = +CA_y + CB_y - W = 0;$$

$$+0.866CA + 0.6CB - 800 \text{ lb.} = 0 \ldots \text{ Eq. (2)}$$

d. Solving the two equations simultaneously:

Equation (1): $-0.5CA + 0.80CB = 0$

$$\therefore CA = +\frac{(.80)}{(.50)}CB = +1.6CB$$

Substituting into Equation (2):

$+0.866(1.6CB) + 0.6CB = +800$ lb.

Simplifying:

$$+1.39CB + .6CB = +800 \text{ lb};$$

$$\therefore CB = +\frac{800 \text{ lb.}}{1.99} = 402 \text{ lb.}$$

Solving for cable force *CA:*

$$CA = 1.6CB = 1.6(402 \text{ lb.}) = 643 \text{ lb.}$$

Perhaps a more convenient method of accounting for force components prior to writing the equations of equilibrium is to construct a table.

Force	F_x	F_y
CA	$CA_x = -CA \cos 60°$ $= -0.5CA$	$CA_y = +CA \sin 60°$ $= +0.866CA$
CB	$CB_x = +\dfrac{4}{5}CB = +.8CB$	$CB_y = +\dfrac{3}{5}CB = +.6CB$
W	0	-800 lb.

The two equations of equilibrium are then written by summing vertically all forces listed under the F_x column, and similarly for the F_y column.

Solution: (Graphical—Figure 3.10)

In the graphical solution, either the tip-to-tail or the parallelogram method may be employed. Generally, the tip-to-tail method is much more direct in solving the problem and will be the method illustrated in the subsequent examples.

Tip-to-tail method: (Scale: $^1/_4$" = 100 lb.)

a. Begin the solution by establishing a reference origin at point C (the point of concurrency) on the $x - y$ coordinate axis.

b. Draw the weight $W = 800$ lb. (or CD) to scale in the given downward direction.

c. To the tip of the first force W, place the tail of the second force CA. Draw CA at a 60° inclination from the horizontal. Since the magnitude of CA is still unknown, we are yet unable to terminate the force. Only the line of action of the force is known.

d. Equilibrium is established in the graphical solution when the tip of the last force closes on the tail of the first force W. Therefore, the tip of force CB must close at the origin point C. Construct the line of action of force CB at a slope of 3:4 from the horizontal. The intersection of lines CA and CB defines the limits of each force. Scale off the magnitudes of CB and CA.

If the drawing is properly done, the same values for cable tensions CA and CB should result.

$CA \cong 640$ lb. and $CB \cong 400$ lb.

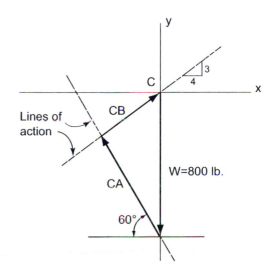

Figure 3.10　Graphical tip-to-tail solution.

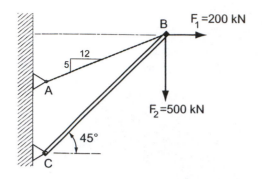

Figure 3.11 *A boom and cable with two loads.*

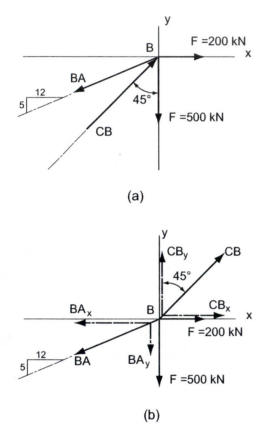

Figure 3.12 *(a) FBD of point B. (b) FBD of point B using the principle of transmissibility for force CB.*

Example Problem 3.2

A strut *CB* is guyed back to a wall support through cable *BA*, as shown in Figure 3.11. Two forces are applied to the concurrent joint at **B**. Determine the tension in cable *BA* and the force in member *CB* for a condition of equilibrium to occur. Check your results graphically.

Solution: (Analytical)

The first step in any of the equilibrium problems is the construction of the free-body diagram. Isolate the concurrent joint and show all of the forces acting at the point. Intuitively, guess at the direction of each unknown force and show the corresponding sense of the force on the FBD. Incorrect assumptions about the unknown force directions will be revealed at the end of the process when numeric answers are found using the equations of equilibrium. Answers yielding a negative sign simply mean that the direction of the force was incorrectly assumed in the initial FBD, however, the magnitude remains unaffected.

In this example, the cable will be assumed in tension and the strut *CB* in compression. An arrowhead shown pulling away from the concurrent point designates a tension force while an arrowhead directed at the point is considered in compression. An illustration of this convention is shown in the FBD of concurrent point **B**. In Figure 3.12, FBD(a), the arrowhead for force *CB* is shown pushing at point **B** to indicate compression. However, in the alternate FBD(b), the force *CB* has been moved along its line of action (the principle of transmissibility) and appears to be a pulling or tension force. Sometimes this technique is used to simplify the FBD when showing all of the component forces acting at the concurrent point. However, if this is too confusing, stick with the original FBD(a).

Step two is the resolution of all angled forces into their respective *x* and *y* components. It is sometimes convenient to record the component forces in a tabular manner as a way of keeping track of all of the forces.

Force	F_x	F_y
BA	$-\dfrac{12}{13}BA$	$-\dfrac{5}{13}BA$
CB	$+CB\cos 45° = +0.707CB$	$+CB\sin 45° = +0.707CB$
F_1	+200 kN	0
F_2	0	−500 kN

Write the two equations of equilibrium and solve for the unknown forces *BA* and *CB*.

$$\sum F_x = -\frac{12}{13}BA + 0.707CB + 200 \text{ kN} = 0 \quad \dots \text{Eq. (1)}$$

$$\sum F_y = -\frac{5}{13}BA + 0.707CB - 500 \text{ kN} = 0 \quad \dots \text{Eq. (2)}$$

Solve both equations simultaneously.

Multiply Eq. (1) × (−1):

$$+\frac{12}{13}BA - 0.707CB - 200\text{ kN} = 0$$

Adding both Equations (1) and (2) together:

$$-\frac{7}{13}BA = -700\text{ kN}$$

$$\therefore BA = \frac{700(13)}{7} = +1300\text{ kN}$$

Note that BA was initially assumed as a tension force and the resulting answer was positive, indicating that the assumption was correct. BA is a tension member with a magnitude of 1300 kiloNewtons.

Substituting the value of BA into Equation (1) or (2):

$$+\frac{12}{13}(1300\text{ kN}) - 0.707CB = +200\text{ kN}$$

$$\therefore CB = \frac{1200\text{ kN} - 200\text{ kN}}{0.707} = 1414\text{ kN}$$

The direction for CB was also assumed correctly and, indeed, CB is in compression.

Solution (Graphical tip-to-tail: Scale: 1 mm = 10 kN):

Begin the graphical solution by drawing the forces that have known magnitudes and directions (see Figure. 3.13). This solution will construct the tip-to-tail diagram following the sequence:

$$F_1 \rightarrow F_2 \rightarrow BA \rightarrow CB.$$

For equilibrium to exist, a closed-force polygon results from the tip-to-tail method with the arrow of the last first ending at the tail of the first force (the beginning point).

Draw the horizontal force F_1 to scale (20 mm in length). To the tip of force F_1 place the tail of the second force F_2. Force F_2 is a vertically downward force that measures 50 mm in length. Force BA has a known direction with a slope of 5:12. Since the magnitude of force BA is still unknown, only the line of action of the force can be drawn. For equilibrium to be established, the last force CB must close at the origin point B. Draw force CB with a line of action at 45° from the horizontal. The intersection of the lines BA and CB defines the limits of the forces. Magnitudes of BA and CB are obtained by scaling the respective force lines.

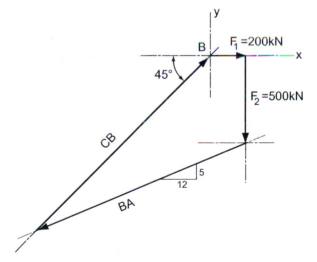

Figure 3.13   Graphical tip-to-tail.

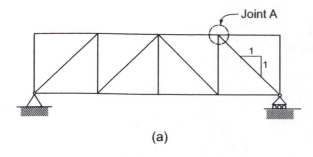

(a)

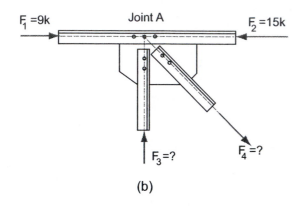

(b)

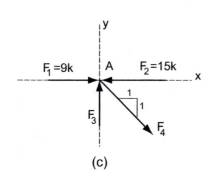

(c)

Figure 3.14   (a) Horizontal roof truss. (b) Detail of joint A. (c) FBD of joint A.

Example Problem 3.3

Two truss member forces of magnitude $F_1 = 9$ kips and $F_2 = 15$ kips are applied to a bolted connection as shown in Figure 3.14b. Knowing that the connection is in equilibrium, determine the magnitudes of member forces F_3 and F_4. If the bolts used for this connection have a capacity of 5 kips each, how many bolts would you specify for connecting members F_3 and F_4? Check the problem graphically.

Solution (Analytical—Figure 3.14):

Begin by constructing the FBD of the concurrent joint at A. Then, resolve all forces into x and y directions.

Force	F_x	F_y
F_1	+9 k	0
F_2	−15 k	0
F_3	0	+F_3
F_4	+0.707 F_4	−0.707 F_4

Writing the two equations of equilibrium:

$$\sum F_x = +9\,k - 15\,k + 0.707F_4 = 0;$$

$$\therefore F_4 = +\frac{6\,k}{0.707} = +8.5\,k$$

$$\sum F_y = +F_3 - 0.707\left(\underset{F_4}{8.5\,k}\right) = 0$$

$$\therefore F_3 = +6\,k$$

Member F_3 is a compression force and F_4 is in tension.

The number of bolts for each member are determined by dividing the member force by the bolt capacity of 5 kips.

For member $F_3 = 6$ k:

$$\text{Number of bolts "}n\text{"} = \frac{6\,k}{5\,^k/_{bolt}}$$

$$= 1.2\ \text{bolts} \to \text{Use 2 bolts}$$

For member $F_4 = 8.5$ k:

$$n = \frac{8.5\,k}{5\,^k/_{bolt}} = 1.7\ \text{bolts} \to \text{Use 2 bolts}$$

Example Problem 3.4

Determine the maximum weight W that can be safely supported by the cable system shown in Figure 3.15 if cables AB and BC have a breaking strength of 500 pounds each.

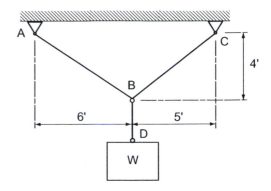

Figure 3.15   Cable system supporting a weight W.

Solution: (Analytical—Figure 3.16)

Directions for cables AB and BC are defined by the dimensions shown. Determine the length of each cable and establish the slope relationship for each cable. Construct the FBD of the concurrent system at B and resolve each force into its respective x and y components.

Write the two equations of equilibrium and solve for the two unknowns W and BA or BC.

Determining the lengths of BA and BC:

$$BA = \sqrt{(6)^2 + (4)^2} = \sqrt{52} = 7.2 \text{ ft.}$$

$$BC = \sqrt{(5)^2 + (4)^2} = \sqrt{41} = 6.4 \text{ ft.}$$

Organizing the components into a tabular format:

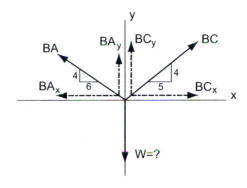

Figure 3.16   FBD of concurrent point B.

Force	F_x	F_y
BA	$-\dfrac{6}{7.2}BA$	$+\dfrac{4}{7.2}BA$
BC	$+\dfrac{5}{6.4}BC$	$+\dfrac{4}{6.4}BC$
W	0	$-W$

$$\sum F_x = \left(\frac{-6}{7.2}\right)BA + \left(\frac{5}{6.4}\right)BC = 0$$

$$\therefore -0.83BA + 0.78BC = 0$$

Simplifying,

$$0.78BC = 0.83\,BA; \quad \therefore BC = 1.06BA$$

This relationship is key in determining the cable that will be at the maximum 500-lb. capacity. Since $BC > BA$, then the 500-lb. capacity must be assigned to BC such that BA remains under the 500-lb. limit. Therefore:

$$BC = 500 \text{ lb.} \quad \text{and} \quad BA = \frac{BC}{1.06}$$

$$\therefore BA = \frac{500 \text{ lb.}}{1.06} = 471.7 \text{ lb.}$$

The weight W can now be determined by using the second equation of equilibrium.

$$\sum F_y = \left(\frac{+4}{7.2}\right)\underbrace{(471.7 \text{ lb.})}_{BA} + \left(\frac{+4}{6.4}\right)\underbrace{(500 \text{ lb.})}_{BC} - W = 0$$

$$0.555(471.7 \text{ lb.}) + 0.625(500 \text{ lb.}) - W = 0$$

$$\therefore W = 262 \text{ lb.} + 312.5 \text{ lb.} = 574.5 \text{ lb.}$$

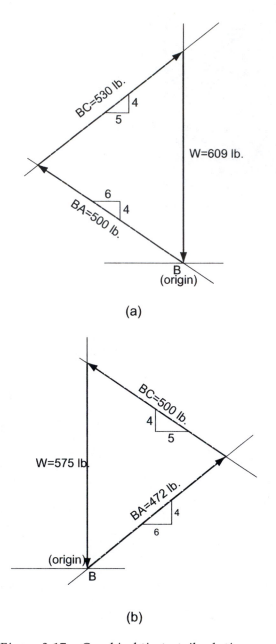

(a)

(b)

Figure 3.17   Graphical tip-to-tail solution.

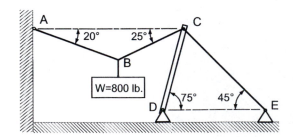

Figure 3.18   Compound cable arrangement.

Solution (Graphical—Figure 3.17: Scale: $\frac{1}{2}$" = 100 lb.):

In using the tip-to-tail method to solve this problem graphically, only two unknowns can exist, consistent with the two equations of equilibrium allowed in the analytical method. The two unknowns are either magnitude and/or direction. In this example, **W** has a known direction but the magnitude is unknown. Directions for *BA* and *BC* are also known but the magnitude of one of them is unknown. One cable will be assigned the 500-lb. capacity while the other cable will develop a magnitude less than the 500-lb. limit.

Begin the procedure by assigning one of the cables to the 500-lb. limit. Assume *BA* = 500 lb. and construct the force polygon as shown in Figure 3.17a. The sequence of construction is: *BA* = 500 lb. → *BC* → *W*.

By scaling the magnitude of force *BC* we find that the magnitude exceeds the maximum capacity of 500 pounds. Therefore, the solution in Figure 3.17a is invalid. In the diagram shown in Figure 3.17b, force *BC* is assigned the 500-lb. limit. This results in the correct solution since *BA* = 472 lb., less than the 500-lb. maximum requirement.

The solution in Figure 3.17b is correct, with *BC* = 500 lb., *BA* = 472 lb., and *W* = 575 lb.

Example Problem 3.5

A compound cable system supports a weight *W* = 800 lb. at point **B**, as shown in Figure 3.18. Cable *BA* is attached to a wall support at *A* and concurrent point *C* is supported by a compression strut *DC*. Determine all of the cable forces and the compression in strut *DC*.

Solution:

This problem involves two distinct concurrent points, one at **B** and the other at **C**. All cable forces and the strut force in *DC* do not have a single, common point of concurrency. Therefore, this problem must be solved as two separate equilibrium of a particle conditions. The solution begins by constructing the FBD of the concurrent point that contains only two unknowns. In this case, begin with point **B** where the direction and magnitude of the weight **W** is known (Figure 3.19a).

Force	F_x	F_y
BA	$-BA \cos 20° = -.94BA$	$+BA \sin 20° = +.342BA$
BC	$+BC \cos 25° = +.906BC$	$+BC \sin 25° = +.423BC$
W	0	-800 lb.

Write the two equations of equilibrium:

$$\sum F_x = -0.94BA + 0.906BC = 0 \dots (1)$$

$$\sum F_y = +0.342BA + 0.423BC - 800 \text{ lb.} = 0 \dots (2)$$

Solving the two equations simultaneously:

Eq. (1) × (.342): $-0.321BA + 0.310BC = 0$

Eq. (2) × (.940): $+0.321BA + 0.398BC - 752 \text{ lb.} = 0$

Adding the two equations together:

$+0.708BC = +752$ lb.;

$$\therefore BC = +\frac{752 \text{ lb.}}{0.708} = +1062 \text{ lb.}$$

Substituting back into Equation (1):

$+0.321BA = +0.310BC;$

$$\therefore BA = +\frac{0.31(1062 \text{ lb.})}{0.321} = +1026 \text{ lb.}$$

Now that cable forces BA and BC are known, move to concurrent point **C** and solve for the two unknowns CE and DC. Refer to the FBD in Figure 3.19b.

Force	F_x	F_y
CE	$+CE \cos 45° = +.707CE$	$-CE \sin 45° = -.707CE$
CB	$-.906\left(\underset{CB}{1062 \text{ lb.}}\right)$ $= -962$ lb.	$-0.423\left(\underset{CB}{1062 \text{ lb.}}\right)$ $= -449$ lb.
DC	$+DC \sin 15° = +.259DC$	$+DC \cos 15° = +.966DC$

Writing the equations of equilibrium:

$$\sum F_x = +0.707CE - 962 \text{ lb.} + 0.259DC = 0 \dots (1)$$

$$\sum F_y = -0.707CE - 449 \text{ lb.} + 0.966DC = 0 \dots (2)$$

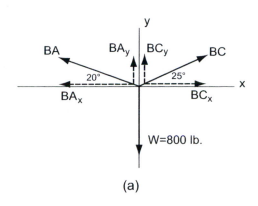

(a)

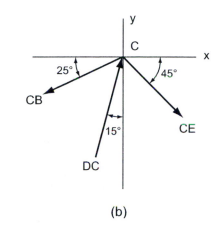

(b)

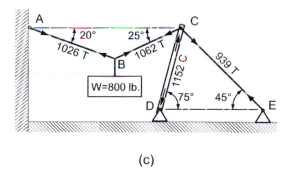

(c)

Figure 3.19   (a) FBD of point B. (b) FBD of point C. (c) Force summation diagram.

Adding the two equations together, we obtain:

$$1.225DC = 1411 \text{ lb.} \quad \therefore DC = +\frac{1411 \text{ lb.}}{1.225} = +1152 \text{ lb.}$$

Solving for CE:

$$+0.707CE + .259(1152 \text{ lb.}) = +962 \text{ lb.}$$

$$+0.707CE = 664 \text{ lb.}$$

$$\therefore CE = +\frac{664 \text{ lb.}}{0.707} = +939 \text{ lb.}$$

3.2   EQUILIBRIUM OF A RIGID BODY—

Nonconcurrent, Coplanar Force System

This section will examine the equilibrium of a rigid body subjected to forces as well as moments (couples). A rigid body is assumed as a system comprised of an infinite number of particles and takes the form of beams, trusses, columns, walls, and other building elements.

A rigid body is considered to be in equilibrium when the resultant of all external forces and moments acting on it is equivalent to zero (Figure 3.20a). Failure to provide equilibrium for a system may result in disastrous consequences (Figure 3.20b). Mathematically, equilibrium of a rigid body may be written as:

$$\sum F_x = 0; \qquad \sum F_y = 0; \qquad \sum M_i = 0$$

where i = any point on the rigid body.

These equations are necessary and sufficient to justify a state of equilibrium. Since only three equations may be written for the coplanar system, no more than three unknowns can be solved.

Alternate sets of equilibrium equations may be written for a rigid body; however, it is required that there always be three equations of equilibrium. One force equation with two moment equations or three moment equations represents alternate sets that are valid.

$$\sum F_x = 0; \qquad \sum M_i = 0; \qquad \sum M_j = 0$$

or

$$\sum F_y = 0; \qquad \sum M_i = 0; \qquad \sum M_j = 0$$

or

$$\sum M_i = 0; \qquad \sum M_j = 0; \qquad \sum M_k = 0$$

where i, j, and k are three different reference points on the rigid body.

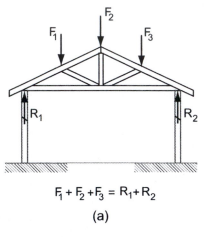

$$F_1 + F_2 + F_3 = R_1 + R_2$$

(a)

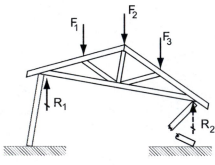

$$F_1 + F_2 + F_3 > R_1 + R_2$$

(b)

Figure 3.20   (a) A structure in equilibrium. (b) Non-equilibrium or failure condition.

Free-Body Diagrams of Rigid Bodies

Free-body diagrams of rigid bodies include a system of forces that no longer have a single point of concurrency. Forces are nonconcurrent but remain coplanar in a two-dimensional system. The magnitudes and directions of the known external forces should be clearly indicated on free-body diagrams (Figure 3.21).

Forces that are normally considered acting on a rigid body are as follows:

- Externally applied forces.
- Weight of the rigid body, if significant.
- Reaction forces or constraints.
- Externally applied moments.
- Moment reactions or constraints.
- Forces developed within a sectioned member.

Unknown external forces, usually the support reactions or constraints, develop on the rigid body to resist translational and rotational tendencies. The type of reaction offered by the support depends on the constraint condition. Some of the most commonly used support constraints are summarized in Table 3.1 and illustrated in Table 3.2.

Note: In the drawing of the FBDs, the author will place a "hash" mark on force arrows that denote reactions. This helps to distinguish the reaction forces from applied forces and loads.

Most material dealt with in this text will be assumed weightless unless otherwise specified. Whenever the rigid body weight is significant in a problem, one can easily include it in the calculations by adding another force passing through the centroid (center of gravity) of the rigid body (Figure 3.22).

When the sense of the reacting force or moment is not apparent, arbitrarily assign a direction to it. If your assumption happens to be incorrect, the calculated answer(s) in the equilibrium equations will result in a negative value. The magnitude of the numerical answer is still correct; only the assumed direction of the force or moment was wrong.

If the negative answer is to be used in further computations, substitute it into equations with the negative value. It is recommended that no vector direction changes be attempted until all computations are completed.

Free-body diagrams (FBDs) should include slopes and critical dimensions since these may be necessary in computing

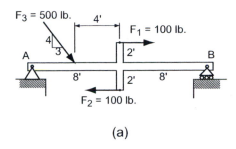

(a)

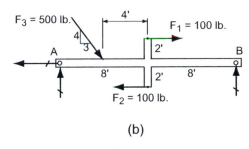

(b)

Figure 3.21   (a) Beam with a hinge and roller support. (b) FBD of the beam—rigid body.

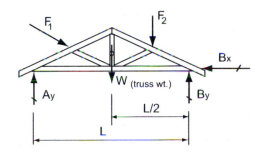

Figure 3.22   FBD of a truss including its own weight.

Table 3.1   Supports Conditions for Coplanar Structures.

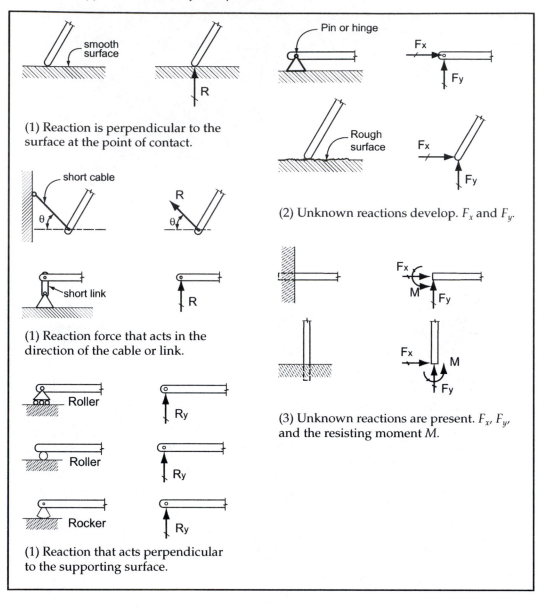

Table 3.2   Connection and Support Examples.

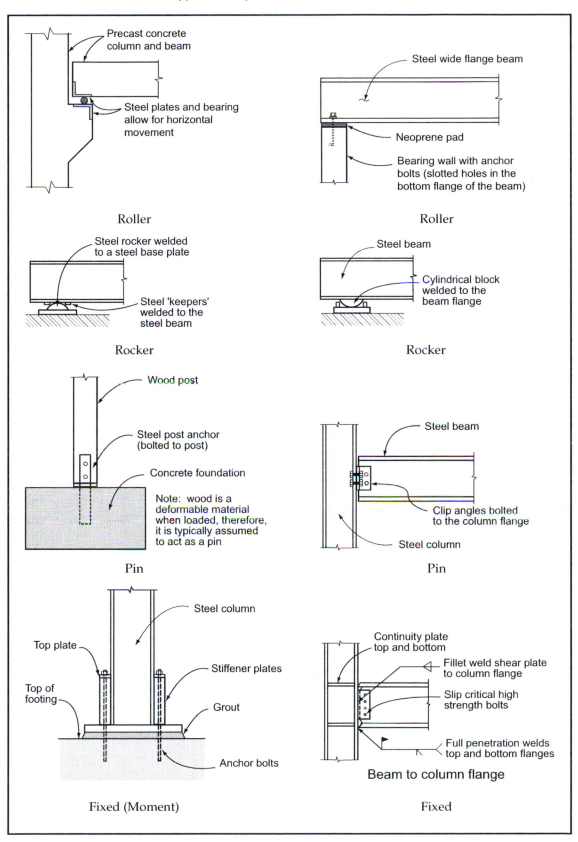

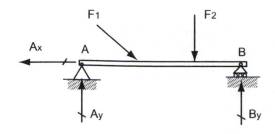

Figure 3.23   Beam with a hinge and roller support.

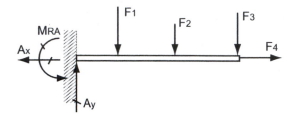

Figure 3.24   A cantilever beam with four loads.

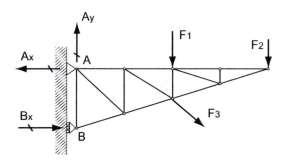

Figure 3.25   A cantilever truss with a hinge and roller support.

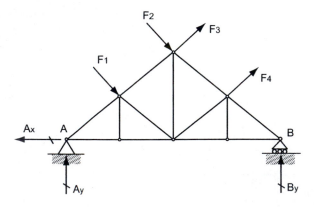

Figure 3.26   FBD of a roof truss with wind loads applied on both slopes.

moments of forces. Figures 3.23 through 3.26 show examples of such FBDs.

Example Problem 3.6—Equilibrium of Rigid Bodies

A bridge spans across a river carrying the applied loads as shown in Figure 3.27. Assuming that the bridge itself weighs 4 kips (located at midspan), determine the support reactions that develop at *A* and *B*.

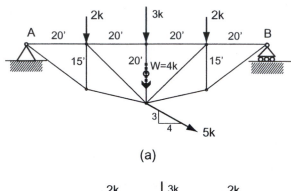

(a)

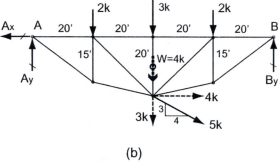

(b)

Figure 3.27   (a) Bridge truss. (b) FBD of the truss.

Solution:

The first step in solving any of these equilibrium problems is the construction of a free-body diagram (FBD). Directions of A_x, A_y, and B are arbitrarily assumed.

a. The pin support at *A* develops two reaction constraints: A_x and A_y. Both forces are independent of each other and constitute two separate unknowns.

b. The roller support at *B* develops only a single reaction component in the vertical direction.

c. Since the force equations of equilibrium ($\Sigma F_x = 0$ and $\Sigma F_y = 0$) are in the x and y reference coordinate system, forces that are inclined should be resolved into x and y components. The 5-kip applied force at the truss midspan is resolved into a 4-kip horizontal and a 3-kip vertical force.

d. Weights of rigid bodies, if significant, must be included in the FBD as part of the loads.

e. Write the three equations of equilibrium and solve for the three unknown support reactions.

$$\sum F_x = -A_x + 4k = 0; \quad \therefore A_x = +4k(\leftarrow)$$

The assumed direction for A_x was assumed correctly.

It is generally convenient to sum moments about a point that contains at least one of the unknown reactions.

$$\sum M_A = -2k(20 \text{ ft.}) - 3k(40 \text{ ft.}) - 2k(60 \text{ ft.})$$

$$- 3k(40 \text{ ft.}) + 4k(20 \text{ ft.})$$

$$- 4k(40 \text{ ft.}) + B_y(80 \text{ ft.}) = 0$$

$$80B_y = +480k; \quad \therefore B_y = +6k(\uparrow)$$

Assumption for the direction of B_y was also correct.

$$\sum F_y = +A_y - 2k - 3k - 2k - 3k - 4k + \underset{(B_y)}{6k} = 0$$

$$\therefore A_y = +8k(\uparrow)$$

Example Problem 3.7

A cantilevered, stadium-type truss supports roof loads as shown in Figure 3.28.

A guying cable CA and a hinge support at B are provided for stability and equilibrium. Draw a FBD of the truss and solve for the support reaction B and the cable tension in CA. Neglect the truss weight.

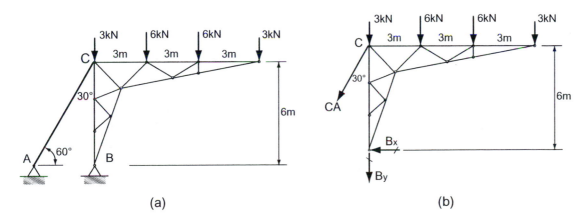

Figure 3.28   (a) Cantilevered stadium truss. (b) FBD of the truss.

Solution:

Again, the construction of the FBD is the first and key step in generating the correct equations of equilibrium.

Cut the cable anywhere between points *C* and *A*, and show the force *CA* as a tension force pulling on the truss at *C*.

Reactions B_x and B_y are assumed directions. Use your intuition in guessing about the reaction directions and don't worry if they are assumed incorrectly. The positive or negative sign that accompanies the final answers will reveal the correct direction. In this example, the direction of B_x will be assumed resisting to the left and B_y is downward.

$$CA_x = CA \sin 30° = 0.5CA$$

$$CA_y = CA \cos 30° = 0.866CA$$

$$\sum M_B = +0.5CA(6 \text{ m}) - 6 \text{ kN}(3 \text{ m})$$

$$- 6 \text{ kN}(6 \text{ m}) - 3 \text{ kN}(9 \text{ m}) = 0$$

$$\therefore CA = \frac{(18 + 36 + 27)\text{kN-m}}{3 \text{ m}} = +27 \text{ kN(tension)}$$

Note that CA_y and the 3-kN force at *C* both pass through reference point *B* and do not enter into the moment equation taken about *B*.

$$\sum F_x = -CA_x - B_x = 0;$$

But $CA_x = 0.5(27 \text{ kN}) = 13.5 \text{ kN}$
$$\therefore B_x = -13.5 \text{ kN}$$

The initial assumption about the direction for B_x was incorrect. Therefore, $B_x = 13.5 \text{ kN} (\rightarrow)$.

$$\sum F_y = -CA_y - 3 \text{ kN} - 6 \text{ kN} - 6 \text{ kN} - 3 \text{ kN} - B_y = 0$$
$$\underset{(23.4 \text{ kN})}{}$$

$$\therefore B_y = -41.4 \text{ kN}$$

Since the result for B_y is negative, the correct direction should be upward.

Example Problem 3.8

A painter, weighing 200 pounds, is 16 feet up the length of a ladder that weighs 50 pounds. The ladder, shown in Figure 3.29, is 24 feet in length and inclined at a 60° angle with the ground. What are the reaction forces developed?

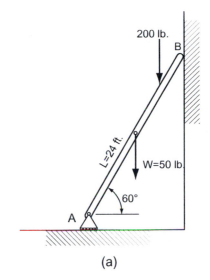

(a)

Solution:

The top of the ladder is assumed to be pushing against a smooth vertical surface which develops a single horizontal support reaction. At the base of the ladder, a hinged foot with a rubber pad (to develop a horizontal, frictional resistance) develops two reactions, horizontal and vertical.

Construct the FBD of the ladder and write the three equations of equilibrium. Determine the horizontal and vertical distances between supports *A* and *B*.

Horizontal distance between *A* and *B*:

$$d_x = 24 \text{ ft.} \times \cos 60° = 24 \text{ ft.} \times (0.5) = 12 \text{ ft.}$$

Vertical distance between *A* and *B*:

$$d_y = 24 \text{ ft.} \times \sin 60° = 24 \text{ ft.} \times (0.866) = 20.8 \text{ ft.}$$

$$\sum F_y = -200 \text{ lb.} - 50 \text{ lb.} + A_y = 0;$$

$$\therefore A_y = +250 \text{ lb.}(\uparrow)$$

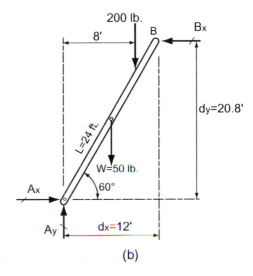

(b)

In writing a moment equation, it is always convenient to select a point that includes at least one of the unknown support reactions.

$$\sum M_A = -(50 \text{ lb.})(6 \text{ ft.}) - (200 \text{ lb.})(16 \text{ ft.} \times \cos 60°)$$
$$+ B_x(20.8 \text{ ft.}) = 0$$
$$\therefore B_x = +91.3 \text{ lb}(\leftarrow)$$

$$\sum F_x = +A_x - \left(\underset{B_x}{91.3 \text{ lb}}\right) = 0$$

$$\therefore A_x = +91.3 \text{ lb.}(\rightarrow)$$

What happens if the ladder shifts to a more vertical position with θ = 75°?

$$\sum M_A = -(50 \text{ lb.})\left(\frac{6.21 \text{ ft.}}{2}\right) - (200 \text{ lb.})(4.14 \text{ ft.})$$
$$+ B_x(23.2 \text{ ft.}) = 0$$

$$\therefore B_x = \frac{(155.3 \text{ lb.-ft.}) + (828 \text{ lb.-ft.})}{23.2 \text{ ft.}}$$

$$= +42.4 \text{ lb}(\leftarrow)$$

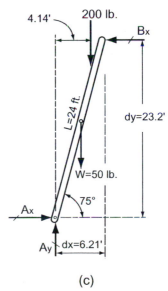

(c)

Figure 3.29 (a) Ladder leaning against a wall. (b) FBD of the ladder at 60°. (c) FBD of the ladder at 75°.

$$\sum F_x = +A_x - \underset{B_x}{(42.4 \text{ lb.})} = 0$$

$$\therefore A_x = +42.4 \text{ lb.}(\rightarrow)$$

$$\sum F_y = +A_y - 50 \text{ lb.} - 200 \text{ lb.} = 0$$

$$\therefore A_y = +250 \text{ lb.}(\uparrow)$$

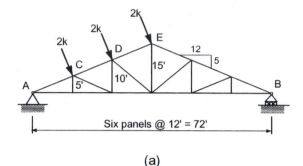

(a)

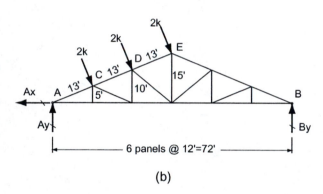

(b)

Figure 3.30   (a) Wind loads applied to the windward slope. (b) FBD of the roof truss.

Example Problem 3.9

The windward slope of a roof, shown in Figure 3.30, is subject to wind forces acting perpendicular to the surface. Assuming a hinge and roller type support system at the truss ends, determine the reactions developed at *A* and *B*.

Solution:

Since the geometry of the roof truss is known, the length of the segments between *AC, CD,* and *DE* can be computed as 13 feet. Use the dimensions along the roof slope for the moments generated by the perpendicular forces acting at *C, D,* and *E*.

$$\sum M_A = -2k(13 \text{ ft.}) - 2k(26 \text{ ft.})$$

$$- 2k(39 \text{ ft.}) + B_y(72 \text{ ft.}) = 0$$

$$\therefore B_y = \frac{(26 + 52 + 78)k\text{-ft.}}{72 \text{ ft.}} = +2.17k(\uparrow)$$

Resolve the applied forces into *x* and *y* components to determine A_x and A_y using the force equations of equilibrium.

$$\sum F_x = -A_x + \left(\tfrac{5}{13}\right)(2k) + \left(\tfrac{5}{13}\right)(2k) + \left(\tfrac{5}{13}\right)(2k) = 0$$

$$\therefore A_x = (3)\left(\tfrac{5}{13}\right)(2k) = +2.3k(\leftarrow)$$

$$\sum F = +A_y - (3) \times \left(\tfrac{12}{13}\right)(2k) + 2.17k = 0$$

$$\therefore A_y = 5.54k - 2.17 = +3.37k(\uparrow)$$

Example Problem 3.10

A utility pole is embedded firmly at the base and supports the two applied loads as shown in Figure 3.31a. Draw a FBD of the pole and determine the support reactions generated at the base E in response to the loading.

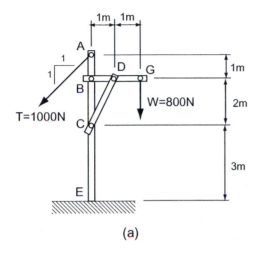

(a)

Figure 3.31 (a) Utility pole with a rigid base support.

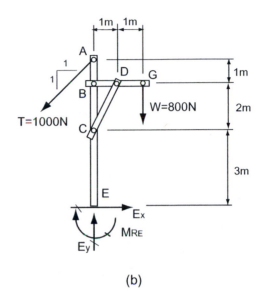

(b)

Figure 3.31 (b) FBD of the utility pole.

Solution:

Resolve the tensile force at A into its respective x and y components.

$$T_x = \frac{-T}{\sqrt{2}} = \frac{-1000 \text{ N}}{\sqrt{2}} = -707 \text{ N}$$

$$T_y = \frac{-T}{\sqrt{2}} = \frac{-1000 \text{ N}}{\sqrt{2}} = -707 \text{ N}$$

$$\sum F_y = \underset{(T_y)}{-707 \text{ N}} - 800 \text{ N} + E_y = 0;$$

$$\therefore E_y = +1507 \text{ N} (\uparrow)$$

$$\sum F_x = \underset{(T_x)}{-707 \text{ N}} + E_x = 0; \quad \therefore E_x = +707 \text{ N} (\rightarrow)$$

$$\sum M_E = +\left(\underset{T_x}{707 \text{ N}}\right)(6 \text{ m})$$

$$-\left(\underset{W}{800 \text{ N}}\right)(2 \text{ m}) - M_{R_E} = 0$$

$$\therefore M_{R_E} = +2642 \text{ N-m}$$

The resisting moment at E was correctly assumed in the clockwise direction.

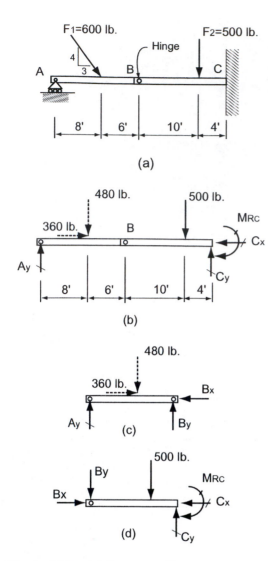

(a)

(b)

(c)

(d)

Figure 3.32   (a) Compound beam.
(b)–(d) FBD of the beam and components.

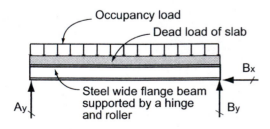

Figure 3.33   Simply supported beam with a
uniform load distribution.

Example Problem 3.11

A simply supported beam *AB*, shown in Figure 3.32, is attached to a cantilever beam *BC* using a hinge at *B*. Construct FBDs and solve for the support reactions at *A* and *C* and the internal forces developed at the hinge at *B*.

Solution:

Since the roller at *A* and the fixed support at *C* develop a total of four unknown support reactions, the three equations of equilibrium are insufficient to solve for all of the unknowns. However, it is possible to separate the compound beam into two additional FBDs that provide the opportunity to write additional equations of equilibrium.

From FBD in Figure 3.32C.

$$\sum M_B = +480 \text{ lb.}(6 \text{ ft}) - A_y(14 \text{ ft.}) = 0$$

$$\therefore A_y = \frac{480 \text{ lb.}(6 \text{ ft.})}{(14 \text{ ft.})} = +205.7 \text{ lb.}(\uparrow)$$

$$\sum F_y = +205.7 \text{ lb.} \underset{(A_y)}{-} 480 \text{ lb.} + B_y = 0$$

$$\therefore B_y = 480 \text{ lb.} - 205.7 \text{ lb.} = +274.3 \text{ lb.}(\uparrow)$$

$$\sum F_x = +360 \text{ lb.} - B_x = 0$$

$$\therefore B_x = +360 \text{ lb.}(\leftarrow)$$

From FBD in Figure 3.32d

Record the values obtained for *Bx* and *By* as known forces.

$$\sum F_x = +360 \text{ lb.} - C_x = 0; \ \therefore C_x = +360 \text{ lb.}(\leftarrow)$$

$$\sum F_y = -274.3 \text{ lb.} - 500 \text{ lb.} + C_y = 0;$$

$$\therefore C_y = +774.3 \text{ lb.}$$

$$\sum M_C = +274.3 \text{ lb.}(14 \text{ ft.})$$

$$+ 500 \text{ lb.}(4 \text{ ft.}) - M_{R_c} = 0;$$

$$\therefore M_{R_C} = (3840 \text{ lb.-ft.}) + (2000 \text{ lb.-ft.}) = +5840 \text{ lb.-ft.}$$

3.3   EQUILIBRIUM INVOLVING DISTRIBUTED LOADS

Beams with Distributed Loads

A simply supported or simple beam is generally defined as a beam supported at both ends with one support being a hinge (pin) and the other a roller (Figure 3.33). Three support reactions develop to produce a condition of equilibrium. Overhang beams (single or double) are supported by a hinge and roller in which one or both of the supports are moved inward from the beam's end (Figure 3.34). Cantilevered beams are supported at one end only and require a fixed (rigid) support condition to satisfy the three conditions of equilibrium (Figure 3.35).

Distributed loads, as the term implies, act on a relatively large area—too large to be considered as a point load without introducing an appreciable error. Examples of distributed loads include:

- Furniture, appliances, people, and so on.
- Wind pressure on a building.
- Fluid pressure on a retaining wall.
- Snow load on a roof.
- The self-weight of a beam.

Point or concentrated loads have a specific point of application, whereas distributed forces are scattered over large surfaces. Most common load conditions on building structures begin as distributed loads (see Figure 3.33). Support reactions for beams and other rigid bodies are calculated in the same manner employed for concentrated loads. The equations of equilibrium,

$$\sum F_x = 0; \quad \sum F_y = 0; \quad \sum M_i = 0$$

where i = any point on the rigid body, are still valid and necessary.

In order to compute the beam reactions, a distributed load is replaced by an equivalent concentrated load, which acts through the geometric center of the distribution, or through what's referred to as the centroid of the load area. Centroids will be discussed in more detail in Chapter 7. The magnitude of the equivalent concentrated load is equal to the area under the load distribution.

It should be noted, however, that the concentrated load is equivalent to the distributed loading only as far as external forces are concerned. The internal stress condition, primarily bending, and the deformation of the beam are very much affected by a change of a uniform load to a concentrated load.

The location of the equivalent concentrated load is based on the centroid of the load area. By geometric construction, the centroids of two primary shapes are shown in

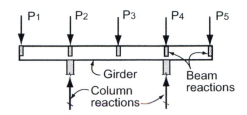

Figure 3.34   Double overhang girder with five concentrated beam reactions.

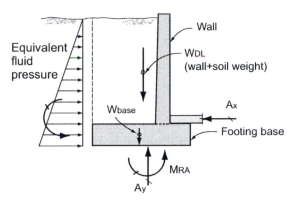

Figure 3.35   Cantilevered retaining wall.

Figures 3.36 and 3.37. Trapezoidal shapes may be thought of as two triangles, or as a rectangle with a triangle. These two combinations are illustrated in Figures 3.38 and 3.39.

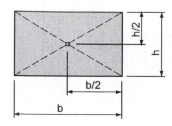

Figure 3.36   Centroid of a rectangle Area = (b × h).

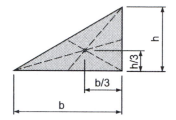

Figure 3.37   Centroid of a triangle Area = (1/2 × b × h).

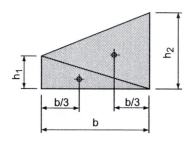

Figure 3.38   Trapezoid as two triangles Area = (1/2 × b × h₁) + (1/2 × b × h₂).

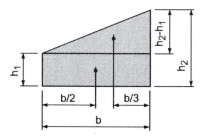

Figure 3.39   Trapezoid as a rectangle and triangle Area = (b × h₁) + (1/2 × b × (h₂ − h₁)).

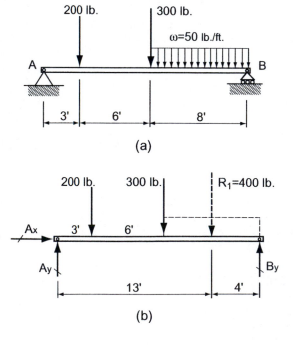

Figure 3.40   (a) Simple beam with a uniform and concentrated loads. (b) FBD of the simple beam.

Example Problem 3.12—Simple Beam

A simple beam supports two concentrated loads and a uniformly distributed load over 8 ft. of the span. See Figure 3.40. Construct a FBD of the beam and solve for the support reactions at *A* and *B*.

Solution (Figure 3.40):

Isolate the beam from its supports and draw the FBD that includes the magnitude and location of the equivalent concentrated load that represents the uniform load distribution.

The magnitude of the equivalent concentrated load is equal to the area under the load curve, in this case, a rectangle.

$$R_1 = \omega \times 8 \text{ ft.} = (50 \text{ }^{lb.}/_{ft.}) \times 8 \text{ ft.} = 400 \text{ lb.}$$

Resultant R_1 is located at the centroid of the rectangle which is 4 ft. from the right support at *B*.

Support reactions are determined by writing the three equations of equilibrium based on the FBD.

$$\sum M_A = -(200 \text{ lb.})(3 \text{ ft.}) - (300 \text{ lb.})(9 \text{ ft.})$$

$$- (400 \text{ lb.})(13 \text{ ft.}) + B_y(17 \text{ ft.}) = 0$$

$$\therefore B_y = \frac{+8500 \text{ lb.-ft.}}{17 \text{ ft.}} = +500 \text{ lb.}(\uparrow)$$

$$\sum M_B = + (400 \text{ lb.})(4 \text{ ft.}) + (300 \text{ lb.})(8 \text{ ft.})$$

$$+ (200 \text{ lb.})(14 \text{ ft.})$$

$$- A_y(17 \text{ ft.}) = +400 \text{ lb.}(\uparrow)$$

$$\sum F_x = 0; \quad \therefore A_x = 0$$

Example Problem 3.13—Overhang Beam (Figure 3.41)

A single overhang beam supports a uniformly distributed load over an 8-ft. section and a triangular distribution over the remaining 12 ft. Draw the appropriate FBD and solve for the support reactions at A and B.

Solution:

Compute the equivalent concentrated loads for the rectangular and triangular distributions.

$$R_1 = \omega \times (8 \text{ ft.}) = \left(300 \text{ }^{\text{lb.}}\!/_{\text{ft.}}\right) \times (8 \text{ ft.}) = 2400 \text{ lb.}$$

$$R_2 = \frac{1}{2}(\omega)(12 \text{ ft.}) = \frac{1}{2}\left(300 \text{ }^{\text{lb.}}\!/_{\text{ft.}}\right)(12 \text{ ft.}) = 1800 \text{ lb.}$$

Writing the three equations of equilibrium:

$$\sum F_x = 0; \quad \therefore A_x = 0$$

$$\sum M_A = -(2400 \text{ lb.})(4 \text{ ft.}) - (1800 \text{ lb.})(12 \text{ ft.})$$

$$+ B_y(16 \text{ ft.}) = 0$$

$$\therefore B_y = \frac{+31{,}200 \text{ lb.-ft}}{16 \text{ ft.}} = +1950 \text{ lb.}(\uparrow)$$

$$\sum F_y = +A_y - 2400 \text{ lb.} - 1800 \text{ lb.} + \underset{B_y}{1950 \text{ lb.}} = 0$$

$$\therefore A_y = +2250 \text{ lb.}(\uparrow)$$

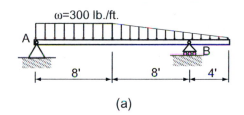

ω=300 lb./ft.

(a)

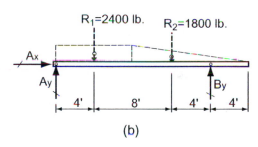

R_1=2400 lb. R_2=1800 lb.

(b)

Figure 3.41   (a) Single overhang beam. (b) FBD of the beam.

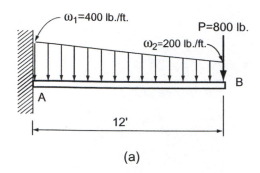

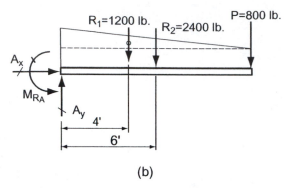

Figure 3.42 (a) Cantilever beam with trapezoidal load. (b) FBD of the cantilever beam.

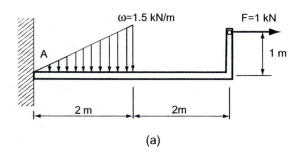

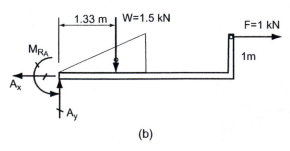

Figure 3.43 (a) Cantilever beam with upturned end. (b) FBD of the cantilever beam.

Example Problem 3.14—Cantilever (Figure 3.42)

Determine the support reactions developed at A for a cantilever beam supporting a trapezoidal load and a point load at the free end.

Solution:

The trapezoidal load is segmented into a triangular load with a maximum $\omega = 200 \; ^{lb.}/_{ft.}$ and a rectangular load of $\omega = 200 \; ^{lb.}/_{ft.}$ over the entire 12-ft. span.

$$R_1 = \frac{1}{2}\left(200 \; ^{lb.}/_{ft.}\right)(12 \text{ ft.}) = 1200 \text{ lb.}$$

$$R_2 = \left(200 \; ^{lb.}/_{ft.}\right)(12 \text{ ft.}) = 2400 \text{ lb.}$$

$$\sum F_x = 0; \qquad \therefore A_x = 0$$

$$\sum F_y = +A_y - 1200 \text{ lb.} - 2400 \text{ lb.} - 800 \text{ lb.} = 0$$

$$\therefore A_y = +4400 \text{ lb.}(\uparrow)$$

$$\sum M_A = +M_{R_A} - (1200 \text{ lb.})(4 \text{ ft.}) - (2400 \text{ lb.})(6 \text{ ft.})$$

$$- (800 \text{ lb.})(12 \text{ ft.}) = 0;$$

$$\therefore M_{R_A} = +28{,}800 \text{ lb.-ft.} \quad \text{(counterclockwise)}$$

Example Problem 3.15—Cantilever Beam

A cantilever beam, shown in Figure 3.43, is subjected to a triangular load over half the span and a horizontal force at the upturned end.

Draw a FBD of the beam and solve for the support reactions at A.

Solution:

$$W_1 = \frac{1}{2}\left(1.5 \; ^{kN}/_m\right)(2 \text{ m}) = 1.5 \text{ kN}$$

$$\sum F_x = -A_x + 1 \text{ kN} = 0; \quad \therefore A_x = +1 \text{ kN}(\leftarrow)$$

$$\sum F_y = +A_y - 1.5 \text{ kN} = 0;$$

$$\therefore A_y = +1.5 \text{ kN}(\uparrow)$$

$$\sum M_A = +M_{R_A} - (1.5 \text{ kN})(1.33 \text{ m})$$

$$-\left(1 \underset{F}{\text{ kN}}\right)(1 \text{ m}) = 0$$

$$\therefore M_{R_A} = (+2 \text{ kN-m}) + 1 \text{ kN-m} = + \underset{\text{(counterclockwise)}}{(3 \text{ kN-m})}$$

Example Problem 3.16 (Figures 3.44 and 3.45)

A compound beam has three supports at *A*, *B* and *D* and an internal hinge at *C*. Two uniformly distributed loads cover the entire length of the beams. Draw the appropriate FBDs and determine the reactions at the supports and the internal pin forces at *C*.

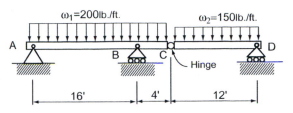

Figure 3.44   Compound beam with a hinge.

Solution:

Since there are more support reactions in the FBD OF Figure 3.45a, of the entire beam system than the three equations of equilibrium permitted, additional FBDs must be drawn. Additional equations of equilibrium can be written for each FBD drawn; three for the FBD in Figure 3.45b and three for the FBD in Figure 3.45c.

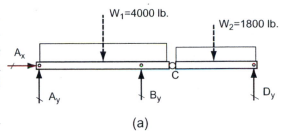

(a)

Calculate the equivalent concentrated load for each uniform distribution shown.

$$W_1 = \left(200\,^{\text{lb.}}\!/_{\text{ft.}}\right) \times (20\text{ ft.})$$

$$= 4000 \text{ lb. located 10 ft. right of } A.$$

$$W_2 = \left(150\,^{\text{lb.}}\!/_{\text{ft.}}\right) \times (12\text{ ft.})$$

$$= 1800 \text{ lb. located 6 ft. left of } D.$$

Begin the solution for the support reactions with the FBD in Figure 3.45b.

$$\sum F_x = 0; \quad \therefore C_x = 0$$

$$\sum M_C = -W_2 \times (6\text{ ft.}) + D_y \times (12\text{ ft.}) = 0;$$

$$\therefore D_y = \frac{(1800\text{ lb.}) \times (6\text{ ft.})}{(12\text{ ft.})} = +900 \text{ lb.}(\uparrow)$$

$$\sum F_y = +C_y - 1800\text{ lb.} + 900\text{ lb.} = 0$$

$$\therefore C_y = +900 \text{ lb.}$$

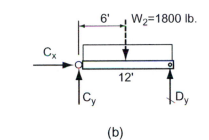

(b)

The vertical pin force C_y is applied as a downward force on the FBD in Figure 3.45c.

Solving for the support reactions in the FBD Figure 3.45c.

$$\sum M_A = -W_1 \times (10\text{ ft.}) + B_y \times (16\text{ ft.})$$

$$- C_y \times (20\text{ ft.}) = 0$$

$$\therefore B_y = \frac{+(4000\text{ lb.}) \times (10\text{ ft.}) + (900\text{ lb.}) \times (20\text{ ft.})}{(16\text{ ft.})}$$

$$B_y = +3625 \text{ lb.}(\uparrow)$$

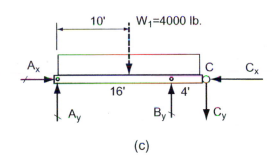

(c)

*Figure 3.45   (a) FBD of the entire beam.
(b) FBD of beam segment CD. (c) FBD of beam segment ABC.*

$$\sum F_y = +A_y - \underset{(W_1)}{4000 \text{ lb.}} + \underset{(B_y)}{3625 \text{ lb.}} - \underset{(C_y)}{900 \text{ lb.}} = 0$$

$$\therefore A_y = +1275 \text{ lb.}(\uparrow)$$

$$\sum F_x = 0; \quad \therefore A_x = 0$$

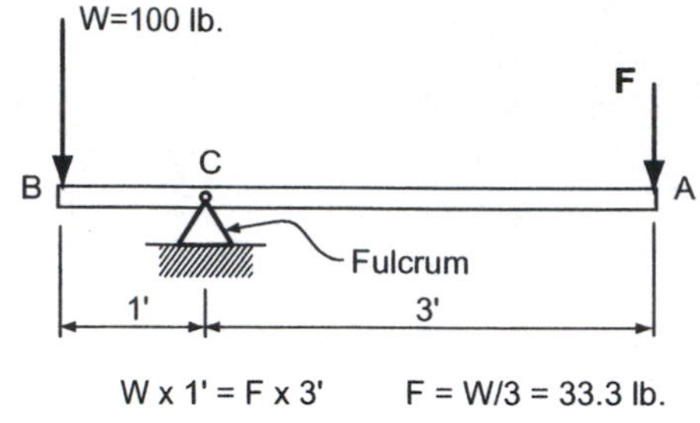

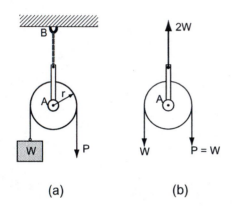

W x 1' = F x 3'   F = W/3 = 33.3 lb.

Figure 3.46   A simple machine lever.

3.4   EQUILIBRIUM INVOLVING PULLEYS

A device that transfers a force from point *A*, where it is applied, to another point *B*, where it is used, is a *machine*. A lever is a machine that enables the effort on one side of the *fulcrum* (a point of support) to lift a load on the other side (Figure 3.46). The functioning of the lever is so simple and straightforward that it is called a *simple machine*. Another simple machine is the pulley. A pulley is a sheave or small wheel with a grooved rim around which a rope or chain is passed to change the direction and point of application of a pulling force (Figures 3.47 and 3.48).

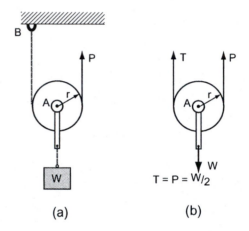

(a)   (b)

Figure 3.47   (a) Fixed pulley. (b) FBD of the pulley.

Figure 3.48   (a) Movable pulley. (b) FBD of the pulley.

The FBD of the pulley in Figure 3.47b shows that the pull **P** is equal to the weight being lifted, **W**. A force of **2W** is exerted on the support at **B**. This pulley arrangement, known as a *fixed pulley*, is used to change the direction of the rope force. In contrast, the pulley in Figure 3.48 is a *moving pulley* in which the pulley moves as a force **P** is applied. The FBD of the moving pulley indicates that the force **P** necessary to lift the weight is **W/2**. In reality, the weight of the pulley should also be added to the applied weight **W**.

The problems to be considered in this section will assume that the rope is perfectly flexible (to wind around the pulley) and there is no friction between the rope and pulley.

Making these assumptions, it is reasonable to assume that the force throughout the length of the rope that goes around the pulleys remains the same. A simple FBD of a pulley about its center axel will result in the rope tensions on either side being equal to preserve rotational equilibrium. Liberal use of FBDs of the pulleys is a simple way to determine the forces throughout the system.

Example Problem 3.17

A pulley system involving two fixed and one movable pulley, shown in Figure 3.49, is arranged to lift a load $W = 1200$ lb. attached at C.

Draw FBDs of each pulley and solve for the force P necessary to just lift the load W. Neglect the pulley weights.

Solution:

Begin with the construction of FBDs of each pulley and note that the force on either side of a pulley is equal. In the FBD of pulley A, the force developed through the center axel at A is:

$$\sum F_y = +A - P - P = 0; \quad \therefore A = 2P$$

At the fixed pulley B:

$$\sum F_y = +B - P - P = 0; \quad \therefore B = 2P$$

Pulley C has three resisting forces for the weight W. Thus:

$$\sum F_y = +P + P + P - W = 0;$$

$$\therefore P = {W}/{3} = {1200\ \text{lb.}}/{3} = 400\ \text{lb.}$$

A Dutch engineer and mathematician, Simon Stevin (1548–1620), studied the pulley and developed a simple analysis based on the idea that the force throughout the length of the rope that goes around the pulleys remains the same. Applying Stevin's method to this example, isolate the movable pulley as a free-body diagram. There are three identical ropes supporting the weight W and, since the pulley is assumed in equilibrium, the three ropes must balance the load. All three rope forces are equal since they are part of the same continuous rope and their magnitudes are equal to $P = {W}/{3} = {1200\ \text{lb.}}/{3} = 400$ lb.

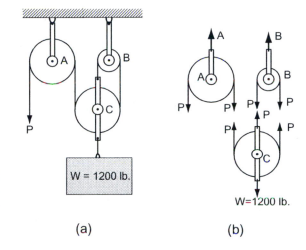

(a)   (b)

Figure 3.49   (a) Pulley arrangement. (b) FBD of pulleys.

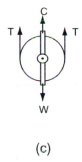

(c)

Figure 3.49   (c) FBD of the movable pulley.

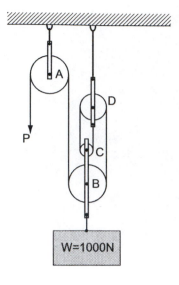

(a)

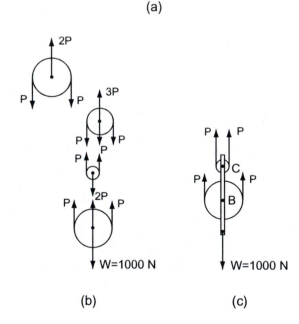

(b)   (c)

Figure 3.50   (a) Pulley arrangement.
(b) FBD of the pulleys. (c) FBD of the movable
pulley block.

Example Problem 3.18

A pulley arrangement, as shown in Figure 3.50, is hoisting a 1000-N weight using two stationary and two movable pulleys in the system. Construct FBDs of all of the pulleys and solve for the required force P, just at the point of movement. Neglect the pulley weights. Check your solution using Stevin's method of isolating the movable pulley block.

Solution:

Free-body diagrams of all of the pulleys, including the equal forces developed on each side of the pulley, reveal a relatively simple vertical equilibrium condition. By summing forces in the y-direction for pulley B:

$$\sum F_y = +P + 2P + P - \underset{(W)}{1000\ \text{N}} = 0$$

$$4P = 1000\ \text{N}; \quad \therefore P = \frac{W}{4} = \frac{1000\ \text{N}}{4} = 250\ \text{N}$$

Stevin's method isolates as a FBD only the movable portion of the pulley system.

Thus:

$$\sum F_y = +P + P + P + P - 1000\ \text{N} = 0$$

$$\therefore 4P = 1000\ \text{N} \quad \text{Then;} \quad P = 250\ \text{N}$$

The addition of more pulleys to the system or combining the pulleys in different combinations can produce any degree of force multiplication as desired, assuming the absence of friction.

Example Problem 3.19

A block and tackle pulley system, as shown in Figure 3.51, is used in hoisting a weight of 800 lb. Determine the pull P necessary to begin lifting the weight, assuming that the pulley weights are neglected.

Solution:

$$\sum F_y = +P + 2P + P - W = 0;$$

$$\therefore 4P = W; \quad \text{Then,} \quad P = 200 \text{ lb.}$$

It is difficult to believe that this force multiplication can continue unabated to a point in which a pulley system could be devised where a one-pound force could lift an 800-lb. weight. Unfortunately, all of this force advantage does not come without consequence. Stevin proposed what he called the *common rule* for pulleys in which he states: *what is gained in force is lost in distance moved.*

In this example, the force multiplication is four times, which means that it will require four feet of rope being pulled to raise the weight one foot in elevation.

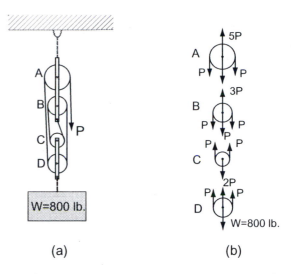

(a)   (b)

Figure 3.51   (a) Pulley arrangement. (b) FBD of pulleys.

Example Problem 3.20

A weight W is being hoisted by the crane system shown in Figure 3.52a. Determine the cable tension in CA and the pin reactions developed at support B.

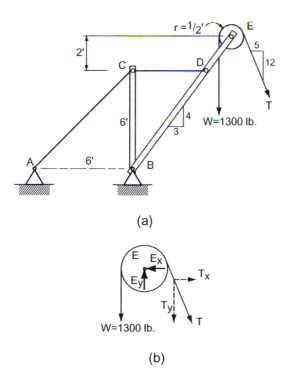

(a)

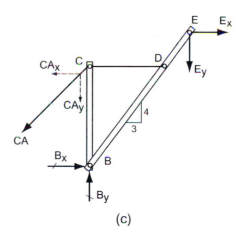

(c)

Figure 3.52   (c) FBD of the crane without the sheave.

(b)

Figure 3.52   (a) Crane hoist assembly. (b) FBD of the sheave.

Solution:

The first step should be the isolation and construction of the FBD of the sheave at E. Assuming that the sheave is just at the point of equilibrium and the pin friction is negligible, force T and the weight W are equal since the summation of moments about the axel at E is zero. The pin forces at E are determined by summing forces in the x and y directions.

$$\sum M_E = +1300 \text{ lb.} \underset{(W)}{\times} r - T \times r = 0;$$

$$\therefore T = W = 1300 \text{ lb.}$$

$$T_x = \frac{5}{13}T = 500 \text{ lb.};$$

$$\text{and} \quad T_y = \frac{12}{13}T = 1200 \text{ lb.}$$

$$\sum F_y = -1300 \underset{(W)}{\text{ lb.}} + E_y - \quad\quad = 0$$

$$\therefore E_y = +1300 \text{ lb.} + 1200 \text{ lb.} = +2500 \text{ lb.}(\uparrow)$$

$$\sum F_x = -E_x + 500 \text{ lb.} = 0;$$

$$\therefore E_x = +500 \text{ lb.}(\leftarrow)$$

Remove the sheave from the crane structure and draw a FBD, applying the pin forces equal and opposite at E. There are three unknown reaction forces, two at B and the cable force CA.

Write the three equations of equilibrium and solve for the unknown reactions.

$$CA_x = \frac{CA}{\sqrt{2}}; \quad \text{and} \quad CA_y = \frac{CA}{\sqrt{2}}$$

$$\sum M_B = +CA_x(6 \text{ ft.}) - 2500 \underset{(E_y)}{\text{ lb.}}(6 \text{ ft.}) - 500 \underset{(E_x)}{\text{ lb.}}(8 \text{ ft.}) = 0$$

$$CA_x = \frac{(2500 \text{ lb.})(6 \text{ ft.}) + (500 \text{ lb.})(8 \text{ ft.})}{6 \text{ ft.}}$$

$$= +3167 \text{ lb.}$$

$$CA_y = 3167 \text{ lb.} \quad \text{and} \quad CA = 3167\sqrt{2} = 4478 \text{ lb.}$$

$$\sum F_x = -3167 \underset{(CA_x)}{\text{ lb.}} + B_x + 500 \underset{(E_x)}{\text{ lb.}} = +2667 \text{ lb.}(\rightarrow)$$

$$\sum F_y = -2500 \underset{(CA_y)}{\text{ lb.}} + B_y - 2500 \underset{(E_y)}{\text{ lb.}} = +5667 \text{ lb.}(\uparrow)$$

Supplementary Problems

Section 3.1—Equilibrium of a Particle

3.1.1 A 500-lb. weight is supported by a tension cable *AB* and a compression strut *CB*. Assuming that the system is in equilibrium and all forces at *B* are concurrent, determine *AB* and *CB*. Solve the problem analytically and check graphically.

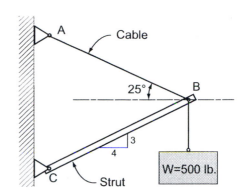

Problem 3.1.1

3.1.2 Two cables *AB* and *BC* support a weight *W* = 200 N as shown. Draw a FBD of the concurrent point *B* and determine the cable forces necessary for equilibrium to exist.

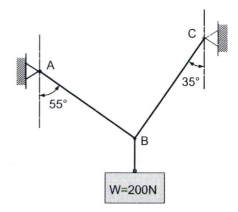

Problem 3.1.2

3.1.3 Construct a FBD of concurrent joint *C* and solve for the member forces *CA* and *BC*. Indicate whether the members are in tension or compression.

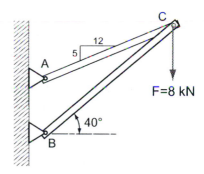

Problem 3.1.3

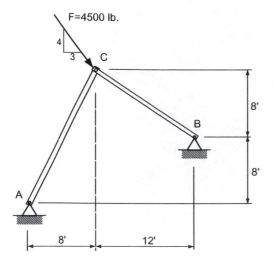

F=4500 lb.

4

3   C

8'

B

8'

A

8'   12'

Problem 3.1.4

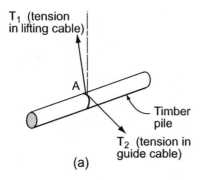

T_1 (tension in lifting cable)

A

Timber pile

T_2 (tension in guide cable)

(a)

Problem 3.1.5 a

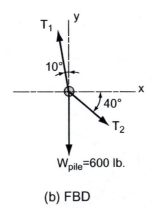

T_1 y

10°

x

40°

T_2

W_{pile}=600 lb.

(b) FBD

Problem 3.1.5b FBD.

3.1.4 Two members, *AC* and *BC*, support an applied load at the concurrent joint *C*. Draw a FBD of joint *C* and solve for the member forces. Are they in tension or compression?

3.1.5 A timber pile is being hoisted into position by a crane. The lifting cable is 10° off from the vertical axis and the guide cable exerts a force T_2 at an angle of 40° from the horizontal axis. Assuming that a state of equilibrium is achieved at point *A*, determine the tension forces T_1 and T_2.

3.1.6 A precast concrete wall panel weighing 8 kips is being lifted into position by using a chain sling. Determine the shortest length of chain *ABC* which may be used if the tension in the chain is not to exceed 10 kips. Assume that a condition of equilibrium is achieved at point *B* and the chain tensions *AB* and *BC* are equal.

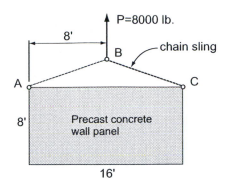

Problem 3.1.6

3.1.7 A weight of 3 kN is held in a stationary position by a system of cables and strut *CB*. Determine the forces developed in the cables *BA*, *DB*, and *DE* and the compression strut *CB*. Draw all appropriate FBDs.

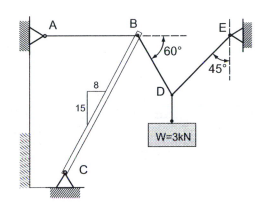

Problem 3.1.7

3.1.8 A 200-lb. weight is supported by cables *DC*, *AC*, and *DE* and by the vertical pole *BC*. Determine all cable forces and the force in the pole *BC*.

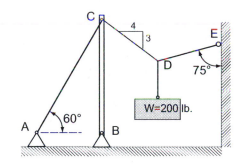

Problem 3.1.8

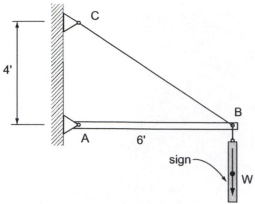

Problem 3.1.9

3.1.9 A sign is suspended from the concurrent point at **B**. Cable *BC* has a maximum capacity of 180 lb. and the strut is able to resist a compression force of up to 125 lb. Determine the maximum weight, *W*, of the sign.

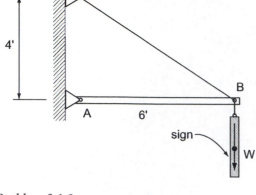

Problem 3.2.1

Section 3.2—Equilibrium of Rigid Bodies
Draw the appropriate FBD for each of the problems in this section.

3.2.1 A 7-m span girder supports the reactions from three roof beams. Determine the support reactions at *A* and *B*.

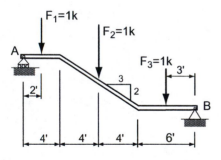

Problem 3.2.2

3.2.2 A bent beam supporting a stair system is loaded with equivalent concentrated loads as shown. Solve for the support reactions at *A* and *B*.

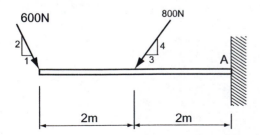

Problem 3.2.3

3.2.3 Determine the support reactions developed at *A* for the cantilevered balcony beam.

3.2.4 Solve for the support reactions at *A* and *B*.

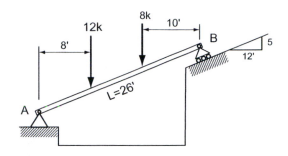

Problem 3.2.4

3.2.5 Determine the support reactions at *C* and the cable force *BA*.

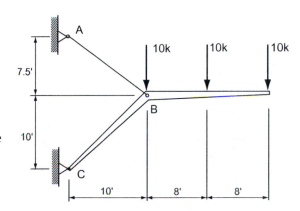

Problem 3.2.5

3.2.6 A balcony railing is bolted to a beam as shown. Draw a FBD of the railing. Which bolt develops a larger horizontal reaction force?

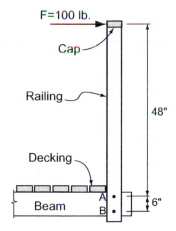

Problem 3.2.6

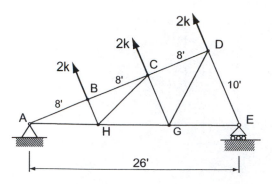

Problem 3.2.7

3.2.7 A saw-tooth roof truss is subjected to uplift wind forces acting perpendicular to the roof surface. Draw a FBD of the roof truss and solve for the support reactions at *A* and *E*.

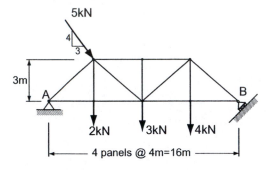

Problem 3.2.8

3.2.8 A truss spanning 16 m over a river is supported by a hinge at *A* and a roller at *B*. The roller rests on a sloping 1:1 surface. Construct a FBD of the truss and solve for the support reactions at *A* and *B*.

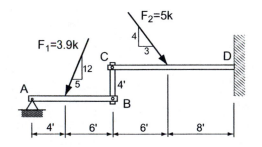

Problem 3.2.9

3.2.9 A short link *BC* joins beam *AB* with the cantilever beam *CD*. Two inclined forces are applied to the compound beam. Draw the appropriate FBDs and solve for the support reactions at *A* and *D* and the link force *BC*.

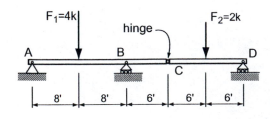

Problem 3.2.10

3.2.10 A glue-laminated timber roof beam supports two concentrated loads as shown. The 12-ft. span *CD* is supported at *C* by a hinge at the end of the overhang of beam *ABC*. Draw FBDs of *ABC* and *CD*, then solve for the support reactions at *A*, *B*, and *D*.

Section 3.3—Equilibrium Involving Distributed Loads
Construct FBDs and solve for the support reactions in each problem.

3.3.1 A double overhang beam is loaded as shown. Solve for the reactions at *A* and *B*.

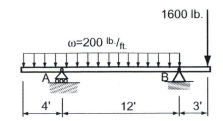

Problem 3.3.1

3.3.2 Snow load on a roof beam creates an unusual load pattern as shown. Determine the reactions at supports *A* and *B*.

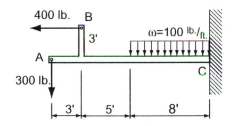

Problem 3.3.2

3.3.3 A cantilever beam has a 3-ft. upturn with a 400-lb. horizontal force applied. Determine the support reactions developed at *C*.

Problem 3.3.3

3.3.4 A beam is subjected to two trapezoidal loads. Solve for the support reactions at *A* and *B*.

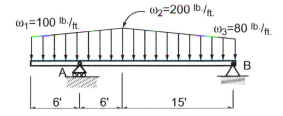

Problem 3.3.4

3.3.5 Determine the support reactions at *A* and *B* for the overhang beam shown.

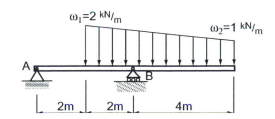

Problem 3.3.5

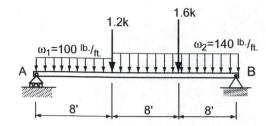

Problem 3.3.6

3.3.6 A steel beam supports the loads as shown. Determine the reactions at *A* and *B*.

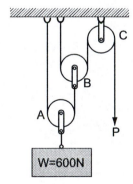

Problem 3.4.1

Section 3.4—Equilibrium Involving Pulleys and Sheaves

3.4.1 What weight, *W*, can be lifted by a pull of *P* = 50 lb. for this arrangement of pulleys?

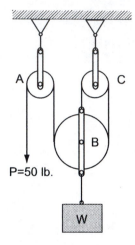

Problem 3.4.2

3.4.2 A 600-N weight is being hoisted by a system of three pulleys. Determine the force *P* necessary to just begin lifting the weight.

3.4.3 Determine the pull, *P*, required to just lift the 800-lb. weight. If the weight is to be moved up 2 feet, how much rope needs to be pulled?

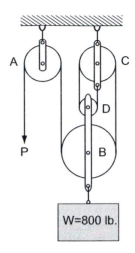

Problem 3.4.3

3.4.4 What is the pull *P* required to lift a weight of 520 pounds? Determine the length of rope that needs to be pulled to raise the weight 3 feet vertically.

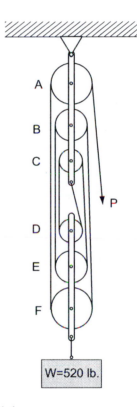

Problem 3.4.4

3.4.5 Isolate and draw a FBD of the sheave at *C*. Then, construct a FBD of the beam *ABC* with the sheave removed. Solve for the support reactions at *A* and *B*.

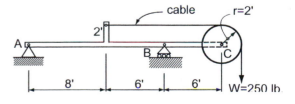

Problem 3.4.5

4 Analysis of Selected Determinate Structural Systems

4.1 PLANAR TRUSSES

A *truss* (Figures 4.1 and 4.2) is a two- or three-dimensional structural framework arranged in triangular patterns using slender straight bars and connected at each end joint using frictionless pins. Triangular configurations, like trusses, are inherently stable, whereas rectangular frameworks connected by pins (Figure 4.3) are unstable. Trusses are often a practical and economical solution to structures spanning long distances, such as bridges and buildings requiring column-free spaces. Truss members are assumed to act as axial (two)-force members in pure tension or compression with the forces at each end directed along the axis of the member. In order for individual truss members to act in tension or compression without bending, several assumptions need to be made: (a) members are linear, (b) members are connected at their ends by pins, (c) member weights are considered negligible in comparison to the applied loads, and (d) loads must be applied at the joints to avoid bending in the members.

Compression members are generally larger in dimension than tension members due to the buckling tendency that results from member instability. Member economy (least material used for the intended design) is generally achieved when diagonals are sloped between 45° to 60° from the horizontal. The perimeter members of a truss are referred to as chords while the interior vertical and diagonal struts connecting the chords are called web members. Generally, horizontal spanning trusses will develop compression forces in the upper chord members and tension in the lower chord while web members can be either in tension or compression.

There are four common ways of calculating the member forces (or joint forces) of a truss. They are the *method of joints*, the *method of sections*, *Maxwell's diagram* (graphical method), and *computer programs*. The first two methods will be illustrated in this book since they are intimately tied to the equations of equilibrium previously introduced. Maxwell's diagram is still a powerful method employing the principles of scaled force polygons done entirely in a graphical manner. Good and relatively inexpensive computer software has

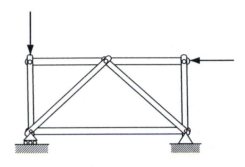

Figure 4.1   Stable, triangulated geometry.

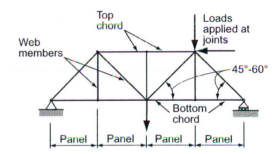

Figure 4.2   Parallel chord truss.

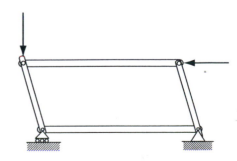

Figure 4.3   Inherently unstable configuration.

now become the industry standard for truss analysis. Nevertheless, a fundamental understanding of trusses and their force mechanisms can easily be achieved through the use of equilibrium equations.

Method of Joints

Trusses are rigid bodies that are optimally in equilibrium under the action of loads; each joint will also be in equilibrium. In this method each joint is considered as an equilibrium problem involving a concurrent force system. A free-body diagram of each joint is constructed and two equations of equilibrium $\left[\sum F_x = 0\right]$ and $\left[\sum F_y = 0\right]$ are written to solve for the two unknown member forces. Remember that in this method, a force analysis of the joint really involves the equilibrium condition of the member forces on the joint (pin). The method of joints is effective when calculating all member forces in a truss. A suggested analysis procedure is as follows: (a) construct a free-body diagram of the entire truss and, if possible, solve for all external support reactions; (b) isolate a joint from the truss that contains no more than two unknowns (generally at the support ends of a spanning truss or at the free end in a cantilever truss) and has at least one known force; (c) draw a FBD of the joint and solve for the two unknown member forces by satisfying the horizontal and vertical conditions of equilibrium; (d) select another joint (usually adjacent to the joint just completed) that contains no more than two unknowns and solve the two conditions of equilibrium; (e) continue this process until all joints have been solved.

Photo 4.1.1   Horizontal and vertical steel trusses used to create a "bent."

Example Problem 4.1 (Method of Joints)

An asymmetrical roof truss, shown in Figure 4.4, supports two vertical roof loads. Determine the support reactions at each end, then, using the method of joints, solve for all member forces. Summarize the results of all member forces on a FBD (this diagram is referred to as a *force summation diagram*).

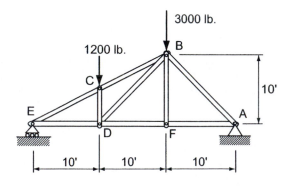

Figure 4.4   Roof truss with two loads.

Solution:

A first step in analyzing this truss is to draw a FBD of the entire truss (Figure 4.5). Next, solve for the support reactions at each end. Since the method of joints involves selecting joints with only two unknowns, joints A and E are the two possible starting points. Hence, the support reactions are necessary since they represent the applied forces at those respective joints.

Solving for the support reactions:

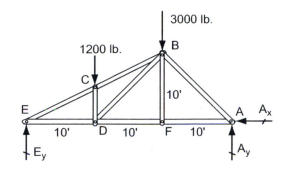

Figure 4.5   Free-body diagram of the truss.

$$\sum M_E = +A_y(30 \text{ ft.}) - 3000 \text{ lb. } (20 \text{ ft.})$$

$$- 1200 \text{ lb. } (10 \text{ ft.}) = 0$$

$$\therefore A_y = \frac{(60{,}000 \text{ lb.-ft.}) + (12{,}000 \text{ lb.-ft.})}{30 \text{ ft.}} = +2400 \text{ lb. } (\uparrow)$$

$$\sum F_x = 0; \quad \therefore A_x = 0$$

$$\sum F_y = +E_y - 1200 \text{ lb. } - 3000 \text{ lb. } + 2400 \text{ lb. } = 0$$

$$\therefore E_y = 1800 \text{ lb.} (\uparrow)$$

Member force analysis proceeds by selecting a joint with no more than two unknowns and at least one known force or reaction. In this example, begin the analysis by drawing a FBD of joints E or A and solve for the two unknown member forces.

Joint A (Figure 4.6):

Write two equations of equilibrium and solve for member forces BA and AF.

$$\sum F_y = -\left(\frac{BA}{\sqrt{2}}\right) + 2400 \text{ lb. } = 0$$

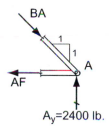

Figure 4.6   FBD of joint A.

Solving for BA; $BA = +2400\sqrt{2}$ lb. $= +3390$ lb. (compression)

$$\sum F_x = \left(+\frac{BA}{\sqrt{2}}\right) - AF = 0;$$

$$AF = \left(+\frac{2400\sqrt{2} \text{ lb.}}{\sqrt{2}}\right) = +2400 \text{ lb. (tension)}$$

Move to another joint that has no more than two unknowns.

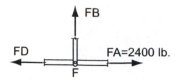

Figure 4.7   FBD of joint F.

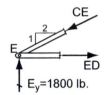

Figure 4.8   FBD of joint E.

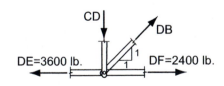

Figure 4.9   FBD of joint D.

Joint *F* **(Figure 4.7):**

Solve for member forces *FB* and *FD:*

$$\sum F_x = -FD + 2400 \text{ lb.} = 0;$$

$$\Delta FD = +2400 \text{ lb. (tension)}$$

$$\sum F_y = 0; \qquad \therefore FB = 0$$

Note: *Although member* **FB** *carries no load in this particular problem, it may well be active if the load condition were changed.*

Joints *B* and *E* both contain two unknowns so they are possible joints to solve next. Joints *C* and *D* still have three unknowns each so they will have to be looked at a bit later. Joint *E* will be selected next since it has only one sloped member (with *x* and *y* components) to resolve.

Joint *E* **(Figure 4.8):**

$$\sum F_y = \left(\frac{-CE}{\sqrt{5}} \right) + 1800 \text{ lb.} = 0;$$

$$\therefore CE = +(1800 \text{ lb.})(\sqrt{5})$$
$$= +4025 \text{ lb. (compression)}$$

$$\sum F_x = \left(\frac{-2CE}{\sqrt{5}} \right) + ED = 0;$$

$$\therefore ED = + \left(\frac{2 \times 4025 \text{ lb.}}{\sqrt{5}} \right) = +3600 \text{ lb. (tension)}$$

Now that member force *ED* is known, joint *D* contains only two unknowns, *CD* and *DF.* Isolate joint *D* and solve for the two unknown member forces.

Joint *D* **(Figure 4.9):**

$$\sum F_x = (-3600 \text{ lb.}) + (2400 \text{ lb.}) + \left(\frac{DB}{\sqrt{2}} \right) = 0$$

$$\therefore DB = (1200\sqrt{2} \text{ lb.}) = +1696 \text{ lb. (tension)}$$

$$\sum F_y = +DB_y - CD = 0$$

$$\therefore CD = \frac{DB}{\sqrt{2}} = \frac{1200\sqrt{2} \text{ lb.}}{\sqrt{2}}$$
$$= 1200 \text{ lb. (compression)}$$

The last unknown member force is *BC.* Construct a FBD of joint *C* (or *B*) and write one equation of equilibrium to solve for *BC.* The second equation of equilibrium for joint *C* will be a check to verify that equilibrium is also established in the other direction.

Joint C (Figure 4.10):

$$\sum F = + EC_x - BC_x = 0$$

But, $BC_x = \dfrac{2BC}{\sqrt{5}}$ and $EC_x = \dfrac{(2 \times 4025 \text{ lb.})}{\sqrt{5}} = 3600 \text{ lb.}$

$\therefore BC = 4025 \text{ lb. (compression)}$

$$\sum F_y = \left(+ \frac{4025 \text{ lb.}}{\sqrt{5}} \right) - 1200 \text{ lb.} + 1200 \text{ lb.}$$

$$- \left(\frac{4025 \text{ lb.}}{\sqrt{5}} \right) = 0$$

$0 = 0;$ checks

All member forces are summarized in the force summation diagram shown in Figure 4.11.

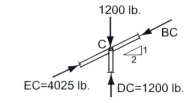

Figure 4.10   *FBD of joint C.*

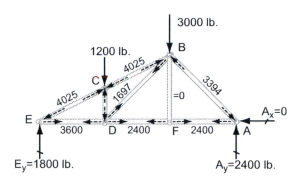

Figure 4.11   *Force summation diagram.*

Example Problem 4.2 (Method of Joints)

A simple stadium truss (Figure 4.12) is loaded with two forces as shown. Solve for the support reactions at A and B, then, using the method of joints, determine all of the member forces.

Summarize the results on a force summation diagram.

Unlike the previous example of a spanning type truss, the member force analysis on this example can commence without solving for the support reactions at A and B. Analysis could begin at joint E and proceed to other joints containing no more than two unknowns until all members are known. The support reactions could be determined as a result of solving joints A and C. However, this example might be well suited for examining the necessity for using two hinges as support reactions rather than the more typical hinge and roller. It would appear intuitively obvious that if a roller support were introduced at A or B, collapse of the truss would result. Two hinge supports are necessary whenever a single member frames into a support, whereby the two hinges serve to create a "closing" member (by use of the ground). Since the two hinges are capable of restraining in the horizontal direction, no movement occurs between the supports. On first glance, a free-body diagram

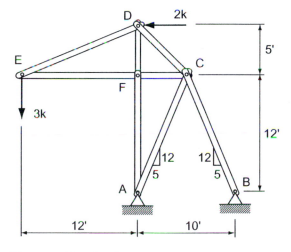

Figure 4.12   *Stadium truss.*

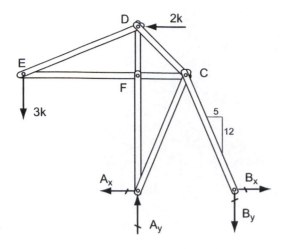

Figure 4.13   FBD of the entire truss.

of the truss, shown in Figure 4.13, would reveal four support reaction unknowns; two at *A* and two at *B*. However, B_x and B_y are related by virtue of the slope of member *BC*. Since *BC* is the only member that frames into support *B*, the resultant reaction *B* must pass through the axis of member *BC*. B_x and B_y are written in terms of the single unknown *B*. Thus, three equations of equilibrium should yield the three support reactions A_x, A_y, and *B*.

Note: $B_x = \dfrac{5B}{13}$ and $B_y = \dfrac{12B}{13}$

$$\sum M_A = +3\,k(12\text{ ft.}) + 2\,k(17\text{ ft.}) - \frac{12B}{13}(10\text{ ft.}) = 0$$

$$B = \left(\frac{13}{12}\right)\left(\frac{36+34}{10}k\right) = 7.6\,k$$

$$B_x = \left(\frac{5}{13}\right)(7.6\,k) = 2.91\,k\ (\rightarrow);\text{ and}$$

$$B_y = \left(\frac{12}{13}\right)(7.6\,k) = 7\,k\ (\downarrow)$$

$$\sum F_x = -2\,k - A_x + 2.91\,k = 0;$$

$$A_x = 0.91\,k\ (\leftarrow)$$

$$\sum F_y = -3\,k + A_y - 7\,k = 0; \qquad A_y = +10\,k\ (\uparrow)$$

Now, with the support reactions known, the method of joints can proceed, beginning at joint *E* or *A*.

Joint *E* (Figure 4.14):

Directions (tension or compression) of members *ED* and *EF* may be assumed initially for the FBD of joint *E*. A negative result from the equation(s) of equilibrium means that the direction is opposite of the initial assumption; however, the magnitude remains unchanged.

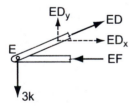

Figure 4.14   FBD of Joint E.

$$ED_x = \frac{12}{13}ED; \quad ED_y = \frac{5}{13}ED$$

$$\sum F_y = +\left(\frac{5}{13}\right)ED - 3\,k = 0; \qquad ED = +7.8\,k\ (T)$$

$$\sum F_x = +\left(\frac{12}{13}\right)(7.8\,k) - EF = 0;$$

$$EF = +7.2\,k\ (C)$$

Joint D (Figure 4.15):

$$DC_x = 0.707DC; \quad DC_y = 0.707DC$$

$$\sum F_x = -\left(\frac{12}{13}\right)(7.8 \text{ k}) - 2 \text{ k} + 0.707DC = 0;$$

$$DC = +13 \text{ k } (T)$$

$$\sum F_y = -\left(\frac{5}{13}\right)(7.8 \text{ k}) - DF - 0.707(13 \text{ k}) = 0$$

$$DF = -3 \text{ k} - 9.2 \text{ k} = -12.2 \text{ k } (C)$$

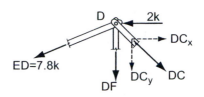

Figure 4.15   FBD of joint D.

The direction of *DF* was assumed wrong for the initial FBD.

Joint F (Figure 4.16):

$$\sum F_x = CF + 7.2 \text{ k} = 0; \qquad CF = -7.2 \text{ k } (C)$$

$$\sum F_y = -12.2 \text{ k} - FA = 0; \qquad FA = -12.2 \text{ k } (C)$$

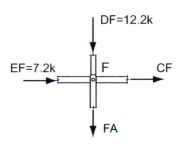

Figure 4.16   FBD of joint F.

Joint C (Figure 4.17):

$$\sum F_x = +7.2 \text{ k} - 0.707(13 \text{ k})$$

$$-\left(\frac{5}{13}\right)CA + \left(\frac{5}{13}\right)CB = 0$$

$$\sum F_y = +0.707(13 \text{ k}) - \left(\frac{12}{13}\right)CA - \left(\frac{12}{13}\right)CB = 0$$

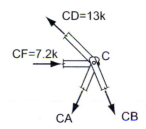

Figure 4.17   FBD of joint C.

Solving the two equations simultaneously:

$$CB = +7.6 \text{ k } (T) \quad \text{and} \quad CA = +2.4 \text{ k } (T)$$

The force summation diagram for the truss is shown in Figure 4.18.

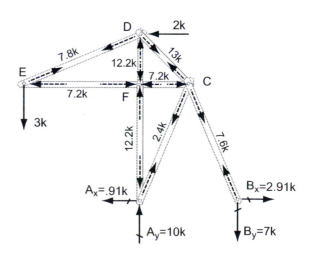

Figure 4.18   Force summation diagram.

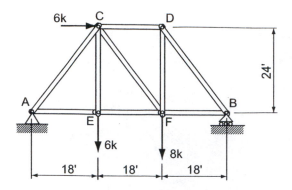

Figure 4.19   Quick method of joints.

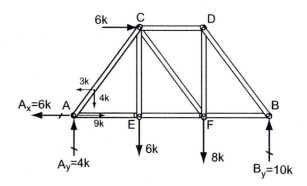

Figure 4.20   FBD-a.

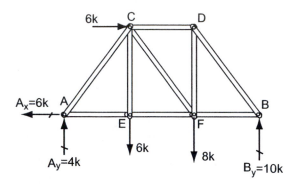

Figure 4.21   FBD-b.

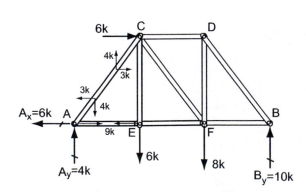

Figure 4.22   FBD-c.

Quick Method of Joints

This example truss (Figure 4.19) will illustrate a "shortcut" version of the method of joints. The principle of equilibrium is still adhered to; however, separate FBDs will not be drawn for each successive joint. Instead, the original FBD drawn for the determination of support equations will be used exclusively. The steps outlined in the earlier examples are still valid and thus will be employed in the "quick" method.

Step 1: *Draw a FBD of the entire truss.*

Step 2: *Solve for the support reactions at A and B. (See Figure 4.20, FBD-a.)*

Step 3: *Isolate a joint with no more than two unknowns. (See Figure 4.21, FBD-b.)*

In this method, it is unnecessary to draw a FBD of the joint. Instead, use the FBD from Steps 1 and 2 and record your solutions directly on the truss. It is usually easier to find the horizontal and vertical components in all the members first and then determine the actual resultant member forces at the very end.

It is common in truss analysis to use slope relationships instead of angle measurements. In this illustration, diagonal members have slopes of 4 vertical to 3 horizontal (3:4:5 triangle). Since all truss members are assumed to be two-force members, the vertical and horizontal force components are related by the slope relationship, 4:3.

Step 4: *Equations of equilibrium for the isolated joint will be done mentally without writing them out. However, occasionally some complex geometry will require equations to be written out and solved simultaneously.*

Let's begin at joint A (joint B is also a possible starting point). Members AC (horizontal and vertical components) and AE (horizontal component only) are unknown. Since member AC has the only vertical component, solve $\Sigma F_y = 0$. This results in a vertical component of 4 k (to balance $A_y = 4$ k). Therefore, with $AC_y = 4$ k, and since AC_x and AC_y are related by virtue of their slope relationship:

$$AC_x = \left(\frac{3}{4}\right)AC_y = 3 \text{ k} (\leftarrow)$$

Next, solve the horizontal equilibrium condition. Reaction $A_x = 6$ k is directed to the left; therefore, AE must be a force of 9 k going to the right. Joint A is now in equilibrium. Record the results (in component form) at the opposite end of each member. (See Figure 4.22, FBD-c.)

Step 5: *Proceed to the next joint with no more than two unknowns. Joint E will be solved next. (See Figure 4.23, FBD-d).*

The two unknowns *EF* (horizontal) and *EC* (vertical) are readily solved since each represents a singular unknown *x* and *y*, respectively. In the horizontal direction, *EF* must resist with 9 k to the right. Member force *EC* = 6 k resists upward to counter the 6 k applied load.

Record member forces *EF* and *EC* at joints *F* and *C*, respectively. (See Figure 4.24, FBD-e.)

Continue to another joint with no more than two unknowns. Try joint *C*. Members *CD* (horizontal) and *CF* (horizontal and vertical) are unknown. Solve the vertical condition of equilibrium first. CA_y = 4 k is upward and *CE* = 6 k is downward, leaving an unbalance of 2 k in the down direction. Therefore, CF_y must resist with a force of 2 k upward. Through the slope relationship:

$$CF_x = \left(\frac{3}{4}\right)CF_y = 1.5 \text{ k}$$

In the horizontal direction, *CD* must develop a resistance of 7.5 k to the left. (See Figure 4.25, FBD-f.)

Step 6: *Repeat Step 5 until all member forces are determined. The final results are shown in the force summation diagram. At this point, resultant forces may be computed by using the known slope relationship for each member. (See Figure 4.26, FBD-g.)*

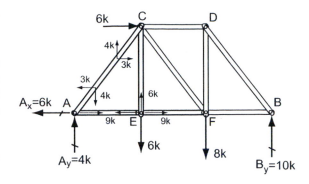

Figure 4.23   FBD-d.

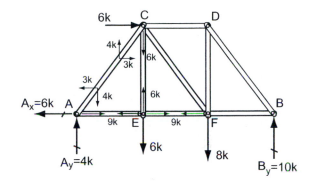

Figure 4.24   FBD-e.

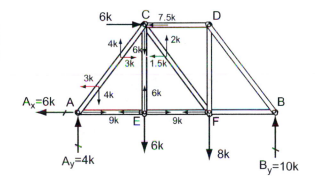

Figure 4.25   FBD-f.

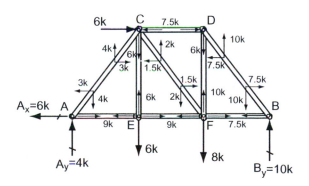

Figure 4.26   FBD-g (force summation diagram).

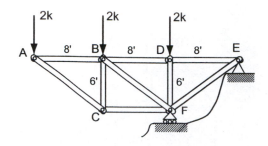

Figure 4.27   Solve for members BD, BF, and CF.

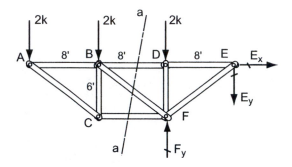

Figure 4.28   FBD with section a-a cutting members BD, BF, and CF.

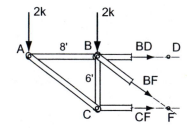

Figure 4.29   FBD-a.

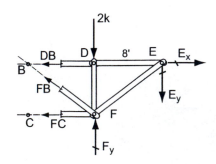

Figure 4.29   FBD-b.

Method of Sections

The *method of sections* is a particularly useful analysis technique when only a few selected member forces in a truss, as shown in Figure 4.27, are desired. Unlike the method of joints which involves the solution of many joints and generally the analysis of the entire truss, the method of sections restricts the solution to selected members. In the method of sections, an imaginary section cut is passed through the truss cutting three members including at least one of the members of interest (Figure 4.28). The truss is now cut into two parts and a free-body diagram is constructed of either section (Figures 4.29a and 4.29b). Since all forces in the free-body diagram are not concurrent at a common point, this constitutes a rigid body type problem and three equations of equilibrium may be written. These three equations of equilibrium can solve for only three unknown member forces; therefore, no more than three unknown members are permitted to be cut for any section passing through the truss. A general procedure for the method of sections is outlined below.

1. Construct a free-body diagram of the entire truss as shown in Figure 4.28.

2. Solve for the support reactions. In most cases support reactions are necessary, but cantilevered and vertical trusses sometimes do not require support reactions for the method of sections to be used.

3. Pass a section through the truss cutting no more than three unknown members. At least one of the cut members should be a desired member.

4. Draw a FBD of that part of the sectioned truss (Figure 4.29a or Figure 4.29b) which seems the least complicated. By inspection, try to assign a tension or compression sense to the unknown cut members. If this seems too difficult, assume unknowns in tension and if the equation of equilibrium yields a negative answer, then the member should be in compression.

5. Three unknown member forces can be solved from this free-body diagram using three equations of equilibrium. It is generally possible to write three equations of equilibrium that avoid simultaneous solutions. Moment equations written about a point (or joint) where two unknowns intersect enables a direct solution of the third force. If two unknown member forces are parallel, an equation is written summing forces in a perpendicular direction to solve for the third member force directly.

Example Problem 4.3 (Method of Sections)

A 64-foot parallel chord truss (Figure 4.30) supports horizontal and vertical loads as shown. Using the method of sections, determine the member forces BC, HG, and GD.

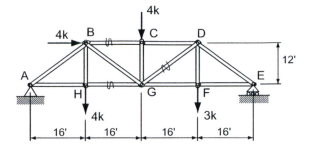

Figure 4.30 Parallel chord truss.

Solution:

The solution begins by constructing a FBD of the entire truss (Figure 4.31) and solving for the support reactions at A and E.

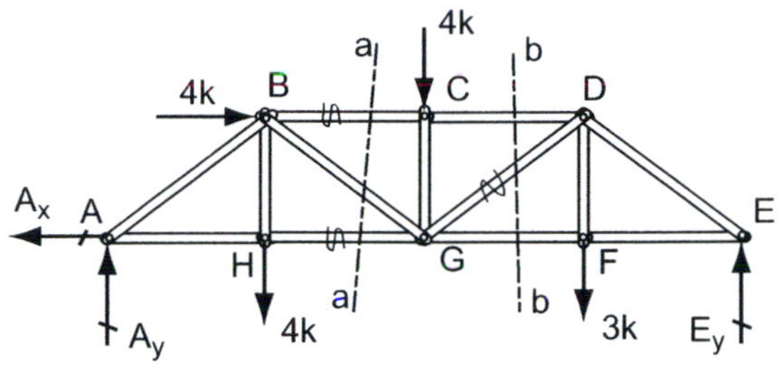

Solving for the support reactions:

$$\sum M_A = -4\,k(12\text{ ft.}) - 4\,k(16\text{ ft.}) - 4\,k(32\text{ ft.})$$

$$- 3\,k(48\text{ ft.}) + E_y(64\text{ ft.}) = 0$$

$$\therefore E_y = +6\,k\,(\uparrow)$$

$$\sum F_y = +A_y - 4\,k - 4\,k - 3\,k + 6\,k = 0;$$

$$\therefore A_y = +5\,k\,(\uparrow)$$

$$\sum F_x = -A_x + 4\,k = 0 \qquad \therefore A_x = +4\,k\,(\leftarrow)$$

Figure 4.31 FBD of the truss with section cuts a-a and b-b.

Members BC and HG can be solved by passing a section cut a-a through the truss and drawing a FBD of either side of the truss cut by the section (Figures 4.32 and 4.33). Individual equations of equilibrium are written to solve for the unknowns BC and HG.

To solve for member BC, using the FBD in Figure 4.32, sum moments about joint G (imaginary) so that unknown member forces BG and HG are not part of the equation. This results in a direct solution for the unknown member force BC.

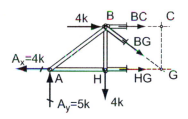

Figure 4.32 FBD of the truss cut by section a-a.

$$\sum M_G = -5\,k(32\text{ ft.}) - 4\,k(12\text{ ft.})$$

$$+ 4\,k(16\text{ ft.}) - BC(12\text{ ft.}) = 0$$

$$\therefore BC = -12\,k\,(\text{compression})$$

Note: *All unknown member forces BC, BG, and HG were initially assumed in tension in the FBD. A negative value obtained by solving an equation of equilibrium simply means that the member is acting in compression.*

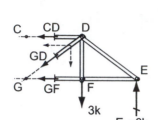

Figure 4.33 FBD of the truss cut by section b-b.

Member *HG* can be solved by summing moments about joint *B*, where members *BC* and *BG* intersect. Since *BC* and *BG* do not appear in the moment equation about *B*, a direct solution for *HG* is possible.

$$\sum M_B = -4\,k(12\text{ ft.}) - 5\,k(16\text{ ft.}) + HG(12\text{ ft.}) = 0$$

$$\therefore HG = +10.67\,k \text{ (tension)}$$

Member force *GD* requires a section cut b-b passing through *GD* (Figure 4.33). A unique equation can be written that solves for *GD* without involving unknown member forces *CD* and *GF*.

Note that for this parallel chord truss, *CD* and *GF* are entirely horizontal in direction; thus, summing forces in the vertical direction will yield a direct solution.

$$GD_x = \left(\frac{4}{5}\right)GD \quad \text{and} \quad GD_y = \left(\frac{3}{5}\right)GD;$$

$$\sum F_y = -\left(\frac{3}{5}\right)GD - 3\,k + 6\,k = 0$$

$$\therefore GD = +5\,k \text{ (tension)}$$

Example Problem 4.4 (Method of Sections)

A steel roof truss, shown in Figure 4.34, supports vertical purlin loads. Utilizing the method of sections, determine the member forces in *BC*, *FC*, and *FG*.

Solution:

Again, since this is a spanning type truss, the support reactions need to be determined before the method of sections can be employed.

$$\sum M_A = -4.8\,k(9\text{ ft.}) - 4.8\,k(18\text{ ft.}) - 4.8\,k(27\text{ ft.})$$

$$- 2.4\,k(36\text{ ft.}) + E_y(36\text{ ft.}) = 0$$

$$\therefore E_y = 9.6\,k \ (\uparrow)$$

$$\sum F_y = -2.4\,k - 4.8\,k - 4.8\,k$$

$$- 2.4\,k + 9.6\,k = 0$$

$$\therefore A_y = +9.6\,k \ (\uparrow)$$

Since there are no horizontally applied forces, no horizontal reaction at E_x develops.

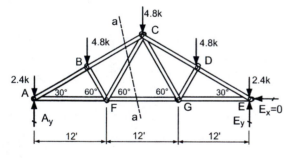

Figure 4.34 Steel roof truss with vertical loads.

To solve for member forces BC, CF, and FG, a section cut a-a is made through the truss and a FBD is drawn of either side of the cut, as shown in Figure 4.35. It is possible to write three distinct equations of equilibrium to solve for each of the unknowns independently.

$$BC_x = BC \cos 30° \quad \text{and} \quad BC_y = BC \sin 30° \text{ and,}$$

$$FC_x = FC \cos 60° \quad \text{and} \quad FC_y = FC \sin 60°$$

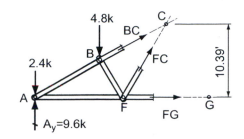

Figure 4.35   FBD of the truss—section a-a.

Solving for member force BC:

$$\sum M_F = -9.6 \text{ k}(12 \text{ ft.}) + 2.4 \text{ k}(12 \text{ ft.})$$
$$+ 4.8 \text{ k}(3 \text{ ft.}) - BC(6 \text{ ft.}) = 0$$
$$\therefore BC = -12 \text{ k } (C)$$

Solving for member FG:

$$\sum M_C = +4.8 \text{ k}(9 \text{ ft.}) + 2.4 \text{ k}(18 \text{ ft.})$$
$$- 9.6 \text{ k}(18 \text{ ft.}) + FG(10.39 \text{ ft.}) = 0$$
$$\therefore FG = +8.32 \text{ k } (T)$$

Solving for member FC:

$$\sum M_A = -4.8 \text{ k}(9 \text{ ft.}) + FC \sin 60°(12 \text{ ft.}) = 0$$
$$\therefore FC = +4.16 \text{ k } (T)$$

Note: *An alternate equation, such as $\sum F_y = 0$, could have been written since member force BC had previously been solved.*

Example Problem 4.5 (Method of Sections)

A bowstring truss (Figure 4.36) supports three loads on its bottom chord joints as shown. Draw a FBD of the truss and solve for the support reactions at A and B. Then, using the method of sections, determine the member forces in CG and CD.

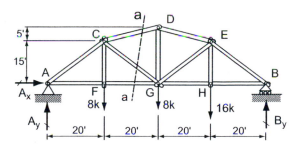

Figure 4.36   Bowstring or camelback truss.

Solution:

Solving for the support reactions (Figure 4.36):

$$\sum M_A = -8 \text{ k}(20 \text{ ft.}) - 8 \text{ k}(40 \text{ ft.})$$
$$- 16 \text{ k}(60 \text{ ft.}) + B_y(80 \text{ ft.}) = 0$$
$$\therefore B_y = +18 \text{ k } (\uparrow)$$

$$\sum M_B = + 16\,k(20\text{ ft.}) + 8\,k(40\text{ ft.})$$
$$+ 8\,k(60\text{ ft.}) - A_y(80\text{ ft.}) = 0$$
$$\therefore A_y = + 14\,k\ (\uparrow)$$
$$\sum F_x = 0; \qquad \therefore A_x = 0$$

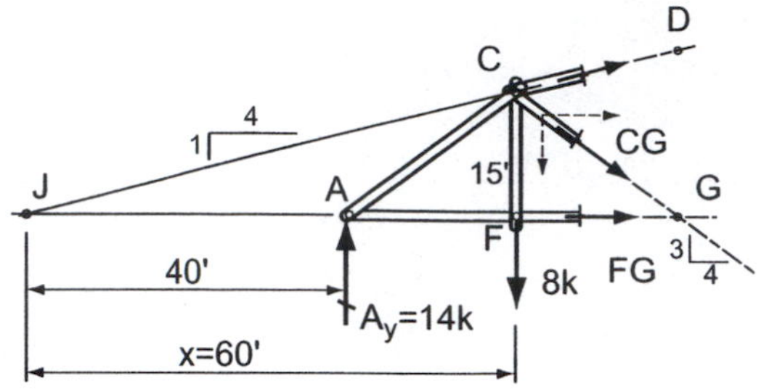

Member force CG can be solved by taking moments about a fictitious joint J (Figure 4.37) where the lines of action of cut members CD and FG intersect. The idea is to use an equation of equilibrium that solves for CG directly without having to determine member forces CD and FG. Fictitious joint J can be located by using the similar triangle relationship of member CD.

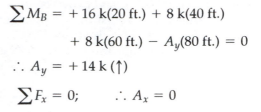

Figure 4.37   *FBD of the truss—section a-a.*

$$\sum M_J = + 14\,k(40\text{ ft.}) - 8\,k(60\text{ ft.})$$
$$- CG_x(15\text{ ft.}) - CG_y(60\text{ ft.}) = 0$$

But: $$CG_x = \left(\frac{4}{5}\right)CG \quad \text{and} \quad CG_y = \left(\frac{3}{5}\right)CG;$$

$$\therefore \left(\frac{4}{5}\right)CG(15\text{ ft.}) + \left(\frac{3}{5}\right)CG(60\text{ ft.})$$
$$= (560\text{ k-ft.}) - (480\text{ k-ft.})$$
$$CG = + 1.67\,k\ (T)$$

$$\sum M_G = - 14\,k(40\text{ ft.}) + 8k(20\text{ ft.})$$
$$- CD_x(15\text{ ft.}) - CD_y(20\text{ ft.}) = 0$$

But: $$CD_x = \frac{4}{\sqrt{17}}CD \quad \text{and} \quad CD_y = \frac{CD}{\sqrt{17}}$$

$$\therefore \left(\frac{4}{\sqrt{17}}CD\right)(15\text{ ft.}) + \left(\frac{CD}{\sqrt{17}}\right)(20\text{ ft.})$$
$$= - (560\text{ k-ft.}) + (160\text{ k-ft.})$$
$$CD = - 20.6\,k\ (C)$$

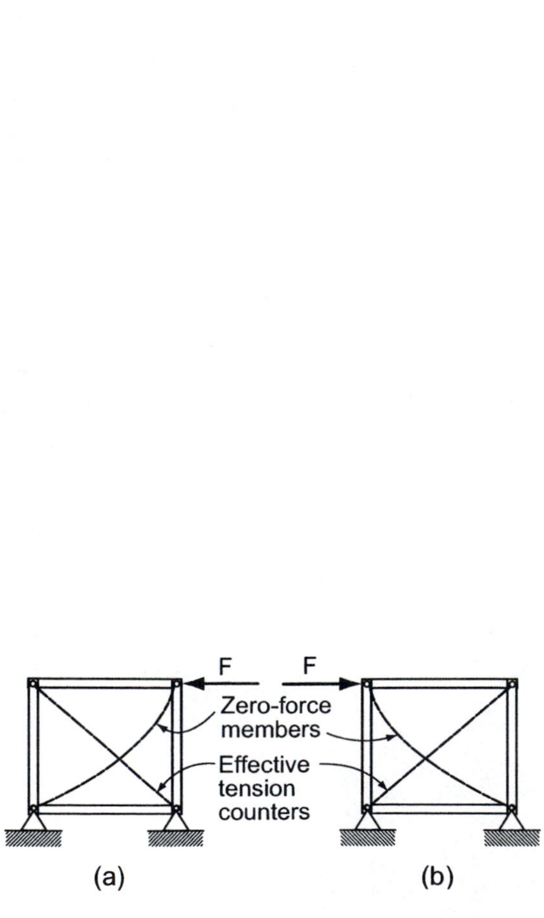

Diagonal Tension Counters

Some trusses subjected to moving loads (freeway, railroad bridges) or lateral forces from wind or earthquake utilize an arrangement of web members within a panel that consists of a pair of crossing diagonals, generally long and slender (Figure 4.38). These diagonals, often referred to as *diagonal tension counters* or *sway bracing*, are usually fabricated of cross-sections such as round rods, flat bars, cables, or light steel angles and tend to be relatively flexible in comparison to other truss members. Diagonal tension counters are designed to carry tensile forces only and are assumed to have

Figure 4.38   *Trusses with diagonal tension counters.*

negligible capacity to resist compressive forces. In a truss containing a tension counter system, one sloping diagonal is subjected to tensile forces while the other is in an apparent compression mode. However, the compressive member buckles and is ineffective due to its slenderness and is thus assumed as a member carrying a zero force.

Example Problem 4.6 (Diagonal Tension Counters)

A parallel chord truss, shown in Figure 4.39, spans 36 ft. from support to support carrying loads at joints E and F. The two internal panels use diagonal tension counters instead of regular web members capable of resisting tension or compression forces. Using the *method of sections*, determine which are the effective tension counters and the tension force developed in each. Show all appropriate free-body diagrams.

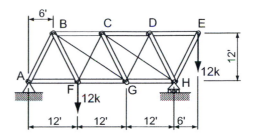

Figure 4.39   Parallel chord truss with diagonal tension counters.

Solution:

Begin the solution by constructing a FBD of the entire truss (Figure 4.40) and solving for the support reactions at A and H. Then, pass a section through the truss cutting the diagonal tension counters within a panel. Note that the section cut will pass through four members, indicating that we have violated the rule stated earlier on cutting no more than three members. Since one of the diagonal tension counters will be considered as a zero-force member (carrying no load), in effect only three members were cut and three equations of equilibrium are available to solve for three unknowns.

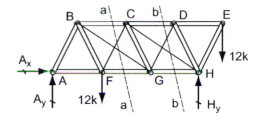

Figure 4.40   FBD of the truss with section cuts a-a and b-b.

$$\sum M_A = -12\ k(12\ ft.) - 12\ k(42\ ft.) + H_y(36\ ft.) = 0$$

$$H_y = \frac{(648\ k\text{-}ft.)}{36\ ft.} = +18\ k$$

$$\sum F = +A_y - 12\ k - 12\ k + 18\ k = 0$$

$$\therefore A_y = +6\ k$$

$$\sum F_x = 0; \quad \therefore A_x = 0$$

Construct a FBD of either half of the truss cut by section a-a, as in Figure 4.41. The cut diagonal tension counters must be shown as tension members since they are only capable of supporting tension forces. One of the diagonals will be ineffective in resisting load and, thus, will be assumed to be zero. Since unknown member forces BC and FG are horizontal, an equation that sums forces in the vertical direction would yield a direct result for the effective tension counter.

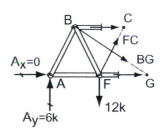

Figure 4.41   FBD of truss cut by section a-a.

$$\sum F_y = +6\ k - 12\ k + FC_y = 0$$

$$\therefore FC_y = 6\ k$$

and by using the slope relationship,

$$FC_x = 3\,k, \quad \text{and} \quad FC = 3\sqrt{5}\,k = 6.71\,k$$

FC is the effective tension counter and $BG = 0$. Note, BG is going in the wrong direction to resist the applied forces in tension.

Again, the effective tension counter can be determined by summing forces (Figure 4.42) in the vertical direction since unknown member forces CD and GH are purely horizontal.

$$\sum F_y = -12\,k + 18\,k - DG_y = 0$$

$$DG_y = 6\,k; \quad \text{and} \quad DG_x = 3\,k$$

$$DG = 3\sqrt{5}\,k = 6.71\,k \text{ (the effective tension counter)}$$

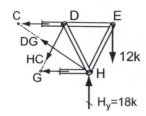

Figure 4.42   FBD of the truss cut by section b-b.

Example Problem 4.7 (Diagonal Tension Counters)

Determine the support reactions at supports A and B. Then, using the method of sections, determine the magnitude of the effective tension counter in panel $DEHG$. Draw a FBD of the right half of the truss cut by section a-a.

Solution (Figure 4.43):

Solving for the support reactions:

$$\sum M_A = -8\,k(20\,\text{ft.}) - 8\,k(40\,\text{ft.})$$

$$-16\,k(60\,\text{ft.}) + B_y(80\,\text{ft.}) = 0;$$

$$\therefore B_y = +18\,k$$

$$\sum M_B = +16\,k(20\,\text{ft.}) + 8\,k(40\,\text{ft.})$$

$$+8\,k(60\,\text{ft.}) - A_y(80\,\text{ft.}) = 0$$

$$\therefore A_y = +14\,k$$

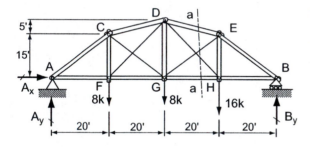

Figure 4.43   Bowstring truss with tension counters.

To determine the effective tension counter in truss panel $DEHG$, pass a section through the truss cutting members DE, GH, and tension counters DH and GE (Figure 4.44). Only one of the tension counters will be effective in resisting loads. Draw a FBD of the cut portion of the truss and solve for the effective counter member.

A direct solution for the unknown effective tension counter can be achieved by writing an equation of equilibrium that excludes the other two unknowns ED and HG. Note that member forces ED and HG have a common point of intersection at imaginary point J, 40 ft. to the right of support B. Point J is determined by extending the line of action of ED along a 1 in 4 slope. By summing moments about point J, the effective tension counter will be the only unknown.

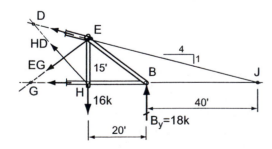

Figure 4.44   FBD of the truss—section a-a.

The 16-k load at joint H, with a moment arm of 60 ft., produces a counterclockwise rotation about J with a magnitude of 960 k-ft. Reaction force $B_y = 18$ k has a 40-ft. moment arm and generates a clockwise rotation of 720 k-ft. about joint J. The resultant moment of these two forces is a 240 k-ft. counterclockwise moment. Tension counter HD is the effective counter capable of generating an opposing clockwise moment.

$$\sum M_J = -18\text{ k}(40\text{ ft.}) + 16\text{ k}(60\text{ ft.})$$

$$-\left(\frac{HD}{\sqrt{2}}\right)(60\text{ ft.}) = 0$$

$$\therefore HD = +4\sqrt{2}\text{ k} = 5.66\text{ k }(T)$$

Member EG is assumed to be a zero-force member.

Maxwell's diagram (graphical truss analysis)

The determination of member forces in a truss can easily be made by utilizing a graphical technique developed in 1864 by James Clerk Maxwell, Professor of Physics and Astronomy at King's College in London. This method is the graphical equivalent of the analytical *method of joints*. A force, as described in Chapter 2, is a vector quantity that can be represented by a scaled arrow. Each truss joint is a system of concurrent, coplanar forces and if the joint is to be in equilibrium, the figure formed by the graphical representation of the force vectors will be a closed polygon.

General procedure for Maxwell's diagram

Step 1: *A necessary first step in the graphical method is the layout of the truss to a convenient, relatively large scale, making certain that all dimensions are scaled accurately. This scaled diagram is also a working free-body diagram (FBD).*

Step 2: *Calculate all support reactions, making sure that the conditions of equilibrium are satisfied.*

$$\sum F_x = 0; \quad \sum F_y = 0; \quad \sum M_i = 0;$$

where i is any point.

Step 3: *In solutions using the graphical force polygon, a special notation system called Bow's notation is utilized. Instead of following the convention of lettering the joints of a truss, Bow's notation uses the lettering or numbering of the spaces between adjacent external forces and in each triangular panel of the truss. Each member and force vector is designated by the two lowercase letters on either side. Labeling generally follows a clockwise order for clarity.*

James Clerk Maxwell (1831–1879)

Maxwell displayed early signs of mathematical talent. When he presented an original paper to the Royal Society of Edinburgh at age 15, many refused to believe that such a young boy could have been the sole author. Later, at Cambridge, he graduated second in his class in mathematics.

In 1864 Maxwell, at the time Professor of Physics and Astronomy at King's College, London, published the well-known graphical solution of the reciprocal stress diagram (Maxwell's diagram). The method enabled engineers to design significant truss structures for over a hundred years. It was a most remarkable contribution to structural mechanics because it was made by a man who had no direct connection to structures or even engineering. Maxwell is best known for his authorship of the theory of electromagnetism. Among other important contributions, he proposed the now-accepted theory that Saturn's rings consist of myriads of small bodies as opposed to being solid or hollow discs.

Maxwell died of cancer before the age of 50. Had he lived a normal life expectancy he would have witnessed much progress in his field of physics, offshooting partially from his work. However, when Einstein's theories, a generation after Maxwell's death, upset all of "classical physics," Maxwell's equations remained virtually as valid as ever.

Step 4: Begin the graphical process by drawing a force diagram for the external forces and reactions that act on the truss. Select the external forces and reactions in a clockwise order around the truss. Make sure you select a convenient scale that results in a large enough diagram to maintain a fair degree of accuracy. Lowercase letters used in denoting the spaces on either side of the external forces or members are changed to capital letters, transforming spaces into points in the force diagram. Since all external forces are in equilibrium, the resulting force polygon must close (the starting point must be the end point as well). This particular force diagram is often called the load diagram.

Step 5: Starting at a joint with no more than two unknown forces (usually at the support end of a spanning truss), draw a force diagram, beginning with the known forces and continuing around a joint until all forces form a closed polygon for the joint. Do not show the sense (arrowhead direction) of each force on the force diagram.

Step 6: Continue this process, joint by joint, until all member forces have been drawn. All spaces lettered using Bow's notation are now represented as points in the force (Maxwell's) diagram. This graphical method is self-checking because, at the last joint, all member forces will already be known and must by necessity form a closed diagram for equilibrium to exist.

Step 7: Scale the magnitudes of each member force and record them on the free-body diagram.

Step 8: The sense (arrowhead direction of each member force at a joint indicating tension or compression) is determined by examining each joint individually and applying the clockwise rule.

This idea is best explained through the use of an example problem.

Example Problem 4.8 (Maxwell's Diagram)

A simple asymmetrical truss as shown in Figure 4.45 is supporting a single 7-kN load at the lower chord joint. Using Maxwell's diagram, determine all member forces.

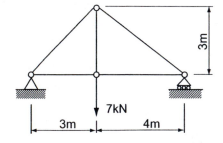

Figure 4.45   Truss analysis—Maxwell's diagram.

Solution:

The solution begins with the scaled drawing of the truss, shown as a free-body diagram (Figure 4.46). Since joints of the truss are not labeled, the spaces around the joint are used to identify the joint. Similarly, members are designated by the spaces on either side. For example, the bottom left joint is called out as *abd* (or *bda* or *dab*). The center, bottom joint is identified as *deca* (using a clockwise order). The two inclined, top chord members are known as members *bd* and *eb* and the vertical member is *de*. Write the equations of equilibrium and determine the support reactions at each end of the truss. Since there are no horizontally applied forces, the horizontal reaction at joint *abd* does not develop.

The scaled FBD of the truss is used in establishing the member force directions in the *load and force* diagrams. Depending on the magnitude of the forces applied to the truss, a convenient scale is selected for drawing the force diagram. In this problem, a scale of 10 mm = 1 kN will be used. The *load diagram* is constructed first, drawing all of the external forces to scale. Forces will be drawn following a clockwise order around the truss. The left reaction *ab* is a vertical line 40-mm long, directed in an upward direction. Moving clockwise around the truss the next external force is the reaction at the right support *bc*. The load diagram should now be a vertical line 70-mm long (or 7 kN), as shown in Figure 4.47.

Next, select a joint that has no more than two unknown members intersecting. In this example, joints *abd* or *bce* are possible starting points. For this illustration, joint *abd* will be selected. Transfer to the load diagram (which will now be referred to as a force diagram) accurately drawn lines of action of members *bd* and *da* (Figure 4.48). *The use of two triangles to transfer lines parallel and perpendicular are particularly useful.*

The intersection of member lines *bd* and *da* defines the point *D*. Note that spaces (indicated by lowercase letters) in the FBD using Bow's notation have now transformed into points (in capital letters) in the force diagram.

Once point *D* is known, member forces *bd* and *da* are known. Move on to a new joint that has no more than two unknowns, such as joint *deca*. Construct member lines *de* and *ec*. The intersection of those two lines (Figure 4.49) locates point *E*. Member forces *de* and *ec* are now known.

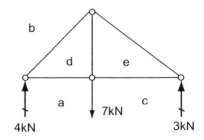

Figure 4.46 FBD with Bow's notation.

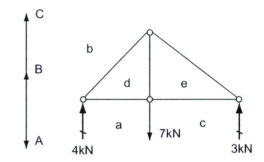

Figure 4.47 Load diagram.

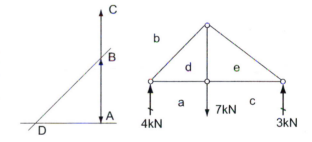

Figure 4.48 Force diagram for BD and DA.

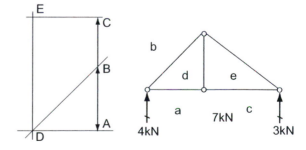

Figure 4.49 Force diagram for DE and EC.

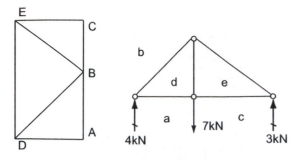

Figure 4.50   Force diagram for BE.

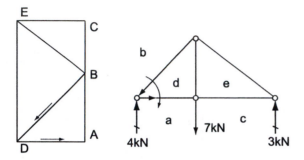

Figure 4.51   Determining the sense for members bd and da.

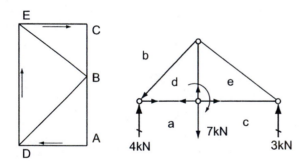

Figure 4.52   Determining the sense for members ad, de, and ec.

The last unknown member force is *eb,* but since all of the points are known, line *EB* can be drawn between *B* and *E* as shown in Figure 4.50.

Determining the sense of each member force (tension or compression) requires the use of both the FBD of the truss and the final force diagram. A joint (any joint) is selected and member forces are called out using a clockwise convention. Going clockwise around joint *bda* in the FBD in Figure 4.51, the inclined member is called out as *bd.*

Looking at the adjacent *force diagram,* the direction of the force from point *B* to point *D* is coming down and toward the left. That same direction is shown on the FBD indicating a compression force, with the arrow pushing on the joint. The horizontal member (continuing clockwise around the joint) is *da.* In the force diagram, the member force *da* is directed to the right, going from point *D* to *A.* In the FBD, the force *da* is also directed to the right, indicating tension since the force is pulling at the joint. Equal and opposite arrows are shown at the joint on the other end for each respective member. Move to another joint (*deca* in Figure 4.52) and repeat the same process. Going in a clockwise order, member *ad* on the force diagram is directed to the left. On the FBD, the same direction for the member force would indicate tension. This confirms the result found earlier at joint *bda.* All member forces and their respective tension or compression are summarized in Figure 4.53.

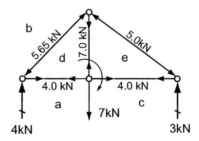

Figure 4.53   Force summation diagram.

Example Problem 4.9 (Maxwell's Diagram)

A triangular roof truss is subject to wind forces acting perpendicular on the windward surface, as shown in Figure 4.54. Solve for the support reactions, then analyze the truss using Maxwell's diagram. Use a scale of $\frac{1}{2}$" = 1 k.

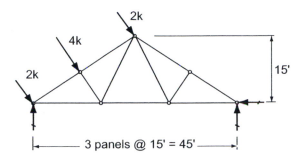

Figure 4.54   Triangular roof truss with wind loading.

Solution:

Draw the truss to scale, then label the truss going in a clockwise order, using Bow's notation. Write the three equations of equilibrium and solve for the support reactions (Figure 4.55).

Using the scaled FBD of the truss to establish the directions of external and member forces, construct the load and force diagrams (Figure 4.56). Magnitudes for the member forces are scaled directly off the force diagram and recorded on a force summation diagram. The sense of each member force (tension or compression) is determined by using the clockwise rule at each joint of the FBD in combination with the force diagram. The results are also summarized on the force summation diagram (Figure 4.57).

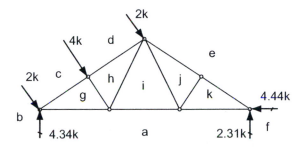

Figure 4.55   FBD of the truss with Bow's notation.

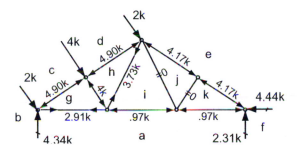

Figure 4.57   Force summation diagram.

Zero-force members

Occasionally a truss analysis will result in some members theoretically carrying zero load. These *zero-force members* are not useless or unnecessary; they simply carry no load for a particular load condition but are necessary for other load conditions and stability purposes. In particular, long members carrying compression forces need to be braced periodically to prevent buckling of the member or to minimize the size of the member needed to resist buckling tendencies for long, unbraced members. Many trusses that appear complicated at first glance can be simplified considerably by recognizing the members that carry no load (zero-force members). A general rule that can be used in determining zero-force members by inspection can be stated as follows:

> At an unloaded joint where three members frame together, if two members are in a straight line, the third is a zero-force member.

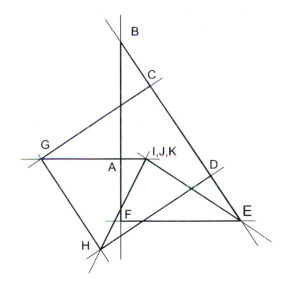

Figure 4.56   Load and force diagram.

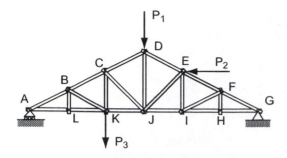

Figure 4.58   Identification of zero-force members in a truss.

Example Problem 4.10 (Zero-Force Members)

Identify, by observation, the zero-force members in the truss shown in Figure 4.58.

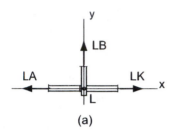

(a)

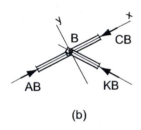

(b)

Figure 4.59   (a) FBD of joint L (or H). (b) FBD of joint B (or F).

Solution:

The zero-force members are: *LB, KB, HF, FI,* and *IE.* Free-body diagrams of joints *L* (or *H*) and *B* (or *F*) (Figures 4.59a, b) would show that *LB* and *HF* are zero; therefore, *KB* and *FI* would also be zero. If *FI* is zero, then *IE* would be zero.

$$\sum F_y = +LB = 0$$

Since no vertical loads are present, $LB = 0$.

Rotate the x and y axes, then:

$$\sum F_y = +KB = 0$$

$$\therefore KB = 0$$

4.2   PINNED FRAMES WITH MULTI-FORCE MEMBERS

The previous section on trusses dealt with straight members, pin connected at the ends with applied loads present only at the pinned joints (Figure 4.60). As a result, all truss members were considered as two-force members shown in Figure 4.61, carrying tension or compression forces along the line of action of the member, with no resulting bending moments.

Pinned frames are structures not limited to two-force members and generally contain multi-force members that are capable of resisting an array of forces (tension, compression, bending, and shear). Many pinned frames resemble trusses in configuration but because forces may be present at points other than at the ends, there is no single resultant force directed along the line of action of the member. When pin forces in a member occur at

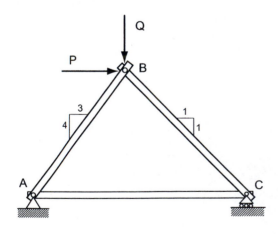

Figure 4.60   A pin-connected truss. External loads are applied at the joint(s) only and members AB, BC, and CA are two-force members.

intermediate points in addition to the ends, bending moments result. Member forces produced by external loads are transmitted through the members to the pins at the joints, and distributed by the pins to other connected members. Pin forces, also referred to as pin reactions, assume friction-less pins and thus disregard the effects of friction.

Member forces in trusses were determined using the method of joints, or for specific members, the method of sections. Both of these analysis techniques depended on the assumption of all members being two-force members with known directions (where the resultant forces at the pinned ends or within the member followed the slope of member) Pinned frame analysis (also referred to as the *method of members*) differs significantly from truss analysis. The determination of internal forces within a multi-force member is more involved and will not be covered in this section. Analysis will be limited to determining the pin forces resulting on each member due to external loads. Forces produced by external loads on pinned frames are transmitted through members to the pins at the joints, and by the pins to other connected members. Forces developed at the pins are called *pin reactions.*

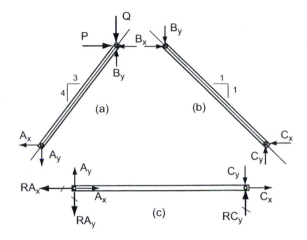

Figure 4.61   FBD of the three truss-component members. Resultants at the ends of each member will pass through the axis of the member.

General Procedure for Determining Pin Reactions (Use Figures 4.62 and 4.63):

- Draw a FBD of the entire frame.
- Solve for the support reactions. In the example of Figure 4.62, all three support reactions can be solved using the three equations of equilibrium. Sometimes only one or two of the support reactions can be solved for at this stage. For example, if the frame in Figure 4.62 had the roller support replaced with another hinge, only vertical support reactions RA_y and RC_y would be solvable. If the frame is statically indeterminate, externally, reactions must be solved by using additional free-body diagrams.

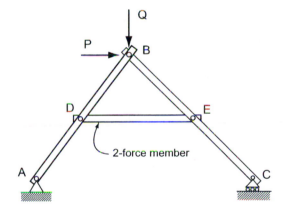

Figure 4.62   A pin-connected frame. External loads are still applied at the joint B. Members ADB and BEC are multi-force members with no single resultant passing through the axis of the respective member.

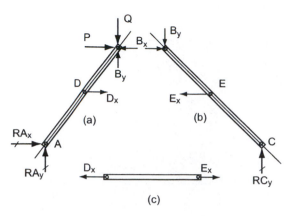

Figure 4.63   FBD of each frame member. Member DE is the only two-force member with a known line of action through the member. Members ADB and BEC develop bending moments due to intermediate pin forces D_x and E_x, respectively.

- To solve for internal pin forces (reactions) and remaining unknown support reactions, separate the individual elements of the frame and construct FBDs of each component member, as shown in Figure 4.63.
- Calculate the internal pin forces by writing equations of equilibrium for the component FBDs. Three equations of equilibrium can be written for each component FBD. Select the most convenient FBD to solve for the desired pin reactions. It is permissible to jump from one FBD to another until all pin reactions are solved. Remember that the direction of the pin reaction on one member must be shown in an equal and opposite sense on the other connected member.
- When external loads are applied at a joint, the load may be arbitrarily assigned to either member. Note, however, that the applied load must appear only once.

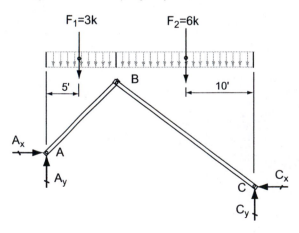

Figure 4.64   Sloping roof beams with snow load.

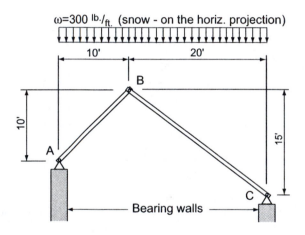

Figure 4.65   FBD of the entire frame.

Example Problem 4.11 (Pinned Frame with Multi-Force Members)

Two inclined roof beams support a snow load equal to $300^{lb.}/_{ft.}$ on the horizontal projection (Figure 4.64). All three joints, A, B, and C, are pins. Supports at A and C are attached to rigid bearing walls. Determine the support reactions at A and C and the pin reactions at B. Draw all of the appropriate free-body diagrams. This particular framework configuration is often referred to as a *three-hinged arch.*

Solution:

Note that since there are four support unknowns and only three equations of equilibrium for the FBD in Figure 4.65, another equation involving the same two unknowns (C_x and C_y) needs to be written. To accomplish this, draw the respective FBDs of the two roof beams and write the equation of equilibrium that includes unknowns C_x and C_y. Then, solve for the unknowns C_x and C_y simultaneously.

Begin with an equation using the FBD of the entire frame as shown in Figure 4.65.

Equation (1):

$$\sum M_A = -(3\,k)(5\,ft.) - (6\,k)(20\,ft.)$$
$$- C_x(5\,ft.) + C_y(30\,ft.) = 0$$

Using the FBD in Figure 4.66b:

Equation (2):

$$\sum M_B = -6\,k(10\text{ ft.}) - C_x(15\text{ ft.}) + C_y(20\text{ ft.}) = 0$$

Simplifying the equations,

$$-C_x + 6\,C_y = 27\,k; \ \ldots\ (\text{Eq. 1})$$
$$+3C_x - 4\,C_y = -12\,k; \ \ldots\ (\text{Eq. 2})$$

Therefore, solving equations (1) and (2) simultaneously,

Eq. (1) $-C_x + 6C_y = 27\,k$

Eq. (2) $\div$ 3: $+C_x - 1.33C_y = -4\,k$

$\therefore C_y = +4.9\,k\,(\uparrow)$; and $C_x = +2.6\,k\,(\leftarrow)$

Now that reaction forces C_x and C_y are known, the reactions A_x and A_y can be solved using the FBD of the entire frame (Figure 4.65).

$$\sum F_x = +A_x - \underset{(2.6\,k)}{C_x} = 0$$

Solving for the single unknown,

$$A_x = +2.6\,k\,(\rightarrow) \quad \text{and}$$

$$\sum F_y = +A_y - 3\,k - 6\,k + \underset{(4.9\,k)}{C_y} = 0;$$

$$\therefore A_y = +4.1\,k\,(\uparrow)$$

The internal pin reactions at B can be solved by using FBDs of either roof beam in Figure 4.67a or b.

If the FBD in Figure 4.67b were used:

$$\sum F_x = +B_x - \underset{(2.6\,k)}{C_x} = 0; \quad \therefore B_x = +2.6\,k$$

$$\sum F_y = -B_y - 6\,k + \underset{(4.9\,k)}{C_y} = 0; \quad \therefore B_y = -1.1\,k$$

The negative sign indicates that the initial assumption about the direction of pin reaction B_y was incorrect. The magnitude of the pin reactions at B are correct; just the direction of B_y needs to be reversed.

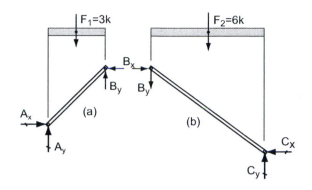

Figure 4.66   FBDs of each inclined roof beam.

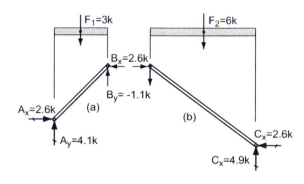

Figure 4.67   Summary of the pin reactions.

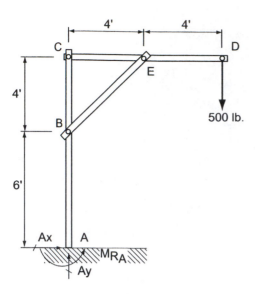

Figure 4.68   Pinned frame with a fixed support.

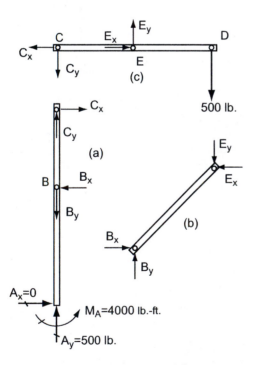

Figure 4.69   FBD of individual frame members.

Example Problem 4.12

A pinned frame with a fixed base at A supports a load at the overhang equal to 500 pounds, as shown in Figure 4.68. Draw free-body diagrams and solve for the support reactions and the pin reactions at B, C, and E.

Solution:

Since the pinned frame is statically determinate externally, all three support equations can be solved by using the requisite three equations of equilibrium.

$$\sum F_x = A_x = 0$$

$$\sum F_y = +A_y - 500 \text{ lb.} = 0; \quad \therefore A_y = +500 \text{ lb. } (\uparrow)$$

$$\sum M_A = +M_A - 500 \text{ lb. } (8 \text{ ft.}) = 0;$$

$$\therefore M_A = +4{,}000 \text{ lb.-ft.}$$

Isolate and construct a free-body diagram of each frame member. Three additional equations of equilibrium, and thus three additional unknowns, can be solved for each member FBD.

Free-body diagrams (a) or (c) in Figure 4.69 may be used to begin the process of determining the pin reaction forces at B, C, and E.

Using FBD (a):

$$\sum M_B = -C_x(4 \text{ ft.}) + \left(\underset{(M_A)}{4000 \text{ lb.-ft.}} \right) = 0;$$

$$\therefore C_x = \frac{(4000 \text{ lb.-ft.})}{4 \text{ ft.}} = +1000 \text{ lb.}$$

$$\sum F_x = +1000 \text{ lb.} - B_x = 0;$$

$$\therefore B_x = +1000 \text{ lb.}$$

Record the results for B_x and C_x as equal and opposite forces on the other corresponding member.

Component pin reactions B_y and C_y cannot be solved with the remaining equation of equilibrium.

It is permissible to jump from one free-body diagram to another in solving for the unknown pin reactions.

Moving to free-body diagram (c);

$$\sum F_x = \underset{(C_x)}{-1000 \text{ lb.}} + E_x = 0; \quad \therefore E_x = +1000 \text{ lb.}$$

$$\sum M_C = +E_y(4 \text{ ft.}) - 500 \text{ lb.}(8 \text{ ft.}) = 0;$$

$$\therefore E_y = +1000 \text{ lb.}$$

$$\sum F_y = -C_y - 1000 \text{ lb.} + \underset{(E_y)}{500 \text{ lb.}} = 0;$$

$$\therefore C_y = +500 \text{ lb.}$$

Note that the (+) sign attached to the answer only indicates that the initial assumption of the component force on the FBD was correct. The actual force is resisting in the downward direction. Record the component force results equal but opposite in direction on the adjoining member.

The last component force B_y can be solved by using FBD(s) in Figures 4.69a or b. Recognize that member BE is a two-force member and the component reactions at either end are related by virtue of the member's slope relationship. For example, member BE has a 1:1 slope relationship. Therefore, the x and y components will be equal and

$$\therefore B_y = +1000 \text{ lb.}$$

It is often useful to record your answers directly on the FBDs to keep track of your solutions.

Example 4.13 (Three-Hinged Arch)

An industrial building is framed using tapered steel sections (haunches) and connected with three hinges (Figure 4.70). Assuming that the loads shown are from gravity loads and wind, determine the support reactions at A and D and the pin reactions at B.

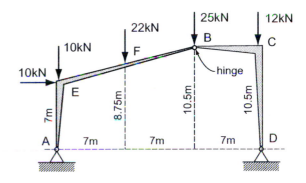

Figure 4.70   Frame as a three-hinged arch.

Solution:

Construct a FBD (Figure 4.71) of the entire three-hinged arch and determine as many of the support reactions as possible using the three available equations of equilibrium.

Since there are four support reactions and only three equations of equilibrium, only A_y and D_y can be solved at this time.

$$\sum M_A = -10 \text{ kN}(7 \text{ m}) - 22 \text{ kN}(7 \text{ m}) - 25 \text{ kN}(14 \text{ m})$$

$$-12 \text{ kN}(21 \text{ m}) + D_y(21 \text{ m}) = 0$$

$$\therefore D_y = 39.3 \text{ kN} (\uparrow)$$

$$\sum F_y = +A_y - 10 \text{ kN} - 22 \text{ kN} - 25 \text{ kN}$$

$$-12 \text{ kN} + 39.3 \text{ kN} = 0$$

$$\therefore A_y = +29.7 \text{ kN} (\uparrow)$$

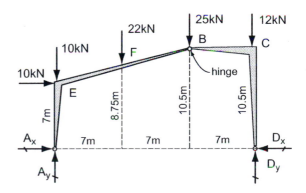

Figure 4.71   FBD of the entire three-hinged arch.

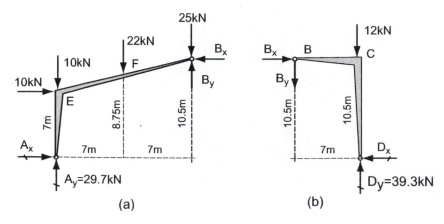

Figure 4.72   FBD of each haunch section.

To determine A_x and D_x and the pin reactions at B, additional free-body diagrams are needed.

Separate the three-hinged arch into its two main components and draw the FBD of each.

Using Figure 4.72a, note that the 25-kN load at B was assigned to member AEB (an assumption).

$$\sum M_B = +22 \text{ kN}(7 \text{ m}) + 10 \text{ kN}(14 \text{ m})$$

$$+ 10 \text{ kN}(3.5 \text{ m}) - 29.7 \text{ kN}(14 \text{ m})$$

$$+ A_x(10.5 \text{ m}) = 0$$

$$\therefore A_x = 8.3 \text{ kN} (\rightarrow)$$

$$\sum F_x = +8.3 \underset{(A_x)}{\text{ kN}} + 10 \text{ kN} - B_x = 0$$

$$\therefore B_x = +18.3 \text{ kN}$$

$$\sum F_y = +29.7 \underset{(A_y)}{\text{ kN}} - 10 \text{ kN} - 22 \text{ kN}$$

$$- 25 \text{ kN} + B_y = 0$$

$$\therefore B_y = +27.3 \text{ kN}$$

The remaining unknown D_x can be solved using free-body diagrams Figure 4.71 or Figure 4.72b.

Using Figure 4.72b:

$$\sum F_x = +18.3 \underset{(B_x)}{\text{ kN}} - D_x = 0$$

$$\therefore D_x = +18.3 \text{ kN} (\leftarrow)$$

As a check, substitute the answer for D_x into an equation for the horizontal condition of equilibrium using Figure 4.71.

Example Problem 4.14

A carport is constructed using a series of identical frames, as shown in Figure 4.73. Wind loads are applied horizontally at joint C and at a distance 10' above support A. All joints are pin connected with *frictionless* pins. Draw a series of free-body diagrams and solve for the support reactions at A and G and the pin reactions at B, C, D, E, and F. Note that members ABC and EFG are multi-force members but BD, CD, DE, and DF are two-force members.

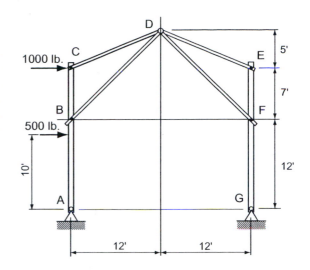

Figure 4.73   Wind loads on a pinned frame.

Solution:

Begin the process by drawing a FBD of the entire frame (Figure 4.74) and solve for as many of the support reactions as permissible.

There are four unknown support reaction components and only three equations of equilibrium; therefore, the external condition is indeterminate to the first degree. However, two of the support reactions, A_y and G_y, can be solved.

$$\sum M_G = -500 \text{ lb. (10 ft.)} - 1000 \text{ lb. (19 ft.)}$$

$$+ A_y(24 \text{ ft.}) = 0$$

$$\therefore A_y = +1000 \text{ lb. } (\downarrow)$$

$$\sum F_y = -1000 \text{ lb.} + G_y = 0$$

$$\therefore G_y = +1000 \text{ lb. } (\uparrow)$$

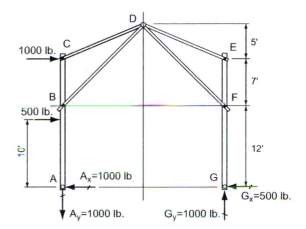

Figure 4.74   FBD of the entire frame.

Draw two additional FBDs separating the frame at the crown joint D. These two new FBDs allow an additional three equations of equilibrium each.

Support reaction force A_x can be solved by using the FBD in Figure 4.75a.

$$\sum M_D = +1000 \text{ lb. (5 ft.)} + 500 \text{ lb. (14 ft.)}$$

$$+ \underset{(A_y)}{1000 \text{ lb. (12 ft.)}} - A_x (24 \text{ ft.}) = 0$$

$$\therefore A_x = +1000 \text{ lb. } (\leftarrow)$$

$$\sum F_y = -1000 \text{ lb.} + D_y = 0$$

$$\therefore D_y = +1000 \text{ lb.}$$

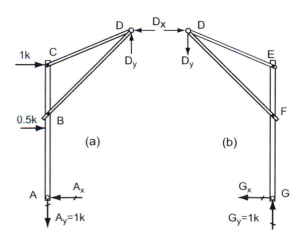

Figure 4.75   FBD of the two subassemblies.

$$\sum F_x = +1000 \text{ lb.} + 500 \text{ lb.} - \underset{(A_x)}{1,000 \text{ lb.}} - D_x = 0$$

$$\therefore D_x = +500 \text{ lb.}$$

Now, using the FBD in Figure 4.75b, solve for the support reaction G_x.

$$\sum M_D = +1000 \text{ lb. (12 ft.)} - \underset{(G_y)}{G_x} (24 \text{ ft.}) = 0$$

$$\therefore G_x = +500 \text{ lb.} (\leftarrow)$$

To solve for the pin reactions at B, C, E, and F, isolate the vertical members ABC and EFG as two additional free-body diagrams.

Writing equations of equilibrium based on member ABC in Figure 4.76a

$$\sum M_C = +DB_x (7 \text{ ft.}) + 500 \text{ lb. (9 ft.)}$$

$$- \underset{(A_x)}{1000 \text{ lb. (19 ft.)}} = 0$$

$$\therefore DB_x = +2070 \text{ lb.}$$

$$\sum F_x = +1000 \text{ lb.} + 500 \text{ lb.} - \underset{(A_x)}{1000 \text{ lb.}}$$

$$+ \underset{(DB_x)}{2070 \text{ lb.}} - CD_x = 0$$

$$\therefore CD_x = +2570 \text{ lb.}$$

Moving to the FBD in Figure 4.76b:

$$\sum F_x = +2570 \text{ lb.} - DC_x = 0$$

$$\therefore DC_x = +2570 \text{ lb.}$$

Since member CD is a two-force member, the component x and y forces are related by virtue of their slope.

Therefore:

$$CD_y = \left(\frac{5}{12}\right)CD_x = \left(\frac{5}{12}\right)(2570 \text{ lb.}) = 1070 \text{ lb.}$$

The same is true for DC_y where:

$$DC_y = \left(\frac{5}{12}\right)DC_x = 1070 \text{ lb.}$$

This same result can be obtained by writing an equation of equilibrium for the horizontal condition.

Using a similar procedure for the right half of the three-hinged arch frame, solve for member and pin forces ED and DF.

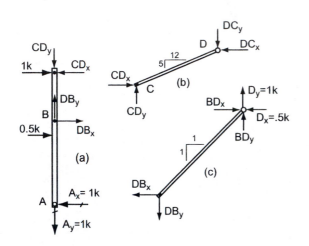

Figure 4.76   FBD of each member of the left subassembly.

Beginning with the FBD in Figure 4.77a:

$$\sum M_E = -500 \text{ lb. (19 ft.)} + \underset{(G_x)}{FD_x} \text{ (7 ft.)} = 0$$

$$\therefore FD_x = +1360 \text{ lb.}$$

$$\sum F_x = -DE_x + \underset{(FD_x)}{1360 \text{ lb.}} - 500 \text{ lb.} = 0$$

$$\therefore DE_x = +860 \text{ lb.}$$

From the FBD in Figure 4.77c:

$$\sum F_x = +DF_x - \underset{(FD_x)}{1360 \text{ lb.}} + 500 \text{ lb.} = 0$$

$$\therefore DF_x = +860 \text{ lb.}$$

From the 1:1 slope relationship for member DF:

$$D_y + DF_y = FD_y = 1360 \text{ lb.}; \therefore DF_y = 360 \text{ lb.}$$

Using the 5:12 slope relationship for member DE:

$$DE_y = ED_y = \left(\frac{5}{12}\right) DE_x = \left(\frac{5}{12}\right)(860 \text{ lb.}) = 360 \text{ lb.}$$

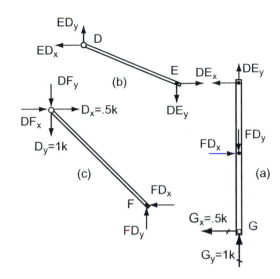

Figure 4.77 FBD of members of the right subassembly.

4.3 FLEXIBLE CABLES WITH CONCENTRATED LOADS

High-strength steel cables are an efficient structural system used for supporting loads over long spans such as in suspension bridges, roof structures, transmission lines, cables, and tramways. The strength-to-weight ratio of steel cables is very high because of their excellent tensile capability. Cables are relatively light and assumed to be flexible and incapable of resisting bending or compression. Since cables support loads only in tension, the system must be kept in tension to maintain its stability. A perfectly flexible cable will change its shape (geometry) when load conditions are changed. The slope that a cable takes under concentrated loads is referred to as a funicular polygon. As the number of applied loads increases, the cable's shape approaches that of a smooth curve. When loads are changed, the cable configuration changes immediately to reestablish a condition of equilibrium.

Cables may be categorized based on whether their loading is concentrated or distributed. This section will examine only cables supporting concentrated loads. The determination of the support reactions and the tension in each segment of the cable between vertical loads can be made by using the equations of equilibrium with corresponding free-body diagrams. In problems involving multiple concentrated loads, it is necessary to know the elevation of the supports and the position or sag of at least one point along the cable (generally at a load point).

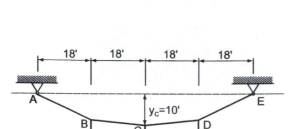

Figure 4.78   Cable with three concentrated loads.

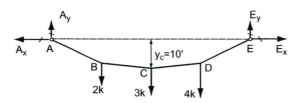

Figure 4.79   FBD of the entire cable.

Any portion of the cable between successive loads is considered as a two-force member and the internal force at any point in the cable follows the line of action of the cable segment. The horizontal component of the cable force remains a constant throughout the cable. Also, since the vertical component of the cable force is dependent upon the cable slope, the largest tension force will be in the portion of the cable with the greatest slope (usually at either support end).

Example Problem 4.15

Three concentrated loads are supported by a steel cable at points B, C, and D. Assume that the sag at point C on the cable, relative to the elevation of the supports, is 10 feet (Figure 4.78). Determine the support reactions at A and E and the maximum cable tension. Also, determine the sag at points B and D.

Solution:

A first step involves the drawing of a free-body diagram of the entire cable system (Figure 4.79) and solving for as many of the support reactions as possible. In this example, since both supports are at the same elevation, the y-component of the support reactions are solvable. If the supports are not at the same elevation, none of the support component reactions can be determined until an additional free-body diagram and equation(s) of equilibrium is written. See Example Problem 4.16 for this latter case.

$$\sum M_A = + E_y\,(72\text{ ft.}) - 2\,\text{k}(18\text{ ft.}) - 3\,\text{k}(36\text{ ft.})$$

$$- 4\,\text{k}\,(54\text{ ft.}) = 0;$$

$$\therefore E_y = \frac{+ 2\,\text{k}(18\text{ ft.}) + 3\,\text{k}(36\text{ ft.}) + 4\,\text{k}(54\text{ ft.})}{72\text{ ft.}}$$

$$= + 5\,\text{k}\,(\uparrow)$$

$$\sum F_y = + A_y - 2\,\text{k} - 3\,\text{k} - 4\,\text{k} + \underset{(E_y)}{5\,\text{k}} = 0$$

$$\therefore A_y = + 4\,\text{k}\,(\uparrow)$$

If the slope of cable segment AB were known, component A_x could be solved using its slope relationship since all cable segments are considered two-force members. The same would be true for the E_x and E_y reactions in relation to the slope of cable segment DE. However, since no slope relationships can be ascertained from the given dimensions, another free-body diagram must be introduced that makes use of the sag dimension of 10 ft. at point C. Assume that the cable is cut at point C and construct a FBD of either half of the section cut. It does not matter to which cable segment

FBD the 3k load at C is assigned. If the section cut is imagined occurring just to the right of point C, the 3k load would appear in the FBD of the left half.

Using the FBD in Figure 4.80, write a moment equation about point C and solve for the unknown support reaction A_x.

$$\sum M_C = +2\,k\,(18\text{ ft.}) - 4\,k\,(36\text{ ft.}) + A_x(10\text{ ft.}) = 0$$

$$\therefore A_x = +10.8\,k\,(\leftarrow)$$

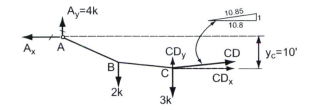

Figure 4.80   Cable segment FBD with the cut just to the right of C.

Then, solve for the component forces in cable segment CD.

$$\sum F_x = -\underset{(A_x)}{10.8\,k} + CD_x = 0 \quad \therefore \Delta CD_x = +10.8\,k$$

$$\sum F_y = +4\,k - 2\,k - 3\,k + CD_y = 0$$

$$\therefore CD_y = +1\,k$$

and $CD = 10.85\,k$ (tension).

The ratio of CD_y to CD_x is: $\dfrac{CD_y}{CD_x} = \dfrac{1}{10.8}$

Knowing the slope of cable segment CD, the sag at point D can be determined. The horizontal distance between C and D is 18 ft. Using the slope relationship of cable segment CD, the vertical difference in elevation between C and D is:

$$\frac{1}{10.8} \times (18\text{ ft.}) = 1.67\text{ ft.}$$

Therefore, the sag at point D, relative to C, is:

$$y_D = 10\text{ ft.} - 1.67\text{ ft} = 8.33\text{ ft.}$$

Returning to the FBD of the entire cable (Figure 4.79), solve for the support reaction component E_x.

$$\sum F_x = -\underset{(A_x)}{10.8\,k} + E_x = 0 \qquad \therefore E_x = 10.8\,k\,(\rightarrow)$$

The reaction components at support E are also the components of the cable forces in segment DE.

$$\therefore DE_x = E_x = 10.8\,k \quad \text{and} \quad DE_y = E_y = 5\,k$$

A verification of the sag y_D at D can be solved using the slope relationship of: $\dfrac{DE_y}{DE_x} = \dfrac{5}{10.8} = \dfrac{1}{2.16}$

If the horizontal distance between D and E is 18 ft., then the sag is:

$$y_D = \left(\frac{1}{2.16}\right) \times (18\text{ ft.}) = 8.33\text{ ft.}$$

This answer checks with the previous solution. Similarly, the cable force components in segment AB are:

$$AB_x = A_x = 10.8\,k \quad \text{and} \quad AB_y = A_y = 4\,k$$

The slope of segment AB is:

$$\frac{AB_y}{AB_x} = \frac{4}{10.8} = \frac{1}{2.7}$$

and the sag at point B relative to A is:

$$\text{sag} = y_B = \left(\frac{1}{2.7}\right) \times (18 \text{ ft.}) = 6.67 \text{ ft.}$$

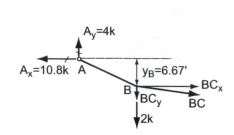

Figure 4.81   FBD of cable segment AB.

Another FBD (Figure 4.81) is required to determine the tension in cable segment BC.

$$\sum F_x = -10.8 \underset{(A_x)}{} k + BC_x = 0$$

$$\therefore BC_x = +10.8 \text{ k}$$

Note that the horizontal component in each cable segment and at the supports remains a constant.

$$\sum F_y = +4 \text{ k} - 2 \text{ k} - BC_y = 0 \quad \therefore BC_y = +2 \text{ k}$$

The slope of cable segment BC is: $\dfrac{BC_y}{BC_x} = \dfrac{2}{10.8} = \dfrac{1}{5.4}$; and the relative vertical distance between B and C equals:

$$y_{B \to C} = \left(\frac{1}{5.4}\right) \times (18 \text{ ft.}) = 3.33 \text{ ft.}$$

A summary of the results of the analysis is shown in Figure 4.82.

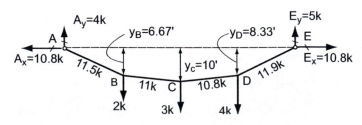

Figure 4.82   Summary of the sag and cable tensions.

Example Problem 4.16

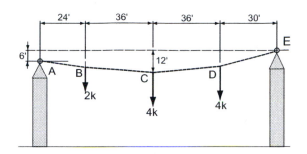

A cable roof system supports three concentrated loads, as shown in Figure 4.83. The ends of the cable differ in elevation by 6 ft. and the sag at C, relative to support E, is 12 ft. Determine the maximum tension in the cable and the sags at B and D.

Figure 4.83   Cable roof system with supports at different elevations.

Solution:

Construct the FBD of the entire cable (Figure 4.84) and write a moment equation of equilibrium about support A or E.

$$\sum M_A = -2\,k(24\,\text{ft.}) - 4\,k(60\,\text{ft.})$$

$$- 4\,k(96\,\text{ft.}) - E_x(6\,\text{ft.}) + E_y(126\,\text{ft.}) = 0$$

$$\therefore \quad -6E_x + 126E_y = 672 \dots\dots(\text{Eq. 1})$$

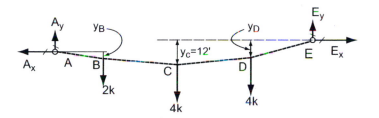

Figure 4.84   FBD of the entire cable system.

Another FBD and equation of equilibrium that involves the same two unknowns E_x and E_y are necessary to solve for the support reactions at E. The FBD will be drawn of the right half of a section cut through C, where the sag is known (Figure 4.85).

$$\sum M_C = -4\,k(36\,\text{ft.}) + E_y(66\,\text{ft.}) - E_x(12\,\text{ft.}) = 0$$

$$\therefore \quad +12E_x - (66)E_y = -144 \dots\dots(\text{Eq. 2})$$

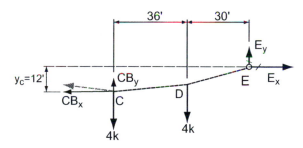

Figure 4.85   FBD of the cable section CDE, cut just to the left of C.

Solving equations (1) and (2) simultaneously:

$$E_x = +23.48k\ (\rightarrow); \quad E_y = +6.45\,k\ (\uparrow); \quad E = 24.35\,k$$

Component forces of the cable segment CB can now be solved with E_x and E_y being known.

$$\sum F_x = -CB_x + \underset{(E_x)}{23.48\,k} = 0 \qquad \therefore CB_x = +23.48\,k$$

$$\sum F_y = +CB_y - 4\,k - 4\,k + \underset{(E_y)}{6.45\,k} = 0$$

$$\therefore CB_y = 1.55 \text{ k}$$

$$\text{and } CB = 23.5 \text{ k}$$

Returning to the FBD of the entire cable (Figure 4.84),

$$\sum F_x = -A_x + \underset{(E_x)}{23.48} \text{ k} = 0 \therefore A_x = +23.48 \text{ k} (\leftarrow)$$

$$\sum F_y = +A_y - 2 \text{ k} - 4 \text{ k} + 4 \text{ k} + \underset{(E_y)}{6.45} \text{ k} = 0$$

$$\therefore A_y = +3.55 \text{ k} (\uparrow) \text{ and } A = 23.75 \text{ k}$$

The force in cable segment AB is equal to the reaction force A.

$$\therefore \Delta AB = 23.75 \text{ k} \quad AB_x = 23.48 \text{ k} \quad \text{and} \quad AB_y = 3.55 \text{ k}$$

The sag at point B can be determined by using the slope relationship of AB_y to AB_x.

$$y_B = \left(\frac{AB_y}{AB_x}\right) \times (24 \text{ ft.}) = \left(\frac{3.55}{23.48}\right) \times (24 \text{ ft.})$$

$$= 3.63 \text{ ft. relative to } A.$$

Similarly, using the FBD in Figure 4.86

$$E = 24.35 \text{ k and } ED_x = 23.48 \text{ k} \quad \text{and} \quad ED_y = 6.45 \text{ k}$$

$$y_E = \left(\frac{ED_y}{ED_x}\right) \times (30 \text{ ft.}) = \left(\frac{6.45}{23.48}\right) \times (30 \text{ ft.})$$

$$= 8.24 \text{ ft. relative to } E.$$

$$\sum F_x = -DC_x + \underset{(E_x)}{23.48} \text{ k} = 0$$

$$\therefore DC_x = +23.48 \text{ k}$$

$$\sum F_y = -4 \text{ k} - DC_y + 6.45 \text{ k} = 0$$

$$\therefore DC_y = 2.45 \text{ k} \quad \text{and} \quad DC = +23.61 \text{ k}$$

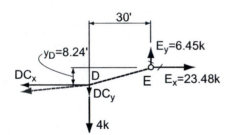

Figure 4.86   FBD of cable segment DE.

4.4   RETAINING WALLS

As its name implies, *retaining walls* are used to hold back (retain) solid or other granular material to maintain a difference in ground elevation. A *dam* is a retaining wall used to resist the lateral pressure of water or other fluids.

There are three general types of retaining walls: the gravity wall (Figure 4.87), the reinforced concrete cantilever retaining wall (Figure 4.88), and the reinforced concrete cantilever retaining wall with counterforts (Figure 4.89).

Gravity retaining walls are generally built of plain concrete or masonry. Height h is generally <4 ft. in height. Gravity walls depend on their mass to give them stability against the horizontal forces from the soil. Sliding resistance (friction) is developed between the concrete and soil at the base. Some major dams are constructed as gravity wall systems but understandably, the base dimensions are immense.

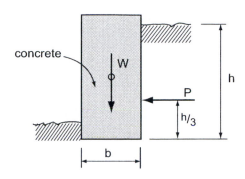

Figure 4.87   Gravity retaining wall.

Reinforced concrete cantilever retaining walls are the most frequently used type of retaining wall with an effectiveness up to about 20 to 25 feet in height h. Stability of this wall type is achieved by the weight of the structure and the weight of the soil on the heel of the slab base. Sometimes, a shear key is included at the bottom of the slab base to increase the wall's resistance to sliding. Retaining walls should have their foundations well below the frost line and adequate drainage (weep) holes near the bottom of the wall should be provided to permit the water accumulation behind the wall to escape.

As the retaining wall height increases, the bending moment in the cantilever wall increases, requiring more thickness. The addition of counterforts (vertical triangular-shaped cross walls) provides the additional depth at the base to absorb the large bending stresses. Counterfort walls behave like one-way slabs spanning horizontally between the counterforts. Counterforts are called buttresses if this same configuration is used when the retained earth is on the flat side of the wall.

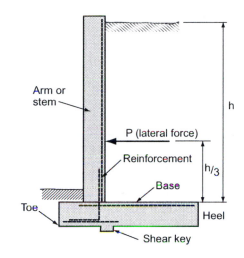

Figure 4.88   Reinforced cantilever retaining wall.

Saturated, loose sand or gravel, granular soil, or mud cause pressures against retaining walls in a manner similar to true fluids (liquids) by exerting a horizontal pressure. In true liquids, like water, the horizontal pressure and vertical pressure are the same at a given depth. However, in soil, the horizontal pressure is less than the vertical pressure, with the ratio dependent on the physical properties of the

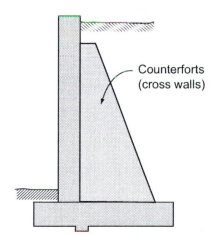

Figure 4.89   Counterfort wall.

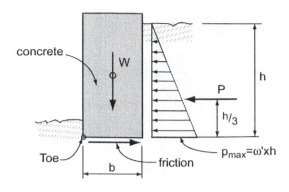

Figure 4.90   FBD of a gravity retaining wall.

soil. Soil pressure, as in liquids, increases proportionately with its depth below grade (Figure 4.90).

Lateral pressure increases linearly from zero at the top to a maximum at the bottom of the footing.

$$p = \omega' \times h$$

where:

   p = the magnitude of the earth pressure in psf or (kN/m^2)

   ω' = the "equivalent" fluid weight (density) of the soil in pounds per cubic feet. Values range from a minimum of 30 pcf (for well graded, clean gravel-sand mixes) to 60 pcf (for clayey sands). SI values are $4.7 \ kN/m^3$ to $9.4 \ kN/m^3$.

   h = soil depth in feet (meters).

$$P = 1/2 \times (p_{max} \times h) \times 1 \text{ ft. or 1 m.}$$

where:

   P = the lateral force (pounds, kips, N or kN) based on the area of the pressure distribution acting on a 1-ft. (1-m) wide strip of wall.

   p_{max} = the maximum pressure at depth h (psf or KN/m^2)

Equivalent fluid pressure against a retaining wall may create conditions of instability. Retaining walls are susceptible to three failure modes: (a) sliding—when the friction at the footing base is insufficient to resist sliding; (b) overturning about the toe—when the lateral force produces an overturning moment greater than the stabilizing moment from the wall's weight, slab base weight, and the soil mass above the heel; and (c) excessive bearing pressure at the toe—when the combination of the vertical downward force and the compression at the toe caused by the horizontal force exceeds the allowable bearing pressure of the soil.

The pressure distribution under the base (Figure 4.91) depends upon the location and magnitude of the resultant (vertical and horizontal) force as it passes through the footing base.

Analysis of a cantilever retaining wall requires that the equilibrium summation of moments about the toe is stable; that is, the weight of the wall plus the backfill on the heel exceeds the overturning moment of the active soil pressure by at least 1.5 (a building-code-imposed safety factor). Once a stable configuration is achieved, the soil pressure distribution on the footing must be calculated to ensure that the bearing pressures are within allowable limits for the soil on site.

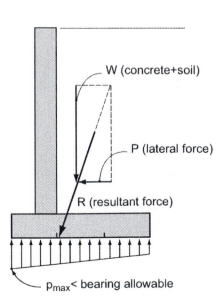

Figure 4.91   Bearing pressure under the wall footing.

Building codes require that a safety factor of 1.5 also be provided to prevent sliding failure. In addition to general instability issues, individual components of a retaining wall (wall thickness, base size, base thickness, and reinforcement steel quantity and location) must be designed to resist the bending moments and shear forces induced by the soil pressure.

Many of the factors concerned with sliding go beyond the scope of statics; therefore, problems in this section will be limited to the investigation of the wall's resistance to overturning and bearing pressure beneath the toe.

Example Problem 4.17 (Overturning Stability)

A small gravity retaining wall (Figure 4.92) is used to accommodate a $4\frac{1}{2}$-ft. drop in elevation. Determine the wall's factor of safety against overturning. $\gamma_{concrete} = 150 \text{ lb./ft.}^3$ and the equivalent fluid density $\omega' = 40 \text{ lb./ft.}^3$.

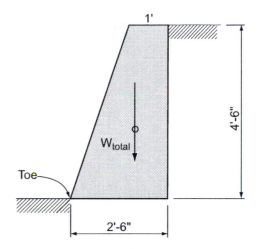

Figure 4.92   Small gravity dam.

Solution (Figure 4.93):

Analyze the retaining wall by assuming a 1-ft. tributary length of the wall as a representation of the entire wall.

$$W_1 = \frac{1}{2}(1.5 \text{ ft.})(4.5 \text{ ft.})(1 \text{ ft.})\left(150^{\text{lb.}}/_{\text{ft.}^3}\right) = 506 \text{ lb.}$$

$$W_2 = (1 \text{ ft.})(4.5 \text{ ft.})(1 \text{ ft.})\left(150^{\text{lb.}}/_{\text{ft.}^3}\right) = 675 \text{ lb.}$$

$$p_{max} = \omega' \times h = \left(40^{\text{lb.}}/_{\text{ft.}^3}\right) \times (4.5 \text{ ft.}) = 180^{\text{lb.}}/_{\text{ft.}^2}$$

$$P_3 = \frac{1}{2}(p_{max}) \times (h) \times (1 \text{ ft.})$$

$$= \frac{1}{2}\left(180^{\text{lb.}}/_{\text{ft.}^2}\right)(4.5 \text{ ft.})(1 \text{ ft.}) = 405 \text{ lb.}$$

The overturning moment M_{OTM} about the *toe* at A is:

$$M_{OTM} = P_3 \times 1.5 \text{ ft.} = (405 \text{ lb.}) \times (1.5 \text{ ft.}) = 608 \text{ lb.-ft.}$$

The *stabilizing* moment or *righting* moment, M_{RM}, is equal to:

$$M_{RM} = W_1 \times (1 \text{ ft.}) + W_2 \times (2 \text{ ft.})$$

$$M_{RM} = (506 \text{ lb.})(1 \text{ ft.}) + (675 \text{ lb.})(2 \text{ ft.}) = 1856 \text{ lb.-ft.}$$

The factor of safety against overturning at A is:

$$SF = \frac{M_{RM}}{M_{OTM}} = \frac{(1856 \text{ lb.-ft.})}{(608 \text{ lb.-ft.})} = 3.05 > 1.5$$

$\therefore$ The wall is stable against overturning.

Force $P_3 = 405$ lb. represents the horizontal sliding force that must be resisted at the base of the footing through frictional resistance.

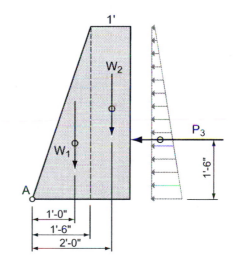

Figure 4.93   Overturning and stabilizing forces acting on the dam.

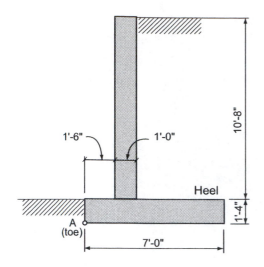

Figure 4.94   Cantilever retaining wall.

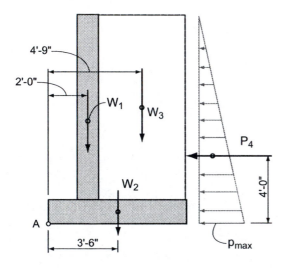

Figure 4.95   Cantilever retaining wall with overturning and stabilizing forces.

Example Problem 4.18 (Overturning Stability)

Determine the overturning stability of the reinforced, cantilever retaining wall shown in Figure 4.94. What is the sliding force that needs to be resisted at the base of the footing?

$\gamma_{conc.} = 150$ pcf; $\gamma_{soil} = 110$ pcf

$\omega' = 35$ pcf *(equivalent fluid density)*

Solution (Figure 4.95):

Assume a 1-ft. wide strip of wall as representative of the entire wall.

$$W_1 = (1 \text{ ft.})(10.67 \text{ ft.})(1 \text{ ft.})\left(150^{lb.}/_{ft.^3}\right) = 1600 \text{ lb.}$$
(wall)

$$W_2 = (1.33 \text{ ft.})(7 \text{ ft.})(1 \text{ ft.})\left(150^{lb.}/_{ft.^3}\right) = 1397 \text{ lb.}$$
(base)

$$W_3 = (4.5 \text{ ft.})(10.67 \text{ ft.})(1 \text{ ft.})\left(110^{lb.}/_{ft.^3}\right) = 5282 \text{ lb.}$$
(soil)

Evaluating the moments about the *toe* at *A:*

$$p_{max} = \omega' \times h = (35^{lb.}/_{ft.^3}) \times (12 \text{ ft.}) = 420^{lb.}/_{ft.^2}$$

$$P_4 = (^1/_2) \times (p_{max}) \times h$$

$$= (^1/_2) \times (420^{lb.}/_{ft.^2}) \times (12 \text{ ft.}) = 2520 \text{ lb.}$$

$$M_{OTM} = P_4 \times (4 \text{ ft.}) = (2520 \text{ lb.}) \times (4 \text{ ft.})$$

$$= 10,080 \text{ lb.-ft.}$$

The righting moment (RM) or the stabilizing moment about *A* is equal to:

$$M_{RM} = W_1 (2 \text{ ft.}) + W_2 (3.5 \text{ ft.}) + W_3 (4.75 \text{ ft.})$$

$$= (1600 \text{ lb.})(2 \text{ ft.}) + (1397 \text{ lb.})(3.5 \text{ ft.})$$
$$+ (5282 \text{ lb.})(4.75 \text{ ft.})$$

$$M_{RM} = 33,180 \text{ lb.-ft.}$$

The safety factor against overturning is computed as:

$$SF = \frac{M_{RM}}{M_{OTM}} = \frac{33,180 \text{ lb.-ft.}}{10,080 \text{ lb.-ft.}} = 3.29 > 1.5$$

The retaining wall is stable against overturning about the toe.

Frictional resistance developed between the bottom of the footing and the soil must be 1.5 (SF) times the horizontal fluid force $P_4 = 2520$ lb. or 3780 lb.

Retaining wall design, in its proportioning and element sizing, must ensure that the bearing pressure under the footing (at the toe) remains below the allowable limit for the soil involved. The pressure distribution varies, depending on the location and magnitude of the resultant (vector sum of the horizontal and vertical forces) as it passes through the base of the footing (Figures 4.96 to 4.98). Not all resultant forces are located beneath the wall section; however, it is generally more economical if the resultant is located within the middle third $(a/3 < x < 2a/3)$ of the base (Figure 4.96) in which the bearing pressures will be in compression throughout the base. If the resultant is located right on the edge of the middle third, where $x = a/3$, the pressure distribution results in a triangle, as shown in Figure 4.97. When the resultant force is located outside the middle third, where $x < a/3$, the pressure would indicate tension at or near the heel. Tension cannot develop between soil and a concrete footing that rests on it. Therefore, the pressure distribution shown in Figure 4.98 results, with the implication that a slight lifting off the soil occurs at the heel. Historical examples exist of some cathedrals built during the middle ages that suffered catastrophic failure when the resultant forces passing through the stone walls and footings fell outside of the middle third of the base. It is generally good practice to proportion the retaining wall to have the resultant fall within the middle third. This practice helps to reduce the magnitude of the maximum pressure and will minimize the variation between the maximum and minimum pressures.

Force diagram in Figure 4.96:

$$P_{max} = \frac{W}{a^2}(4a - 6x) \qquad p_{min.} = \frac{W}{a^2}(6x - 2a)$$

Force diagram in Figure 4.97:

$$p_{max} = \frac{2W}{a} \qquad p_{min.} = 0$$

Force diagram in Figure 4.98:

$$p_{max}\frac{2W}{3x} \qquad p_{min.} = 0$$

If a wall were to be constructed on highly compressible soil, such as some types of clay, a pressure distribution as found in Figures 4.97 and 4.98 would result. The larger settlement of the toe, as compared to the heel, would result in a tilting of the wall. Foundations constructed on compressible soils should have resultants falling at or near the center of the footing. Resultants can fall outside the middle third if the foundation is constructed on very incompressible soil, such as well-compacted gravel or rock.

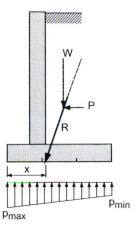

Figure 4.96   Trapezoidal pressure distribution.

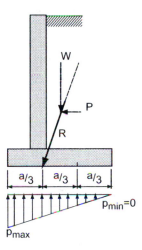

Figure 4.97   Resultant at the third point—a triangular pressure distribution.

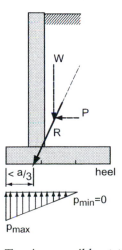

Figure 4.98   Tension possible at the heel.

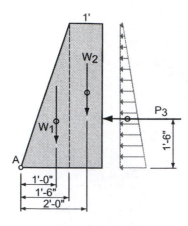

Figure 4.99   Forces on the retaining wall.

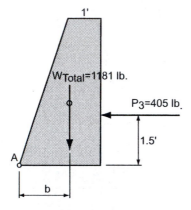

Figure 4.100   Total wall weight at its centroid.

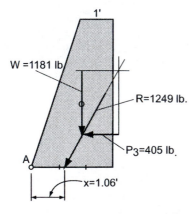

Figure 4.101   Resultant within the middle third.

Example Problem 4.19 (Bearing Pressure Check)

The bearing pressure will be checked for the gravity retaining wall found in Example Problem 4.17. Assume that the allowable bearing pressure is 2000 psf.

Wall weights and the horizontal fluid force P_3 are as shown in Figure 4.99. The location of the resultant vertical force W_{Total} is obtained by writing a moment equation in which:

$$M_A = W_1 \times (1 \text{ ft.}) + W_2 \times (2 \text{ ft.})$$

$$M_A = (506 \text{ lb.})(1 \text{ ft.}) + (675 \text{ lb.})(2 \text{ ft.}) = 1856 \text{ lb.-ft.}$$

From Figure 4.100, a moment equation due to the wall weight can also be written as:

$$M_A = (W_{total}) \times (b);$$

where:

$$W_{total} = W_1 + W_2 = (506 \text{ lb.}) + (675 \text{ lb.}) = 1181 \text{ lb.}$$

Equating the two equations,

$$M_A = W_{total} \times (b) = (1181 \text{ lb.}) \times (b) = 1856 \text{ lb.-ft.}$$

$$\therefore b = 1.57 \text{ ft.}$$

The resultant force R of the vertical and horizontal components is equal to:

$$R = \sqrt{(W_{Tot.}^2 + P_3^2)} = \sqrt{(1181^2 + 405^2)} = 1249 \text{ lb.}$$

Locating the point where the resultant intersects the base of the footing utilizes Varignon's theorem in which the moment caused by the vertical and horizontal forces about the toe at A is the same as the moment resulting from R_y (which is equal to W_{Total}) times the distance x from point A (Figure 4.101).

$$x = \frac{M_{RM} - M_{OTM}}{W_{total}}$$

Values for the overturning moment (M_{OTM}) and the righting moment (M_{RM}) are obtained from work done in Example Problem 4.17.

$$x = \frac{(1856 \text{ lb.-ft.}) - (608 \text{ lb.-ft.})}{(1181 \text{ lb.})} = 1.06 \text{ ft.}$$

The dimension $x = 1.06$ ft. falls within the middle third of the base dimension.

$$\frac{a}{3} \leq x = 1.06 \text{ ft.} \leq \frac{2a}{3}$$

Since the resultant is within the middle third of the base dimension, the equations for the maximum and minimum bearing pressures are:

$$p_{max} = \frac{W}{a^2}(4a - 6x)$$

$$= \frac{(1181 \text{ lb.})}{(2.5^2)}(4 \times 2.5 \text{ ft.} - 6 \times 1.06 \text{ ft.})$$

$$p_{max} = (189 \text{ } ^{lb.}/_{ft.^2})(10 - 6.36) = 688 \text{ } ^{lb.}/_{ft.^2}$$

$$p_{min.} = \frac{W}{a^2}(6x - 2a)$$

$$= \left(\frac{1181 \text{ lb.}}{2.5^2}\right)(6 \times 1.06 - 2 \times 2.5) = 257 \text{ } ^{lb.}/_{ft.^2}$$

The maximum bearing pressure is well within the allowable bearing stress limit of 2000 psf.

Example Problem 4.20 (Bearing Pressure Check)

Perform a check of the bearing pressure beneath the wall footing in Example Problem 4.18. Assume an allowable bearing pressure of 3000 psf.

Solution (Figure 4.102):

The total downward force (W_{Total}) and its location can be found as follows:

$$W_{Tot.} = W_1 + W_2 + W_3$$

$$W_{Tot.} = (1600 \text{ lb.}) + (1400 \text{ lb.}) + (5280 \text{ lb.})$$

$$= 8280 \text{ lb.}$$

$$M_A = (W_1 \times 2 \text{ ft.}) + (W_2 \times 3.5 \text{ ft.})$$

$$+ (W_3 \times 4.75 \text{ ft.})$$

$$M_A = (1600 \text{ lb.} \times 2 \text{ ft.}) + (1400 \text{ lb.} \times 3.5 \text{ ft.})$$

$$+ (5280 \text{ lb.} \times 4.75 \text{ ft.}) = 33,180 \text{ lb.–ft.}$$

Also, $M_A = W_{Total} \times b$

Equating both equations:

$$M_A = (8280 \text{ lb.}) \times (b) = 33,180 \text{ lb.-ft.}$$

$$\therefore b = 4.0 \text{ ft.}$$

The resultant force R is computed as:

$$R = \sqrt{(W_{Tot.}^2 + P_4^2)} = \sqrt{(8280^2) + (2520^2)}$$

$$= 8655 \text{ lb.}$$

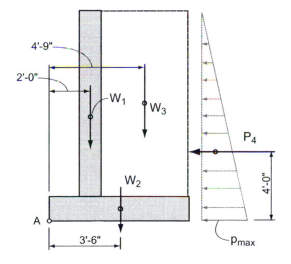

Figure 4.102   Retaining wall forces.

Using the moment values obtained in the solution of Example Problem 4.18, the distance x from the toe at A can be solved.

$$x = \frac{M_{RM} - M_{OTM}}{W_{Tot.}}$$

$$= \frac{(33,180 \text{ lb.-ft.}) - (10,080 \text{ lb.-ft.})}{8280 \text{ lb.}} = 2.8 \text{ ft.}$$

$$a/3 = 2.33 \text{ ft.} < x = 2.8 \text{ ft.} < {}^{2a}/3 = 6.67 \text{ ft.}$$

The resultant is within the middle third of the base; as shown in Figure 4.103, therefore, the entire bearing pressure will be in compression. A trapezoidal distribution results in which:

$$p_{max} = \frac{W}{a^2}(4a - 6x)$$

$$= \left(\frac{8280}{7^2}\right)(4 \times 7 \text{ ft.} - 6 \times 2.8 \text{ ft.}) = 1893 \text{ psf}$$

$$p_{max} = 1893 \text{ psf} < 3000 \text{ psf; OK}$$

$$p_{min.} = \frac{W}{a^2}(6x - 2a)$$

$$= \left(\frac{8280}{7^2}\right)(6 \times 2.8 - 2 \times 7) = 473 \text{ psf}$$

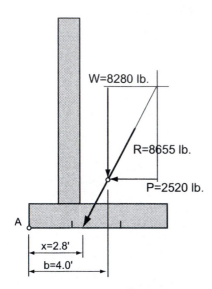

Figure 4.103 *Resultant force passing through the base.*

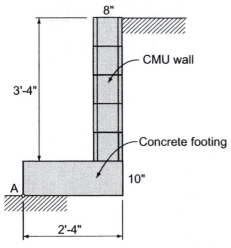

Figure 4.104 *Reinforced CMU retaining wall.*

Example Problem 4.21

Eight-inch nominal concrete masonry (CMU) blocks are used to retain soil as shown in Figure 4.104. Determine the wall's stability against overturning, then check the bearing pressure under the footing base at the toe (A). Assume that the allowable bearing pressure for the soil is limited to 2500 psf. Assume: $\gamma_{CMU} = 125 \, {}^{lb.}\!/_{ft.^3}$;

$\gamma_{conc.} = 150 \, {}^{lb.}\!/_{ft.^3}$; equivalent design fluid pressure

$\omega' = 40 \, {}^{lb.}\!/_{ft.^3}$.

Solution (Figure 4.105):

Assume a 1-ft. wide width of wall as representative of the entire wall.

$$W_1 = W_{CMU} = (1 \text{ ft.})(3.33 \text{ ft.})(^8/_{12} \text{ ft.})(125^{lb.}/_{ft.^3})$$

$$= 279 \text{ lb.}$$

$$W_2 = W_{ftg} = (1 \text{ ft.})(^{10}/_{12}\text{ft.})(2.33 \text{ ft.})(150^{lb.}/_{ft.^3})$$

$$= 291 \text{ lb.}$$

$$W_{Tot.} = W_1 + W_2 = 279 \text{ lb.} + 291 \text{ lb.} = 570 \text{ lb.}$$

The horizontal force on the wall due to the fluid pressure is equal to:

$$p_{max} = (40^{lb.}/_{ft.^3})(4.17 \text{ ft.}) = 167 \text{ psf}$$

$$P = {}^1\!/_2(p_{max})(h)(1 \text{ ft.})$$

$$= {}^1\!/_2(167^{lb.}/_{ft.^2})(4.17 \text{ ft.})(1 \text{ ft.}) = 348 \text{ lb.}$$

The location of the resultant weight W_{Total} is determined by summing moments about the toe at A.

$$M_A = W_1 \times (2 \text{ ft.}) + W_2 \times (1.17 \text{ ft.})$$

$$M_A = (279 \text{ lb.})(2 \text{ ft.}) + (291 \text{ lb.})(1.17 \text{ ft.}) = 899 \text{ lb.-ft.}$$

$$M_A = W_{Tot.} \times (b) = (570 \text{ lb.}) \times (b) = 899 \text{ lb.-ft.}$$

$$\therefore b = 1.58 \text{ ft.} \cong 1.6 \text{ ft.}$$

The resultant force on the retaining wall is:

$$R = \sqrt{(570^2 + 348^2)} = 668 \text{ lb.}$$

The overturning moment caused by the horizontal force about the toe at A is:

$$M_{OTM} = (348 \text{ lb.}) \times \left(\frac{4.14 \text{ ft.}}{3}\right) = 484 \text{ lb.-ft.}$$

The righting or stabilizing moment due to the total weight of the retaining wall is calculated as:

$$M_{RM} = W_{Tot.} \times b = (570 \text{ lb.}) \times (1.6 \text{ ft.}) = 912 \text{ lb.-ft.}$$

Checking the safety factor against overturning:

$$SF = \frac{M_{RM}}{M_{OTM}} = \frac{912 \text{ lb.-ft.}}{484 \text{ lb.-ft.}} = 1.9 > 1.5 \quad \therefore \text{ OK}$$

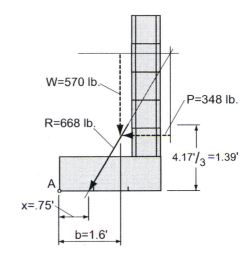

Figure 4.105   Resultant force outside the middle third.

Resultant force **R** intersects the footing base at:

$$x = \frac{M_{RM} - M_{OTM}}{W_{Tot.}}$$

$$= \frac{(912 - 484)\text{lb.-ft.}}{570 \text{ lb.}} = 0.75 \text{ ft.}$$

$$\frac{a}{3} = 0.78 \text{ ft.} > x = 0.75 \text{ ft.}$$

∴ The resultant falls outside the middle third section of the footing base.

$$p_{max} = \frac{2W}{3x} = \frac{2(570)}{3(0.75)} = 507 \text{ psf} < 2500 \text{ psf}$$

∴ OK

Supplementary Problems

Section 4.1—Planar Trusses Method of Joints:
Solve Problems 4.1.1–4.1.2 using the analytical method of joints. Begin with a FBD of the entire truss, then continue with FBDs of individual joints, solving the equations of equilibrium for the unknown member forces.

4.1.1

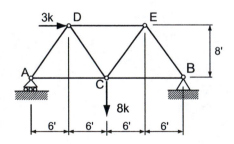

Problem 4.1.1

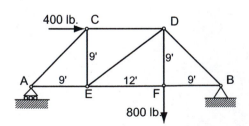

Problem 4.1.2

4.1.2

4.1.3 A cantilever truss, attached to a wall, supports the load from a storefront canopy. Determine the support reactions and the member forces using the method of joints.

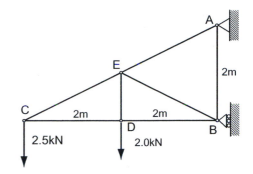

Problem 4.1.3

4.1.4 A roof truss is subjected to wind forces acting perpendicular to the slope of the roof. Using the method of joints, determine the support reactions and member forces.

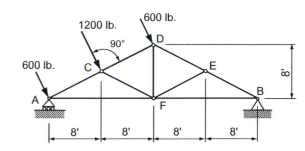

Problem 4.1.4

4.1.5 A roof truss is subjected to vertical gravity loads as shown. Solve for the support reactions and determine the member forces using the method of joints.

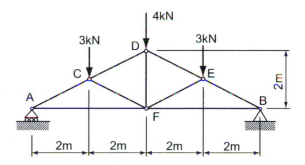

Problem 4.1.5

4.1.6 A parallel chord truss is loaded as shown. Determine all member forces using the method of joints.

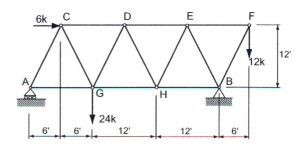

Problem 4.1.6

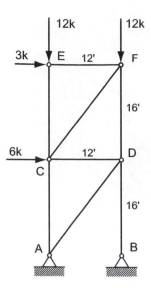

Problem 4.1.7

4.1.7 A vertical truss supports vertical gravity loads at the roof level and horizontal wind forces at the second floor and roof. Using the method of joints, determine all member forces.

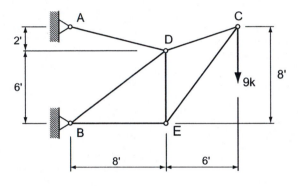

Problem 4.1.8

4.1.8 A cantilever truss supports a single load of 9 k at the free end. Solve for the support reactions and determine all member forces using the method of joints.

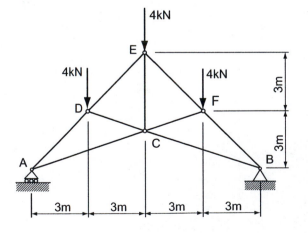

Problem 4.1.9

4.1.9 A roof scissors truss supports three gravity load forces from dead and snow load. Determine all member forces using the method of joints.

4.1.10 An overhang truss supported on a sloping surface supports a single 2k load at the free end. Determine the support reactions and solve for all member forces.

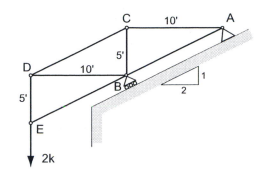

Problem 4.1.10

Method of Sections
Solve Problems 4.1.11–4.1.17 using the *method of sections* to determine the forces in the specified members.

4.1.11 Determine the support reactions and solve for member forces *DE, DC,* and *AC.*

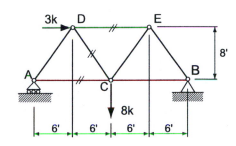

Problem 4.1.11

4.1.12 Using the method of sections, solve for member forces *DC, CF,* and *AF.*

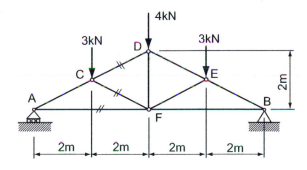

Problem 4.1.12

4.1.13 Solve for member forces *DE, DH,* and *GH.*

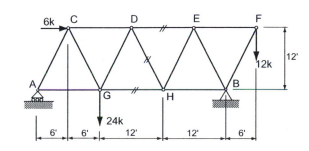

Problem 4.1.13

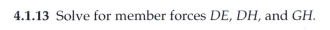

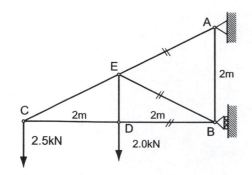

Problem 4.1.14

4.1.14 Determine member forces *AE, EB,* and *DB* for the cantilever truss shown.

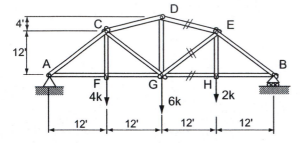

Problem 4.1.15

4.1.15 A bowstring or crescent truss is loaded as shown. Determine the member forces in *DE, EG,* and *GH.*

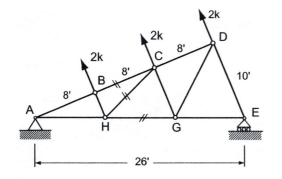

Problem 4.1.16

4.1.16 A saw-tooth roof truss is subjected to uplifting wind forces acting perpendicular to the roof plane. Members *BH, CG,* and *DE* are framed perpendicular to the roof slope *ABCD.* Solve for member forces *BC, CH,* and *HG.*

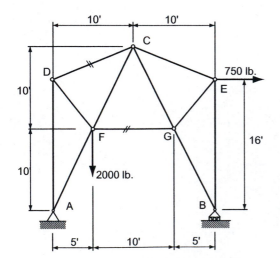

Problem 4.1.17

4.1.17 Determine member forces *DC* and *FG.*

Method of Sections—Diagonal Tension Counters
Solve Problems 4.1.18–4.1.22 using the method of sections.
Double crossing lines in any truss panel indicates a set of
diagonal tension counters. Only one of the counters is effective in tension, while the other counter is assumed as a
zero-force member.

4.1.18

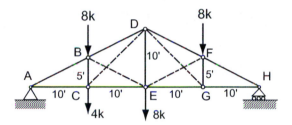

Problem 4.1.18

4.1.19

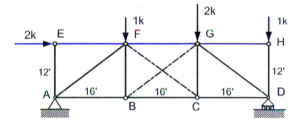

Problem 4.1.19

4.1.20

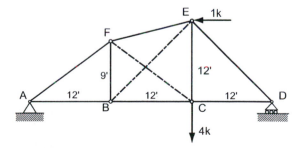

Problem 4.1.20

4.1.21

Problem 4.1.21

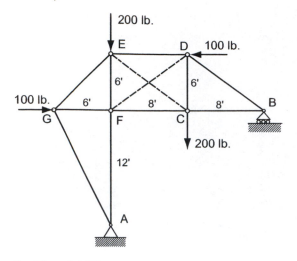

Problem 4.1.22

4.1.22

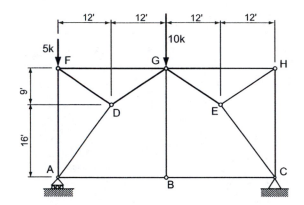

Problem 4.1.23

4.1.23 Determine the support reactions at *A* and *C* and all other member forces for the truss shown. *Hint: Use the method of sections and cut through members* FG, GD, *and* AB. *Solve for all three unknowns, then continue with the method of joints in obtaining other member forces.*

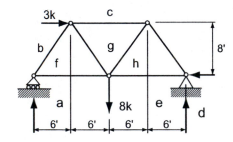

Problem 4.1.24

Maxwell's Diagram—Graphical Truss Analysis
Determine the member forces for each respective truss in Problems 4.1.24–4.1.31 using the graphical method (Maxwell's diagram).

4.1.24 Scale: $\frac{1}{2}$" = 1 k

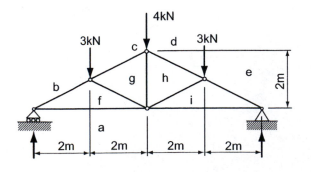

Problem 4.1.25

4.1.25 Scale: 10 mm = 1 kN

4.1.26 Scale: $\frac{1}{4}$" = 100 lb.

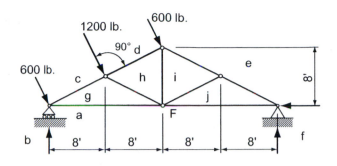

Problem 4.1.26

4.1.27 Scale: $\frac{1}{4}$" = 1 k

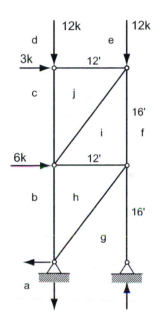

Problem 4.1.27

4.1.28 Scale: $\frac{1}{4}$" = 1 k

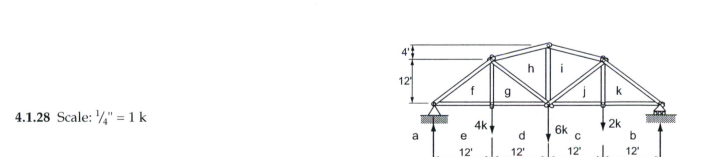

Problem 4.1.28

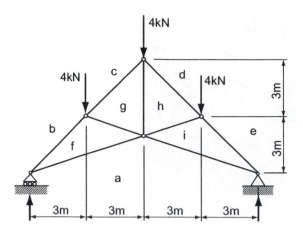

Problem 4.1.29

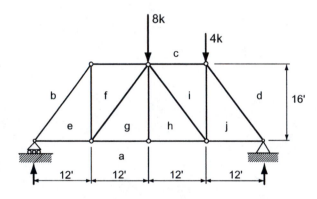

4.1.30 Scale: $\frac{1}{2}$" = 1 k

Problem 4.1.30

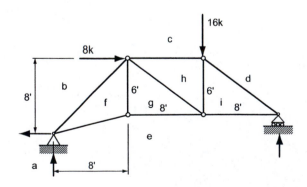
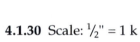

4.1.31 Scale: $\frac{1}{4}$" = 1 k

Problem 4.1.31

Supplementary Problems

Section 4.2—Pinned Frames with Multi-Force Members
4.2.1 Two members, *AC* and *BCD*, are loaded as shown. Determine the support reactions at *A* and *B* and the internal pin forces at *C*. Draw all appropriate free-body diagrams.

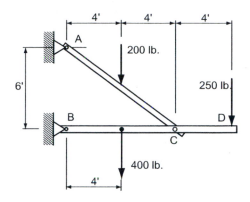

Problem 4.2.1

4.2.2 A bent steel member supports a 12-k load at the free end and is also subjected to a horizontal wind force as shown. Draw FBDs of the two elements and solve for the pin forces at *C*. What are the support reactions at *A* and *B*?

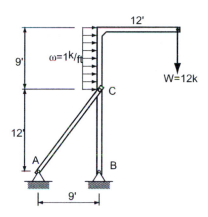

Problem 4.2.2

4.2.3 Determine the support reactions at *A* and *B* and the internal pin forces at *C*. Draw all appropriate FBDs.

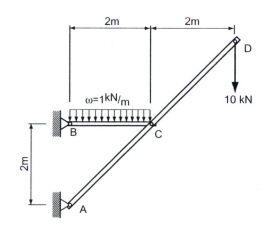

Problem 4.2.3

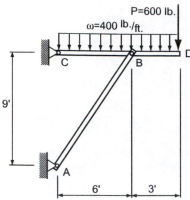

Problem 4.2.4

4.2.4 Determine the support reactions at A and C and the internal pin forces at B. Draw FBDs.

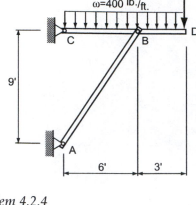

Problem 4.2.5

4.2.5 A frame is loaded by three forces as shown. Construct appropriate FBDs and solve for the support reactions at A and B. Also, determine the internal pin forces at D.

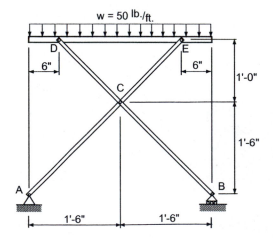

Problem 4.2.6

4.2.6 A picnic table is fully loaded along the entire top surface. Solve for the support reactions and the internal pin forces at C. Construct all appropriate FBDs.

Three-Hinged Arches: Problems 4.2.7–4.2.10

4.2.7 A three-hinged gabled frame supports two unequal roof loads as shown. Determine the support reactions and the internal pin forces at *B*.

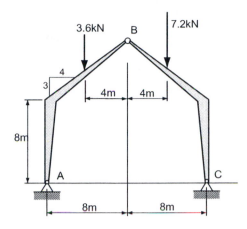

Problem 4.2.7

4.2.8 A wind force is applied to the vertical projection of a building. Determine the support reactions and the pin forces developed at the crown at *B*.

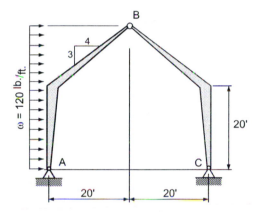

Problem 4.2.8

4.2.9 A glu-laminated three-hinged arch is subjected to a suction (uplifting) force on the leeward roof slope. The suction force acts perpendicular to the roof surface. Determine the resulting support reactions and the internal pin forces at *B*.

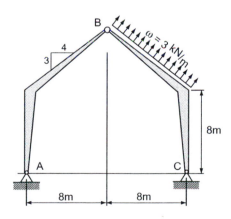

Problem 4.2.9

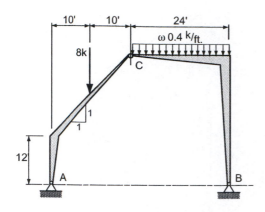

Problem 4.2.10

4.2.10 A three-hinged arch system is used to enclose a space as shown. Two load conditions are being resisted by the two members connected by a pin at C and the support hinges at A and B. Solve for the internal pin forces and the support reactions. Draw FBDs to assist in the solution.

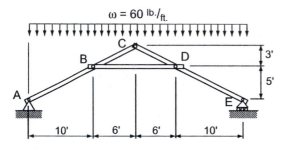

Problem 4.2.11

4.2.11 A pitched roof with a collar tie is subjected to a snow load across the entire horizontal projection of the roof surface. Determine the support reactions at A and E and the pin forces developed at B, C, and D.

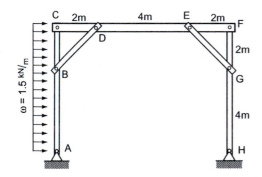

Problem 4.2.12

4.2.12 A knee-braced frame is used in a carport structure and is subjected to a wind force of 1.5 kN per meter acting on the vertical projection of the surface. Assume that the horizontal support reactions at A and H are equal (without this assumption, the problem becomes an indeterminate problem and cannot be solved by the equations of equilibrium alone). Construct the appropriate FBDs and solve for the support reactions and the pin forces at B, C, D, E, F, and G.

Supplementary Problems

Section 4.3—Cables with Concentrated Loads (Neglect the cable weight in the following problems)

4.3.1 Determine the tensions developed in cable segments AC, CD, and DB.

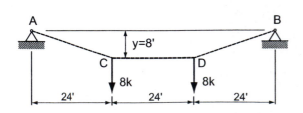

Problem 4.3.1

4.3.2 Three loads are suspended from the cable as shown. If the sag at B and D is equal to 2.5 m, determine the support reaction components at E and the sag at point C.

4.3.3 Using the same diagram as in Problem 4.3.2, determine the sag at point C if the maximum tension in the cable is 12 kN.

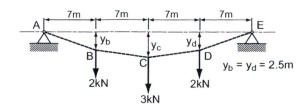

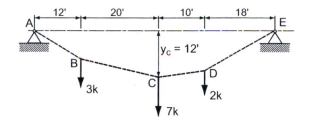

Problem 4.3.2

4.3.4 Determine the cable tension between each force and solve for the length of the cable for the entire system. The sag at C is 12 ft.

4.3.5 If the cable in the previous problem (4.3.4) had a maximum tensile capacity of 20 k, determine the minimum sag permitted at C.

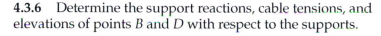

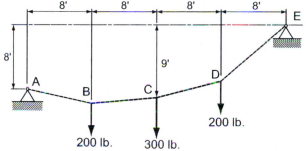

Problem 4.3.4

4.3.6 Determine the support reactions, cable tensions, and elevations of points B and D with respect to the supports.

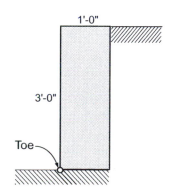

Problem 4.3.6

Supplementary Problems

Section 4.4—Retaining Walls
Assume an allowable bearing pressure of 3000 psf for all problems in this section.

4.4.1 A gravity retaining wall as shown is subjected to a lateral soil pressure as a result of an *equivalent fluid density* of 35 pcf. Calculate the resultant horizontal pressure against the wall and the wall's factor of safety against overturning. Assume that concrete has a density of 150 pcf. Check the bearing pressure under the footing.

Problem 4.4.1

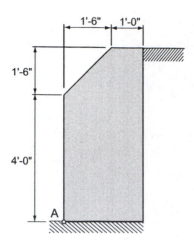

Problem 4.4.2

4.4.2 A 5'-6" high gravity wall is used to contain a soil with an equivalent fluid density of 30 pcf. If concrete has a density of 150 pcf, determine the wall's factor of safety against overturning. Does the bearing pressure remain within the allowable limit?

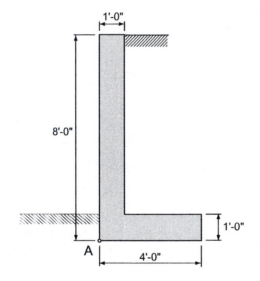

Problem 4.4.3

4.4.3 An L-shaped cantilever retaining wall constructed of reinforced concrete, with a density of 150 pcf, is 8'-0" tall from the base to the top of the soil. Assuming a soil density of 120 pcf and an equivalent fluid density of 40 pcf, determine the overturning stability of the wall. Check the bearing pressure beneath the toe at point A.

4.4.4 Using the same wall configuration as in Problem 4.4.3, determine the wall stability if the wall stem is projected to the opposite side with no soil above it. Check the bearing pressure.

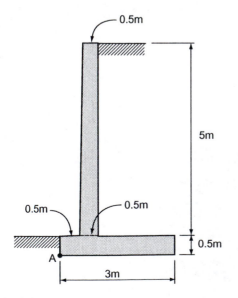

Problem 4.4.5

4.4.5 Determine the overturning stability of the cantilever retaining wall shown. The equivalent fluid density is 5.5 kN per cubic meter, soil density is 18 kN per cubic meter, and the concrete weighs 23.5 kN per cubic meter. Does the bearing pressure beneath the toe remain within the allowable bearing pressure limit?

4.4.6 Check on the overturning stability of the wall shown if the equivalent fluid pressure is 40 pcf and the soil density equals 115 pcf. Use a concrete density of 150 pcf. Evaluate the bearing pressure developed at the base of the footing.

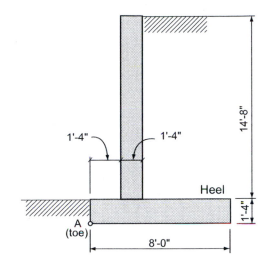

Problem 4.4.6

5 Load Tracing

Early in a project's structural design phase, an initial assumption is made by the designer about the path across which forces must travel as they move throughout the structure to the foundation (ground). Loads (forces) travel along *load paths*, and the analysis method is often referred to as *load tracing*.

Engineers often view structures as interdependent mechanisms by which loads are distributed to their individual members, such as roof sheathing, floor slabs, rafters, joists, beams, and columns. The structural designer makes judgments on the amount of load assigned to each member and the manner in which loads travel throughout the structure (load path).

Load tracing involves the systematic process of determining loads and support reactions of individual structural members as they in turn affect the loading of other structural elements (Figure 5.1). Simple determinate structures can be thoroughly analyzed using free-body diagrams (FBDs) in conjunction with the basic equations of equilibrium studied previously.

Usually, the process begins with the very uppermost member or level, tracing loads down layer by layer, until the last affected member under investigation is solved. In other words, start from the uppermost roof element and work your way down through the structure until you reach the foundation.

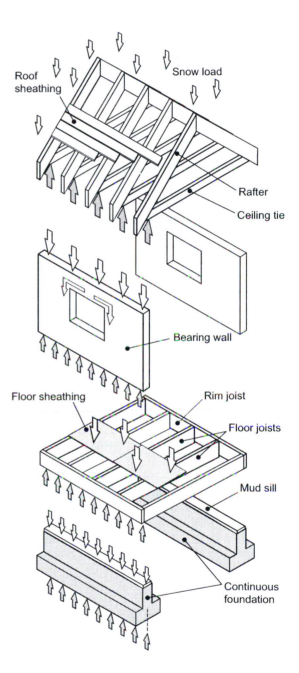

Figure 5.1 Load paths through a simple wood-framed structure.

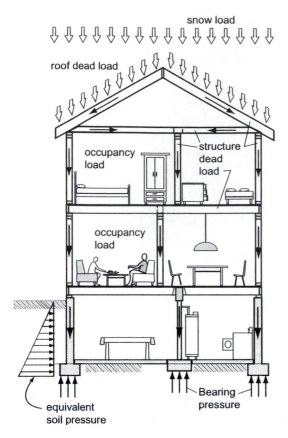

snow load

roof dead load

occupancy load

structure dead load

occupancy load

Bearing pressure

equivalent soil pressure

Figure 5.2   Dead and live loads on a simple structure.

Dead loads: Fin. flr., sheathing, joists, beams, column
Live loads: occupancy (uniform)

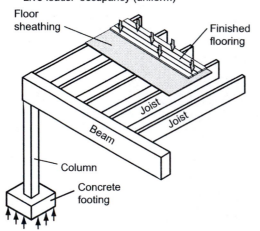

Floor sheathing

Finished flooring

Joist

Joist

Beam

Column

Concrete footing

Figure 5.3   Dead and live loads on a floor system.

5.1   LOADS ON STRUCTURES

Before beginning a more detailed discussion on load tracing, a brief description of structural loads is useful in understanding the types and sources of loads buildings must resist (Figure 5.2).

All structural systems are subjected to a variety of load conditions. Forces on various parts of the structure induce stresses and deformations on the elements within the framework. Aside from their form-defining function, structural systems essentially exist to resist forces that result from two general classifications of loads:

> *Static.* This classification refers to gravity-type forces. Loads or forces are assumed to be applied gradually over a relatively long interval of time.

> *Dynamic.* Dynamic loads are due to inertial forces (momentum of the mass of the structure) caused by sources like earthquakes and explosions. The more sudden the starting or stopping of the structure, the greater the resulting force will be.

Note: *Other dynamic forces are produced by wave action, landslides, falling objects, vibration from heavy machinery, and so forth.*

A light, steel frame building may be very strong in resisting static forces, but a dynamic force may cause large distortions to occur because of the frame's flexible nature. However, a heavily reinforced concrete building may be as strong as the steel building in carrying static loads but may have considerable stiffness and sheer dead weight, which may absorb the energy of dynamic forces with less distortion (deformation).

All of the following forces must be considered in the design of a building structure.

Dead loads are static, fixed loads that include the building structure weight, exterior and interior cladding, flooring, and fixed equipment that generally remains on the structure throughout its lifetime (Figure 5.3). The weight of the structural elements has to be estimated at the beginning of the structural design process and revised as the design proceeds to completion. Dead loads are computed on a conservative basis, which often involves the use of estimates on the higher side to negate redesign for minor changes in the final proportions. Final dead loads can be computed quite accurately using standard material weight tables of common building materials (Table 5.1) once the design is finalized. When activated by earthquake, static dead loads take on a dynamic nature in the form of horizontal inertial forces.

Live loads are transient and moving loads that include occupancy loads, furnishings, and storage. Live loads are extremely variable by nature and normally change during a structure's lifetime as occupancy changes. Live loads are generally static loads applied as a uniform load over floor and roof areas. Some live loads might be considered as moving loads or dynamic in nature, such as automobiles in parking garages or crane loads in industrial buildings, and are applied to particular parts of a structure so as to produce the maximum stress condition.

Building codes specify minimum uniform live loads for the design of roof and floor systems based on a history of many buildings and types of occupancy conditions (Table 5.2). They incorporate safety provisions for overload protection, allowance for construction loads, and serviceability considerations (such as vibration and deflection behavior). The occupancy loads specified in building codes are generally conservative enough to account for the increased stresses caused by the vibration of the structure (people dancing, bouncing down a flight of stairs).

Except for a building's dead load, which is fixed, the other forces just listed can vary in duration, magnitude, and point of application. A building structure must nevertheless be designed for these possibilities. Unfortunately, a large portion of a building structure exists for loads that may never occur or will be present at much lower magnitudes. Dead loading represents the weight of materials required in the building. The structural efficiency of a building is measured by its dead load weight in comparison to the live load carried.

Building designers have always strived to reduce the ratio of dead to live load. New methods of design, new and lighter materials, and old materials used in new ways have contributed to the dead/live load reduction. The size of the structure has an influence on the ratio of dead to live load. A small bridge over a creek, for example, can carry a heavy vehicle—a live load representing a large portion of the dead/live load ratio. The San Francisco Golden Gate Bridge, however, spans a long distance and the material of which it is composed is used chiefly in carrying its own weight. The live load of the vehicular traffic, in comparison to the dead load, has a relatively small effect on the bridge's internal stresses.

Table 5.1 Selected building material weights.

Assembly	lb./ft.²	kN/m²
Roofs:		
3-ply and gravel	5.5	0.26
5-ply and gravel	6.5	0.31
Wood shingles	2	0.10
Asphalt shingles	2	0.10
Corrugated metal	1–2.5	0.05–0.12
Plywood	3/inch	0.0057/mm
Insulation —fiberglass batt	0.5	0.0025
Insulation—rigid	1.5	0.075
Floors:		
Concrete plank	6.5	0.31
Concrete slab	12.5/in.	0.59/mm
Steel decking w/concrete	35–45	1.68–2.16
Wood joists	2–3.5	0.10–0.17
Hardwood floors	4/in.	0.19/mm
Ceramic tile w/thin set	15	0.71
Lightweight concrete	8/in.	0.38/mm
Timber decking	2.5/in.	0.08/mm
Walls:		
Wood studs (ave)	2.5	0.012
Steel studs	4	0.20
Gypsum drywall	3.6/in.	0.17/mm
Partitions (studs w/drywall)	6	0.29

Table 5.2 *Selected live load requirements.*[*]

Occupancy/Use (Uniform load)	lb./ft.²	kN/m²
Apartments:		
Private dwellings	40	1.92
Corridors and public rooms	100	4.79
Assembly areas/ theaters:		
Fixed seats	60	2.87
Stage area	100	4.79
Hospitals:		
Private rooms and wards	40	1.92
Laboratories /operating rooms	60	2.87
Hotels:		
Private guest rooms	40	1.92
Corridors /public rooms	100	4.79
Offices:		
General floor area	50	2.40
Lobbies /first floor corridor	100	4.79
Residential (Private):		
Basic floor area and decks	40	1.92
Uninhabited attics	20	0.96
Habitable attics /sleeping areas	30	1.44
Schools:		
Classrooms	40	1.92
Corridors	80–100	3.83–4.79
Stairs and exits:		
Single family /duplex dwellings	40	1.92
All other	100	4.79

[*] Loads are adapted from various code sources and are listed here for illustrative purposes only. Consult the governing code in your local jurisdiction for actual design values.

The use of modern materials and construction methods has resulted in the smaller rather than the larger buildings showing a higher dead-to-live load ratio. In a traditional house, the live load (LL) is low, and much of the dead load (DL) not only supports itself but also serves as weather protection and space-defining systems. This represents a high DL/LL ratio. In contrast, in a large factory building, the dead load is nearly all structurally effective, and the DL/LL ratio is low.

As spans increase, so do the bending effects caused by dead and live loads; therefore, more material must be introduced into the beam in order to resist the increased bending effects. This added material weight itself adds further dead load and pronounced bending effects as spans increase. The DL/LL ratio not only increases but may eventually become extremely large. The dead/live load ratio has considerable influence on the choice of structure, especially the choice of beam types.

Snow load is often considered a special type of live load because it is so variable. Local building officials or building codes applicable for a specific geographical jurisdiction will dictate the design snow load. Generally, snow loads are determined from a zone map reporting 50-year recurrence intervals of an extreme snow depth. Snow weights can vary from approximately 8 pounds per cubic foot for dry powder snow to 12 pounds per cubic foot for wet snow. Design loads can vary from 10 psf in areas with little snow to over 400 psf in some specific mountainous regions. In many urban areas in the United States, design loads can range from 20 to 35 psf.

The accumulation depth of the snow depends on the slope of the roof. The steeper the slope, the smaller the accumulation. Minimum roof live loads include allowance for minor snowfall and construction loads.

Wind loads on buildings are dynamic in nature. When buildings and structures become obstacles in the path of wind flow, the wind's kinetic energy is converted into potential energy of pressure on various parts of the building. Wind pressures, directions, and duration are constantly changing. However, for calculation purposes, most wind design assumes a static force condition. The fluctuating pressure caused by a constantly blowing wind is approximated by a *mean pressure* that acts on the structure's windward and leeward sides. The "static" or non-varying external forces are applied to the building structure, which simulates the actual varying wind forces.

Direct wind pressures, also referred to as *stagnation pressures*, depend on several variables: wind velocity, height of the wind above ground (wind velocities are lower near the ground), and the nature of the building's surroundings. Wind pressure on a building varies as the square of the velocity (in miles per hour) as expressed in the Bernoulli equation (see Table 5.3):

$$p = 0.00256V^2$$

for flat-sided objects (flat walls).

Table 5.3   *Wind velocity and corresponding wind pressure.*

Wind Velocity in Miles/Hour	Wind Pressure in Lbs./Sq. Ft.	Wind Type
30	2	Stiff breeze
60	9	Wind storm
150	58	Strong hurricane
250	160	Tornado

Photo 5.1.1   *Damage on a carport roof from a small snow load in Seattle.*

(Photo by author.)

Buildings respond to wind forces in a variety of complex and dynamic ways. The wind creates a negative pressure, or suction, on both the leeward side of the building (the side opposite the windward side), and the side walls parallel to the wind direction. Uplift pressure occurs on horizontal or sloping roof surfaces. The corners, edges, and eave overhangs of a building are subjected to complicated forces as the wind passes these obstructions, causing higher localized suction forces than generally encountered on the building as a whole.

Wind is a fluid and acts like other fluids, where a rough surface causes friction and slows the wind velocity near the ground and increases with height. Wind speeds are measured at a standard height of 10 meters (33 feet) above the ground and adjustments are made when calculating wind pressures at higher elevations.

Other buildings, trees, and topography affect how the wind will strike the building. Buildings in vast, open areas are subject to larger wind forces than those in sheltered areas or where the building is surrounded by other buildings. The size, shape, and surface texture of the building also impact the design wind forces.

Resulting wind pressures are treated as lateral loading on walls and in downward pressure or uplift forces (suction) on roof planes (see Figures 2.2, 2.6, and 2.41).

Photo 5.1.2   *Collapse, during construction, of a light-gauge, steel-framed warehouse structure from a wind storm.*

(Photo by Philip Lust.)

Photo 5.1.3   Detail of the collapsed framing where the steel stud is attached to the foundation wall.

(Photo by Philip Lust.)

Earthquake loads (seismic) are *inertial forces* that develop in the structure due to its weight, configuration, building type, and geographic location. Inertial forces are the product of mass and acceleration (Newton's second law: $F = m \times a$). Heavy, massive buildings will result in larger inertial forces; hence, there is a distinct advantage in using a lighter-weight construction when seismic considerations are a key part of the design strategy.

Earthquake, like wind, produces a dynamic force on a building. Lateral forces developed in the structure are a function of the building's mass, configuration, building type, height, and geographic location. During an earthquake, the ground mass moves suddenly both vertically and laterally. The lateral movements are of particular concern to building designers.

In actual earthquakes, there are continuous ground motions that cause the building structure to vibrate.

All objects, including buildings, have a *natural or fundamental period of vibration*. It represents the time it takes an object or building to vibrate through one cycle of vibration (or sway) when subjected to an applied force. When an earthquake ground motion begins a building vibrating, the building begins to displace (sway) back and forth at its natural period of vibration. Shorter, lower buildings have very short periods of vibration (less than 1 second), while tall high rises can have periods of vibration that last several seconds. Fundamental periods are a function of the building's height. An approximate estimate of a building's period is equal to:

$$T = 0.1N$$

where:

N = number of stories
T = period of vibration in seconds

The ground also vibrates at its own natural period of vibration. Many of the soils in the United States have periods of vibration in the range of 0.4 seconds to 1.5 seconds. Short periods are more characteristic of hard soils (rock), while soft ground (some clays) may have periods of up to 2 seconds.

Many common buildings can have periods within the range of the supporting soils, making it possible for the ground motion to transmit at the same natural frequency as that of the building. This may create a condition of resonance (where the vibrations increase dramatically), in which the inertial forces might become extremely large.

For some tall buildings or structures with complex configurations or unusual massing, a dynamic structural analysis is required. Computers are used to simulate earthquakes on the building to study how the forces are developed and the response of the structure to these forces. While the dynamic analysis method may be used for any building, in most cases building codes allow an equivalent static analysis of the loads produced, greatly simplifying the structural design for more conventional structures. Building codes are intended to safeguard against major failures and loss of life, not explicitly for the protection of property.

5.2 LOAD PATHS, TRIBUTARY AREAS, AND FRAMING SYSTEMS

Load Paths

In general, the shorter the *load path* to its foundation and the fewer elements involved in doing so, the greater the economy and efficiency of the structure. The most efficient load paths also involve the unique and inherent strengths of the structural materials used: tension in steel, compression in concrete or masonry. Bending, however, is a relatively inefficient way to resist loads and, as a result, beams become relatively large as loads and spans increase. Sketches of structural members in the form of free-body diagrams are used extensively to clarify the force conditions of individual elements as well as other interconnected members. As long as each element is determinate, the equations of equilibrium are sufficient for the determination of all supporting reactions. Load tracing requires an initial assessment of the general structural framework to determine where the analysis should begin (see Figure 5.1). Each time the load path is redirected, a support condition is created and the loads and reactions at each transfer must be analyzed (using FBDs) and solved using equilibrium equations.

Loads uniformly distributed over an area of roof or floor are assigned to individual members (rafters, joists, beams, girders) based on the concept of *distributive area, tributary area, or contributory area*. This concept typically considers the area that a member must support as being halfway between the adjacent similar members.

Example Problem 5.1

A section of a wood floor framing system (Figure 5.4) will be used to illustrate this concept. Assume that the general load over the entire deck area is a uniform $50\left(^{\text{lb.}}/_{\text{ft.}^2}\right)$.

- The tributary width contributing to the load on beam B is $\frac{1}{2}$ the distance (plank span) between A and B plus $\frac{1}{2}$ the distance between B and C (see Figure 5.5).

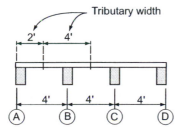

Figure 5.5   *Section cut through the deck and beams.*

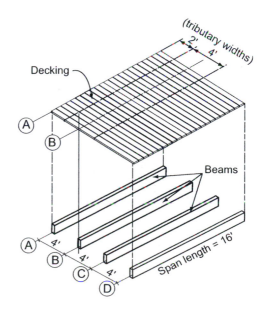

Figure 5.4   *Wood floor framing system.*

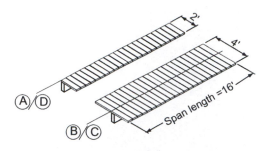

Figure 5.6   Tributary widths for beams A/D *and* B/C.

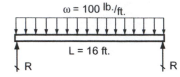

Figure 5.7   FBD of beams A *and* D.

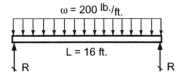

Figure 5.8   FBD of beams B *and* C.

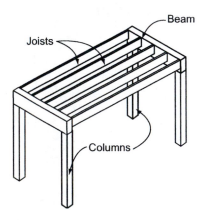

Figure 5.9   Long, lightly loaded joists bearing on shorter beams create a more uniform structural depth. Space can be conserved if the joists and beams are flush framed.

- The tributary width of load on beam $B = 2$ ft. $+ 2$ ft. $=$ 4 ft. The same is true for beam C (see Figure 5.6).
- Similarly, the tributary width for edge beams A and D is 2 ft.

Beam loads resulting from a uniformly applied load condition are determined by multiplying the load in pounds per square foot $\left(^{lb.}/_{ft.^2}\right)$ times the tributary width of load:

$$\omega = (\text{load in } {}^{lb.}/_{ft.^2}) \times (\text{tributary width})$$

The load on each beam may be expressed as (see Figures 5.7 and 5.8):

$$\omega_{A,D} = \left(50\,{}^{lb.}/_{ft.^2}\right) \times (2\text{ ft.}) = 100\,{}^{lb.}/_{ft.}$$

$$\omega_{B,C} = \left(50\,{}^{lb.}/_{ft.^2}\right) \times (4\text{ ft.}) = 200\,{}^{lb.}/_{ft.}$$

Framing Design Criteria: Direction of Span

Architectural character

The structural framing, if exposed, can contribute significantly to the architectural expression of buildings. Short joists loading relatively long beams yield shallow joists and deep beams. The individual structural bays are more clearly expressed. Figures 5.9 to 5.11 illustrate some of the common ways of framing a rectangular floor plan.

Structural efficiency and economy

Considerations should include the materials selected for the structural system, the span capability, and the availability of material and skilled labor. Standard sections and repetitive spacing of uniform members are generally more economical.

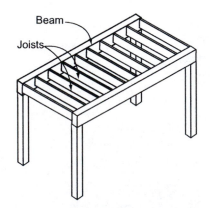

Figure 5.10   Short joists loading relatively long beams yield shallow joists and deep beams. The individual structural bays are more clearly expressed.

Mechanical and electrical system requirements

The location and direction of mechanical systems should be coordinated with the intended structural system. Layering the structural system provides space for ducts and pipes to cross-structural members, eliminating the need to cut openings in the beams. Flush or butt framing saves space in situations where floor-to-floor dimensions are limited by height restrictions.

Openings for stairs and vertical penetrations

While most framing systems will accommodate openings, it is generally more economical to make openings parallel to the dominant spanning direction. Additional headers and connections create point loads on members that are typically designed for light, uniform loads, increasing their size.

Load Paths: Pitched Roof Systems

Single-level framing

Rafters and ceiling joists combine to form a simple truss spanning from wall to wall. In addition to the truss action (rafters pick up compression forces, and ceiling joists develop tension to resist the horizontal thrust), rafters experience bending due to the uniform load along their length, as shown in Figure 5.12.

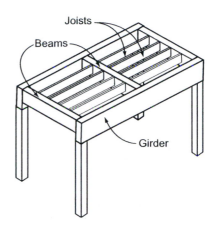

Figure 5.11   Loads can be reduced on selected beams by introducing an intermediate beam(s).

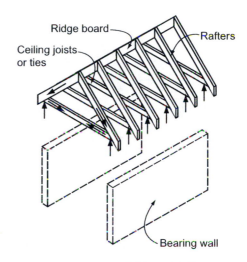

Figure 5.12   Pitched roof: single-level framing.

Double-level framing

Roof joists or beams are supported by a ridge beam on one end and a bearing wall or header beam at the other. No ceiling ties are used since this arrangement does not develop a horizontal thrust (as in the previous example). Notice that each level of structural framing spans in a perpendicular direction to the next layer, as shown in Figure 5.13.

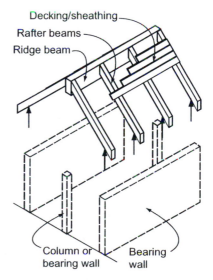

Figure 5.13   Pitched roof: double-level framing.

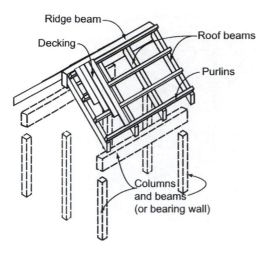

Figure 5.14   Pitched roof: three-level framing.

Three-level framing

The load path sequence in this arrangement starts with the loads transferred from the sheathing (decking) onto the purlins, which distribute concentrated loads onto the roof beams. The roof beams in turn transmit load to the ridge beam at one end and a bearing wall or wall beam at the other. Columns or wall framing support the ridge beam at either end, as shown in Figure 5.14.

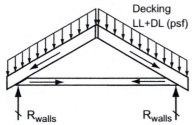

Figure 5.15   Single-level framing: FBD—rafters.

Construction: Pitched Roof Systems

Single-level framing

A common roof system for residential structures is a rafter/ceiling joist arrangement. Loads onto the roof are initially supported by the sheathing (plywood or other structural panels or skip sheathing, usually 1" × 4" boards spaced some distance apart), which in turn loads the rafters (see Figure 5.15).

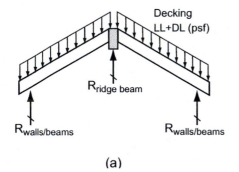

(a)

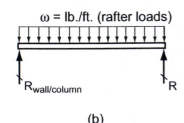

(b)

Figure 5.16   Double-level framing (a) FBD—rafters. (b) FBD—ridge beam.

Double-level framing

Another common roof framing arrangement involves roof joists or beams that are supported by a ridge beam on one end and a bearing wall or header beam at the other. The ridge beam must be supported at each end by a column or bearing wall (see Figure 5.16).

Three-level framing

A method used to achieve a heavier beam appearance is spacing the roof beams (rather than rafters) farther apart, typically 4 ft. to 12 ft. o.c. Perpendicular to the roof beams are purlins, spaced from 1'-6" to 4'-0" o.c., supporting sheathing, decking, or a metal roof. In both the two- and three-level framing systems, the ceiling plane can follow the roof slope (see Figure 5.17).

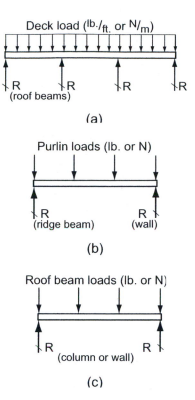

Figure 5.17 *Three-level framing: (a) FBD—purlins. (b) FBD—roof beams. (c) FBD—ridge beam.*

Load Paths: Wall Systems

A bearing wall is a vertical support system that transmits compressive forces through the wall plane and on to the foundation. Uniform compressive forces along the length of the wall result in a relatively uniform distribution of force. Concentrated loads or disruptions in the structural continuity of the wall, such as large window or door openings, will result in a non-uniform distribution of compressive forces on the footing. Bearing wall systems can be constructed with masonry, cast-in-place concrete, site-cast tilt-up concrete, or studs (wood or light-gauge metal framing).

Uniform slab loads are distributed along the top of the bearing wall as ω. A masonry or concrete wall footing will be required to support ω plus the additional wall weight. The load $\omega_2 = (\omega_1 + \text{wall weight})$ remains a uniform load (see Figure 5.18).

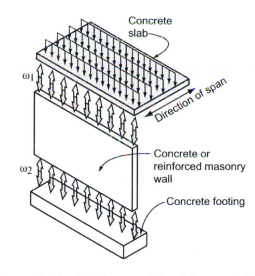

Figure 5.18 *Uniform wall load from a slab.*

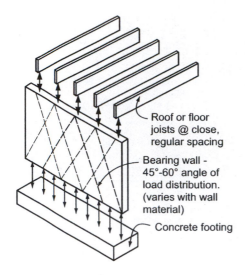

Figure 5.19   Uniform wall load from rafters and joists.

Uniform distribution

Roof or floor joists (in typical light-wood framing) are spaced 16" or 24" on center. This regular, close spacing is assumed as a uniform load along the top of the wall. If there are no openings to disrupt the load path from the top of the wall, a uniform load will result on top of the footing (see Figure 5.19).

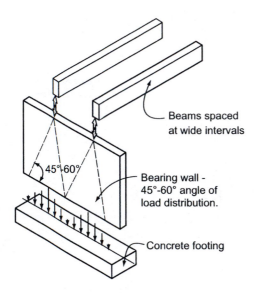

Figure 5.20   Concentrated loads from widely space beams.

Non-uniform distribution

Concentrated loads develop at the top of a wall when beams are spaced at wide intervals. Depending on the wall material, the concentrated load distributes along an angle of 45° to 60° as it moves down the wall. The resulting footing load will be non-uniform, with the largest forces directly under the applied load (see Figure 5.20).

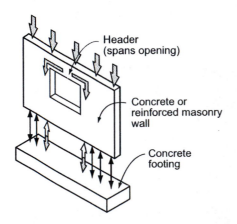

Figure 5.21   Arching over wall openings.

"Arching action" over opening

Openings in walls also redirect the loads to either side of the opening. The natural stiffness of a concrete wall under compression produces an "arching action" that contributes to the lateral distribution of the loads (see Figure 5.21).

Opening in a stud wall

Stud walls (wood and metal) are generally idealized as monolithic walls (except for openings) when loaded uniformly from above. Openings require the use of headers (beams) that redirect the loads to either side. Concentrated loads from the header reactions must be supported by a buildup of studs resembling a column (see Figure 5.22).

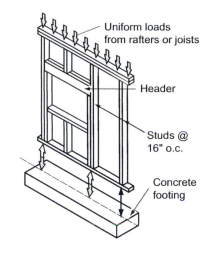

Figure 5.22   Stud wall with a window opening.

Concentrated loads—pilasters

In special cases where the concentrated loads are very large, walls may need to be reinforced with pilasters directly under the beam. Pilasters are essentially columns and carry the large concentrated loads directly to the footing. The walls between the pilasters are now considered as non-bearing walls except for carrying their own weight (see Figure 5.23).

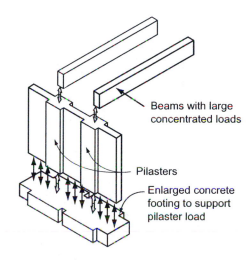

Figure 5.23   Pilasters supporting concentrated loads from widely spaced beams.

Load Paths: Roof and Floor Systems

One-level framing

Although it is not a common framing system, relatively long-spanning decking materials may transmit roof or floor loads directly to bearing walls (see Figure 5.24). Precast hollow-core concrete planks or heavy-timber-plank decking can be used to span between closely spaced bearing walls or beams. Spacing of the supports (the distance between bearing walls) is based on the span capability of the concrete planks or timber decking (see Figures 5.25 and 5.26).

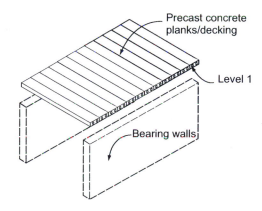

Figure 5.24   One-level framing (roof and floor).

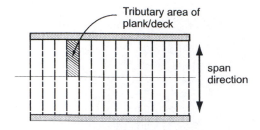

Figure 5.25   Framing plan.

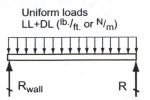

Figure 5.26   FBD—plank: One-level framing.

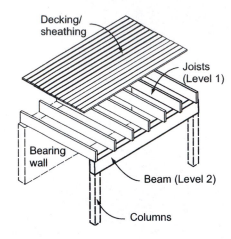

Figure 5.27   Two-level framing (roof and floor).

Two-level framing

This is a very common floor system that uses a regular, relatively closely spaced series of secondary beams (called joists) to support a deck. The decking is laid perpendicular to the joist framing. Span distances between bearing walls and beams affect the size and spacing of the joists (see Figure 5.27).

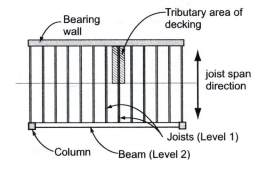

Figure 5.28   Framing plan.

Efficient structural sections in wood and steel joists allow relatively long spans between bearing walls. Lighter deck materials such as plywood panels can be used to span between the closely spaced joists (see Figures 5.28 and 5.29).

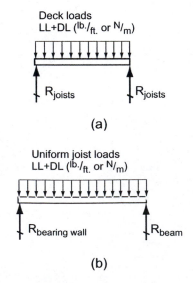

Figure 5.29   (a) FBD—joists: level 1. (b) FBD—beam: level 2.

Three-level framing

When bearing walls are replaced by girders or trusses spanning between columns, the framing involves three levels. Joist loads are supported by beams, which transmit their reactions to girders or trusses. Each level of framing is arranged perpendicular to the level directly above it (see Figure 5.30).

Buildings requiring large open floor areas, free of bearing walls and with a minimum number of columns, typically rely on the span capability of joists supported by beams, girders, and/or trusses. The spacing of the columns and the layering of the beams and girders establish the regular bays that subdivide the space (see Figures 5.31 and 5.32).

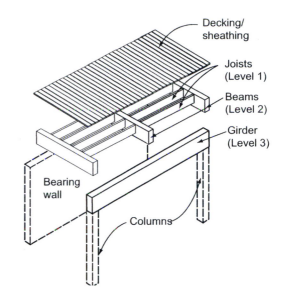

Figure 5.30   Three-level framing.

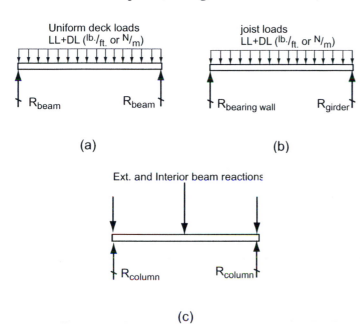

(a)   (b)

(c)

Figure 5.32   (a) FBD—joists (level 1). (b) FBD—beams (level 2). (c) FBD—girder (level 3).

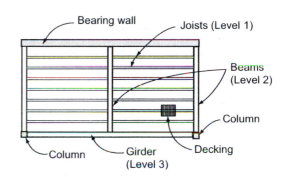

Figure 5.31   Framing plan.

Load Paths: Foundation Systems

The foundation system for a particular structure or building depends on the size, use of the structure, subsurface conditions at the site, and the cost of the foundation system to be used.

A large building with heavy loads can often be supported on relatively inexpensive shallow footings if the subsurface soils are dense and stable. However, the same building constructed at a site containing soft soils or expansive clay may require pile or caisson foundations. Foundations are generally subdivided into two major categories: shallow foundations and deep foundations.

Shallow foundations

Shallow foundations essentially obtain their support on soil or rock just below the bottom of the structure in direct bearing. Vertical loads are transmitted from walls or columns to a footing that distributes the load over a large enough area that the allowable load-carrying capacity of the soil is not exceeded and/or settlement is minimized. Shallow foundations are of three basic types: (a) spread footings—individual column footing, (b) continuous strip footings—supporting a bearing wall, and (c) mat foundations that cover the entire plan area of the building.

Spread Footing. This footing type is usually square, or sometimes circular, in plan and is generally simple and economical for moderate to high soil-bearing capacities. The purpose of this footing is to distribute the load over a large area of soil. Pedestal and footing are reinforced with steel (see Figure 5.33).

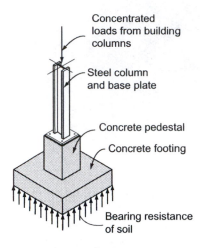

Figure 5.33   Spread footing.

Wall Footing. Wall footings are one of the most common footing types that support relatively uniform bearing wall loads through a continuous foundation wall. The wall footing width remains constant throughout its length if no large concentrated loads occur (see Figure 5.34).

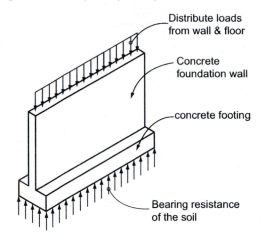

Figure 5.34   Wall footing.

Mat or Raft Foundation. Mat foundations are used when soil bearing is relatively low or where loads are heavy in relation to soil-bearing capacities. This foundation type is essentially one large footing under the entire building, which distributes the load over the entire mat. A mat is called a raft foundation when it is placed deep enough in the soil that the soil removed during excavation equals most or all of the building's weight (see Figure 5.35).

Deep Foundations

The function of a deep foundation is to carry building loads beneath a layer of unsatisfactory soil to a satisfactory bearing stratum. Deep foundations are generally piles, piers, or caissons installed in a variety of ways. There is normally no difference between a drilled caisson and a drilled pier and, most often, only a modest difference in diameter between

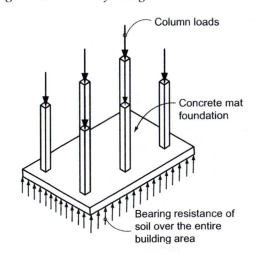

Figure 5.35   Mat or raft foundation.

them. Piles, the most common deep-foundation system, are driven into the earth by pile-driving hammers powered with drop hammers, compressed air, or diesel engines. Building loads are distributed to the soil in contact with the surface area of the pile through skin friction (friction piles), in direct bearing (bearing piles) at the bottom of the pile on a sound stratum of earth or rock, or a combination of skin friction and direct bearing.

Pile Foundations. Timber piles are normally used as friction piles, whereas concrete and steel piles are generally used as bearing piles. When bearing piles must be driven to great depths to reach suitable bearing, a combination of steel and concrete is used. Hollow steel shells are driven into the ground to a predetermined bearing point, and then the casings are filled with concrete (see Figure 5.36).

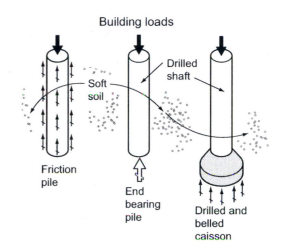

Figure 5.36   Pile foundations.

Pile Caps. Individual building columns are generally supported by a group (cluster) of piles. A thick, reinforced cap is poured on top of the pile group, which distributes the column load to all the piles in the cluster (see Figure 5.37).

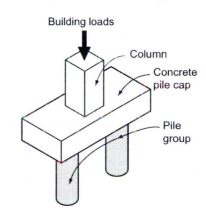

Figure 5.37   Pile cap on one pile group.

Grade Beams. Piles or piers supporting bearing walls are generally spaced at regular intervals and connected with a continuous reinforced concrete grade beam. The grade beam is intended to transfer the loads from the building wall to the piles (see Figure 5.38).

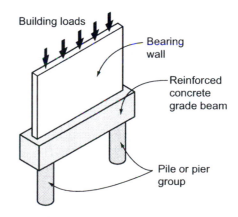

Figure 5.38   Grade beam supporting a bearing wall.

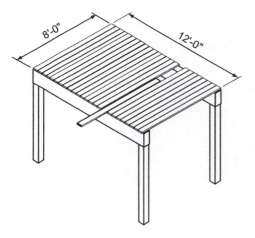

Figure 5.39   Single bay, deck-beam-post system.

5.3 LOAD TRACING OF DETERMINATE FLOOR AND ROOF SYSTEMS— INCLUDING SLOPING ROOFS

The example problems that follow will illustrate the load-tracing methodology as it applies to a variety of structural frameworks and arrangements. Note that the predominance of the examples illustrated are wood-frame structures such as those in Figures 5.39 and 5.45. Wood framing is the one structural material type that generally results in a determinate framing system whereas steel, and in particular cast-in-place concrete, are often designed to capitalize on the advantages of indeterminacy through the use of rigid connections and continuity.

Example Problem 5.2

In the single-bay, post-and-beam deck illustrated, planks typically are available in nominal widths of 4" or 6", but for the purposes of analysis it is permissible to assume a unit width equal to one foot. Determine the plank, beam, and column reactions.

Solution:

Load on the deck (live load) = 60 psf

<u>Deck weight (dead load)   = 8 psf</u>

Total load (LL + DL)   = 68 psf

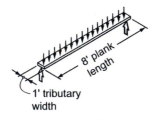

Figure 5.40   Load on a 1' width plank.

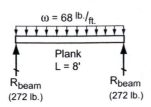

Figure 5.41   FBD of the plank.

Plank Reaction (Figures 5.40 and 5.41):

Looking at the deck in elevation, the load ω is determined by multiplying the pounds per square foot load by the tributary width of the plank.

Therefore:

$$\omega = \left(68\,{}^{lb.}\!/_{ft.^2}\right) \times (1\ \text{ft.}) = 68\,{}^{lb.}\!/_{ft.}$$

Beam Reaction (Figures 5.42 and 5.43):

The planks load the beams with a load of $\omega = 68\,^{lb.}/_{ft.}$ of the plank span. Half of the plank load is transferred to each beam. The beams are loaded by the planks with a load of $272\,^{lb.}/_{ft.}$ of the beam span.

In addition, the beam has a self-weight equivalent to $10\,^{lb.}/_{ft.}$.

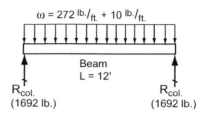

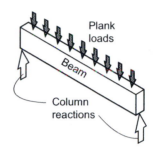

Figure 5.42   FBD of the beam.

Figure 5.43   FBD of the beam.

Column Reaction (Figure 5.44):

Half of each beam load is transferred to the column at each corner of the deck. The columns are loaded by the beams with loads of 1692 lb. at each column. Assume each column has a self-weight of 100 lb.

Ground Reaction (Figure 5.44):

The load at each column is resisted by an equivalent ground reaction of 1792 lb.

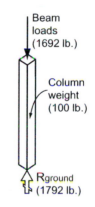

Figure 5.44   FBD of the column.

Example Problem 5.3a

This problem represents an expansion of Example Problem 5.2 where the decking has an additional 6 feet to span and the beams are extended another bay (see Figure 5.45). Loads on the structural system remain the same.

Determine the loads developed in each column support. Assume that columns are located at grids 1-A, 2-A, 3-A, 1-B, 2-B, 3-B, 1-C, 2-C, and 3-C.

Solution:

Deck DL	= 8 psf
Live load	= 60 psf
Total load	= 68 psf
Beam self-weight	= 10 lb./ft.
Column self-weight	= 100 lb.

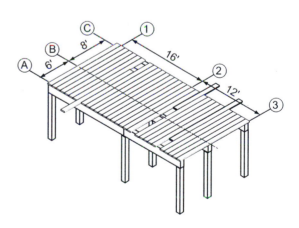

Figure 5.45   Decking/beam/column arrangement.

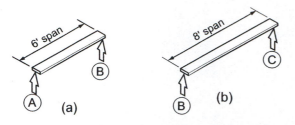

Figure 5.46   Planks: 1' tributary widths.

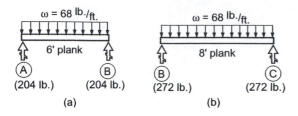

Figure 5.47   FBD of planks.

Plank Reaction (Figures 5.46 and 5.47):

$$\omega = \left(68\,^{\text{lb.}}\!/_{\text{ft.}^2}\right) \times (1 \text{ ft.}) = 68\,^{\text{lb.}}\!/_{\text{ft.}}$$

Beam Reaction (Figure 5.47):

First, analyze the planks that span 6 feet between grid lines *A* and *B*.

$$R = \frac{\omega L}{2} = \frac{\left(68\,^{\text{lb.}}\!/_{\text{ft.}}\right) \times (6 \text{ ft.})}{2}$$

$$= 204 \text{ lb. (Beam reaction)}$$

Next, analyze the plank loads and beam support for the 8-foot span between grids *B* and *C*.

$$R = \frac{\omega L}{2} = \frac{\left(68\,^{\text{lb.}}\!/_{\text{ft.}}\right) \times (8 \text{ ft.})}{2}$$

$$= 272 \text{ lb. (Beam reaction)}$$

Column Reaction:

All of the following beam cases represent uniformly loaded conditions with simply supported ends, which re sult in reactions that are $R = \dfrac{\omega L}{2}$. The resulting reactions of the beams represent the loads present in each column.

First, analyze the beams along grid line *A* (Figure 5.48):

$$\omega = 204\,^{\text{lb.}}\!/_{\text{ft.}} + 10\,^{\text{lb.}}\!/_{\text{ft.}} = 214\,^{\text{lb.}}\!/_{\text{ft.}}$$

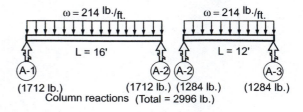

Figure 5.48   FBD of beams along grid line A.

Next, analyze the beams along grid line *B* (Figure 5.49):

$$\omega = 204\,{}^{lb.}/_{ft.} + 272\,{}^{lb.}/_{ft.} + 10\,{}^{lb.}/_{ft.} = 486\,{}^{lb.}/_{ft.}$$

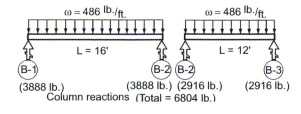

Figure 5.49   FBD of beams along grid line B.

Then, analyze the beams along grid line *C* (Figure 5.50):

$$\omega = 272\,{}^{lb.}/_{ft.} + 10\,{}^{lb.}/_{ft.} = 282\,{}^{lb.}/_{ft.}$$

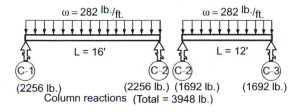

Figure 5.50   FBD of beams along grid line C.

Column Loads and Reactions (Figures 5.51 and 5.52):

The perimeter columns along grid lines 1 and 3 receive half the load of each beam. The interior columns along grid line 2 receive loads from two beams, which are added together to calculate the column loads.

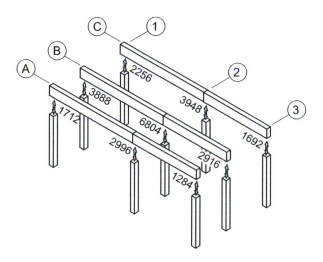

Figure 5.51   Column loads (beam reactions).

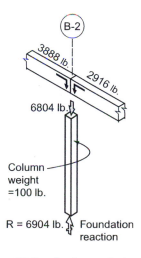

Figure 5.52   FBD of column B-2.

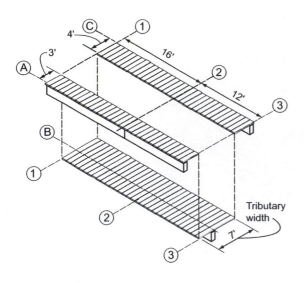

Figure 5.53   Beam tributary widths.

Example Problem 5.3b (Alternate Method)

Another technique may be employed in determining the beam reactions without going through an analysis of the planks. This may be accomplished by evaluating the tributary widths of load for each beam and directly calculating the ω for each beam (Figure 5.53). For example, in the following figure, the tributary width of load assigned to the beams along grid line A is 3'. Therefore:

$$\omega = \left(68\,{}^{\text{lb.}}\!/_{\text{ft.}^2}\right) \times (3\ \text{ft.}) + 10\,{}^{\text{lb.}}\!/_{\text{ft.}} = 214\,{}^{\text{lb.}}\!/_{\text{ft.}}$$

This ω value corresponds to the result obtained in the previous method.

And along grid line C, tributary width = 4 ft.

$$\omega = \left(68\,{}^{\text{lb.}}\!/_{\text{ft.}^2}\right) \times (4\ \text{ft.}) + 10\,{}^{\text{lb.}}\!/_{\text{ft.}}$$

$$= 282\,{}^{\text{lb.}}\!/_{\text{ft.}}\quad\text{(checks)}$$

Similarly, for beams along grid line B:

Tributary width = 3 ft. + 4 ft. = 7 ft.

$$\therefore\ \omega = \left(68\,{}^{\text{lb.}}\!/_{\text{ft.}^2}\right) \times (7\ \text{ft.}) + 10\,{}^{\text{lb.}}\!/_{\text{ft.}}$$

$$= 486\,{}^{\text{lb.}}\!/_{\text{ft.}}\quad\text{(checks)}$$

Example Problem 5.4

A steel-framed floor for an office building, as shown in Figures 5.54 to 5.56, was designed to support a load condition as follows:

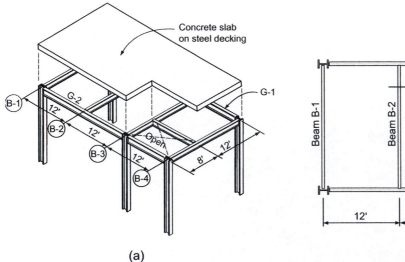

(a)

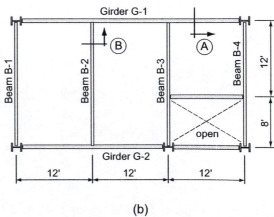

(b)

Figure 5.54   (a) Isometric view of partial steel framing arrangement. (b) Partial floor framing—office structure.

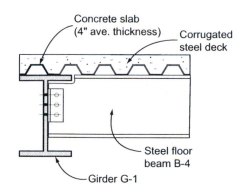

Figure 5.55   Section A at girder G-1.

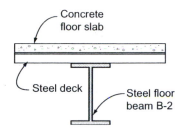

Figure 5.56   Section B at beam B-2.

Loads:

Live load	= 50 psf
Dead loads:	
Concrete	$= 150 \; \#/\text{ft.}^3$
Steel decking	= 5 psf
Mechanical equipment	= 10 psf
Suspended ceiling	= 5 psf
Steel beams	$= 25 \; \#/\text{ft.}$
Steel girders	$= 35 \; \#/\text{ft.}$

Using appropriate FBDs, determine the reaction forces for beams B-1, B-2, and B-3, and girder G-1.

Solution:

Loads:

$$\text{Slab load} = \left(\frac{4 \text{ in.}}{12 \, ^{\text{in.}}/_{\text{ft.}}} \right) \times \left(150 \, ^{\text{lb.}}/_{\text{ft.}^3} \right) = 50 \, ^{\text{lb.}}/_{\text{ft.}^2}$$

Dead loads: = 50 psf (slab)
                + 5 psf (decking)
                + 10 psf (mech. equip.)
                + 5 psf (ceiling)
_____
Total DL   = 70 psf

Dead load + Live load = 70 psf + 50 psf + 120 psf

Beam B-1 (Figures 5.57 and 5.58):

(Tributary width of load is 6')

$$\omega_1 = \left(120\,^{lb.}/_{ft.^2}\right) \times (6\ ft.) + \underset{\text{(beam wt.)}}{25\,^{lb.}/_{ft.}} = 745\,^{lb.}/_{ft.}$$

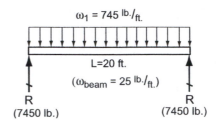

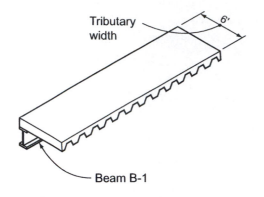

Tributary width

Beam B-1

Figure 5.57   Tributary width for beam B-1.

Figure 5.58   FBD of beam B-1.

Beam B-2 (Figures 5.59 and 5.60):

(Tributary width of load is 6' + 6' = 12')

$$\omega_2 = \left(120\,^{lb.}/_{ft.^2}\right) \times (12\ ft.) + 25\,^{lb.}/_{ft.} = 1465\,^{lb.}/_{ft.}$$

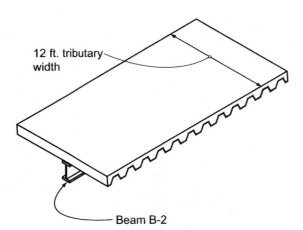

12 ft. tributary width

Beam B-2

Figure 5.59   Tributary width for beam B-2.

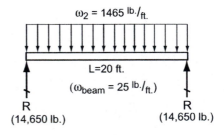

Figure 5.60   FBD of beam B-2.

Beam B-3 (Figures 5.61 to 5.62): This beam has two different load conditions due to the changing tributary width created by the opening.

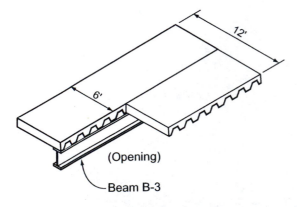

(Opening)

Beam B-3

Figure 5.61   Tributary widths for beam B-3.

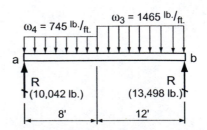

Figure 5.62   FBD of beam B-3.

For 12' of span:

$$\omega_3 = \left(120\,^{\text{lb.}}\!/_{\text{ft.}^2}\right) \times (12\text{ ft.}) + 25\,^{\text{lb.}}\!/_{\text{ft.}} = 1465\,^{\text{lb.}}\!/_{\text{ft.}}$$

For 6' of span:

$$\omega_4 = \left(120\,^{\text{lb.}}\!/_{\text{ft.}^2}\right) \times (6\text{ ft.}) + \underset{\text{(beam wt.)}}{25\,^{\text{lb.}}\!/_{\text{ft.}}} = 745\,^{\text{lb.}}\!/_{\text{ft.}}$$

$$\sum M_a = 0$$

$$- \left(745\,^{\text{lb.}}\!/_{\text{ft.}}\right)(8\text{ ft.})(4\text{ ft.}) - \left(1465\,^{\text{lb.}}\!/_{\text{ft.}}\right)(12\text{ ft.})(14\text{ ft.})$$

$$+ B_y(20\text{ ft.}) = 0$$

$$\therefore B_y = 13{,}498\text{ lb.}$$

$$\sum F_y = 0$$

$$- \left(745\,^{\text{lb.}}\!/_{\text{ft.}}\right)(8\text{ ft.}) - \left(1465\,^{\text{lb.}}\!/_{\text{ft.}}\right)(12\text{ ft.})$$

$$+ 13{,}498\text{ lb.} + A_y = 0$$

$$\therefore A_y = 10{,}042\text{ lb.}$$

Girder G-1 (Figures 5.63 and 5.64): Girder G-1 supports reactions from beams B-2 and B-3. Beam B-1 sends its reaction directly to the column and causes no load to appear in girder G-1.

$$\left[\sum M_a = 0\right]$$

$$- (14{,}650\text{ lb.})(12\text{ft.}) - (13{,}498\text{ lb.})(24\text{ ft.})$$

$$- \left(35\,^{\text{lb.}}\!/_{\text{ft.}}\right)(36\text{ ft.})(18\text{ ft.}) + B_y(36\text{ ft.}) = 0$$

$$\therefore B_y = 14{,}512\text{ lb.}$$

$$\left[\sum F_y = 0\right] - 14{,}650\text{ lb.} - 13{,}498\text{ lb.}$$

$$+ 14{,}512\text{ lb.} + A_y = 0$$

$$\therefore A_y = 14{,}896\text{ lb.}$$

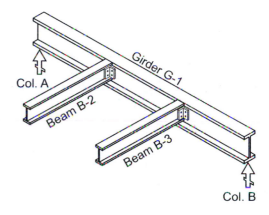

Figure 5.63   Girder G-1 (partial framing).

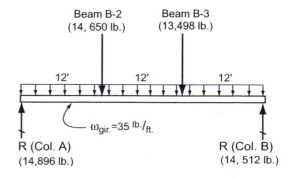

Figure 5.64   FBD of girder G-1.

Example Problem 5.5

In this example, the load trace will involve the framing for a small deck addition to a residence as shown in Figures 5.65 and 5.66. Once post reactions have been determined, a preliminary footing size will be designed assuming the soil capacity of 3000 psf is known from a geotechnical investigation.

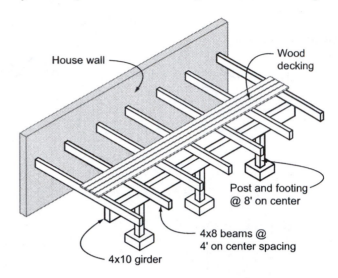

Figure 5.65   Isometric of a residential deck.

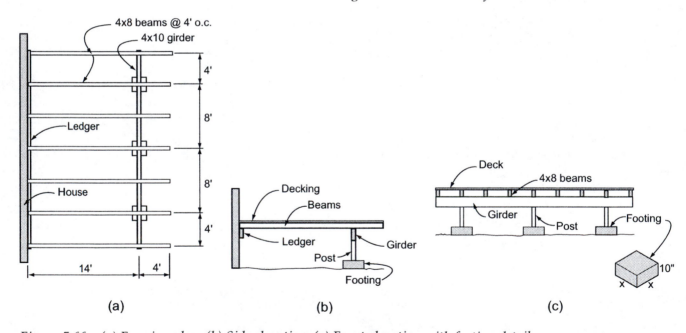

Figure 5.66   (a) Framing plan. (b) Side elevation. (c) Front elevation with footing detail.

Loads:

$$\text{Live load} \qquad = 60\,^{\text{lb.}}/_{\text{ft.}^2}$$

Dead loads:

$$\text{Decking} \qquad = 5\,^{\text{lb.}}/_{\text{ft.}^2}$$
$$\text{Beams} \qquad = 5\,^{\text{lb.}}/_{\text{ft.}}$$
$$\text{Girder} \qquad = 10\,^{\text{lb.}}/_{\text{ft.}}$$
$$\gamma_{\text{concrete}} \qquad = 150\,\#\,/\,\text{ft.}^3\ (\text{density})$$

For this load-trace problem, we will investigate the following:

1. Draw a FBD of the typical beam showing its load condition.
2. Draw a FBD of the girder with its load conditions shown.
3. Determine the load in each post.
4. Determine the size x of the critical pier footing (account for the weight of the concrete).

Solution:

1. **Beam, typical interior** (Figures 5.67 and 5.68):

   DL: 5 psf (4 ft.) $= 20^{\,\text{lb.}}/_{\text{ft.}}$

         Beam wt. $= 5^{\,\text{lb.}}/_{\text{ft.}}$

         ω_{DL} $= 25^{\,\text{lb.}}/_{\text{ft.}}$

   LL: 60 psf (4 ft.) $= 240^{\,\text{lb.}}/_{\text{ft.}}$

         $\omega_{DL\,+\,LL}$ $= 265^{\,\text{lb.}}/_{\text{ft.}}$

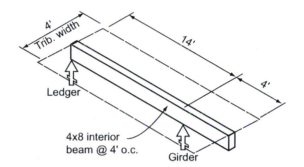

Figure 5.67 Tributary width for an interior beam.

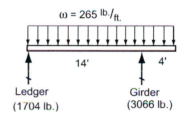

Figure 5.68 FBD of an interior beam.

2. **Beam, typical exterior** (Figures 5.69 and 5.70):

   DL: 5 psf (2 ft.) $= 10^{\,\text{lb.}}/_{\text{ft.}}$

         Beam wt. $= 5^{\,\text{lb.}}/_{\text{ft.}}$

         ω_{DL} $= 15^{\,\text{lb.}}/_{\text{ft.}}$

   LL: 60 psf (2 ft.) $= 120^{\,\text{lb.}}/_{\text{ft.}}$

         $\omega_{DL\,+\,LL}$ $= 135^{\,\text{lb.}}/_{\text{ft.}}$

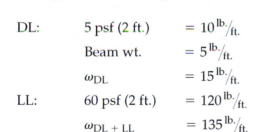

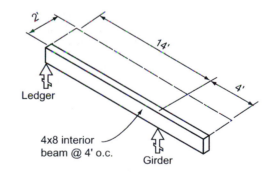

Figure 5.69 Tributary width for an exterior beam.

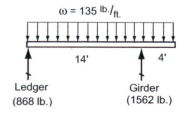

Figure 5.70 FBD of an exterior beam.

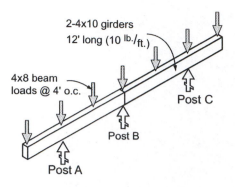

Figure 5.71 *Isometric of the girder with beam loads.*

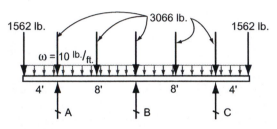

Figure 5.72 *FBD of the girder.*

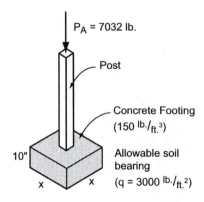

Figure 5.73 *Load on the pier footing.*

3. and 4. **Girder and post** (Figures 5.71 and 5.72):

$$\sum M_B = +(3066\ \text{lb.})(4\ \text{ft.}) + (3066\ \text{lb.})(8\ \text{ft.})$$
$$+ (1562\ \text{lb.})(12\ \text{ft.})$$
$$+ \left(10^{\text{lb.}}/_{\text{ft.}}\right)(12\ \text{ft.})(6\ \text{ft.}) - A_y(8\ \text{ft.}) = 0$$
$$\therefore\ A_y = 7032\ \text{lb.}$$

$$\sum F_y = -\left(\frac{3066\ \text{lb.}}{2}\right) - 3066\ \text{lb.}$$
$$- 3066\ \text{lb.} - 1562\ \text{lb.}$$
$$- \left(10^{\text{lb.}}/_{\text{ft.}}\right)(12\ \text{ft.}) + \underset{\text{(right side)}}{B_y} + 7032\ \text{lb.} = 0$$

$$\therefore\ \underset{\text{(right)}}{B_y} = 2315\ \text{lb.}$$

The total reaction at post B is the sum of the reactions:

$$\therefore\ \underset{\text{(right)}}{B_y} + \underset{\text{(left)}}{B_y} = 4630\ \text{lb.}$$

5. **Critical Footing** (Figure 5.73):

The soil is capable of resisting a total bearing pressure of 3000 pounds per square foot.

Note: pressure $= \dfrac{\text{load}}{\text{area}}$; $q = \dfrac{P}{A}$

where:

q = allowable soil bearing capacity $\left(^{\text{lb.}}/_{\text{ft.}^2}\ \text{or}\ ^{\text{k}}/_{\text{ft.}^2}\right)$
P = load on the footing (pounds or kips)
A = base area of the footing (ft.^{2})

By setting $q = 3000^{\text{lb.}}/_{\text{ft.}^2}$ (allowable capacity of the soil), we need to deduct the weight of the footing itself to determine the footing's capacity to resist applied load from above.

Therefore:

$q_{\text{net}} = q -$ footing weight (as pressure in psf)

The footing weight can be solved for by converting the density of concrete $\gamma_{\text{concrete}} = 150^{\text{lb.}}/_{\text{ft.}^3}$ into equivalent pounds per square foot units by multiplying:

footing wt. (psf)
$$= (\gamma_{\text{concrete}}) \times (\text{thickness of concrete in feet})$$

$$\text{footing wt.} = \left(150^{\text{lb.}}/_{\text{ft.}^3}\right) \times \left(\frac{10\ \text{in.}}{12^{\text{in.}}/_{\text{ft.}}}\right) = 125^{\text{lb.}}/_{\text{ft.}^2}$$

The remaining soil capacity to resist point loads is expressed as:

$$q_{net} = 3000\,{}^{lb.}\!/_{ft.^2} - 125\,{}^{lb.}\!/_{ft.^2} = 2875\,{}^{lb.}\!/_{ft.^2}$$

$$q_{net} = \frac{P}{A} = \frac{P}{x^2}$$

$$\therefore x^2 = \frac{P}{q_{net}} = \frac{7032\ lb.}{2875\,{}^{lb.}\!/_{ft.^2}} = 2.45\ ft.^2$$

$$\therefore x = 1.57\ ft. = 1\ ft. - 7\ in.$$

(minimum square footing size)

Example Problem 5.6

Calculate the load trace for the sloped roof structure shown in Figures 5.74 to 5.77.

Roofing	= 5 psf
Roof sheathing	= 3 psf
Rafters	= 4 plf

Dead loads are assumed applied along the length of the rafters.

Snow (SL): 40 psf} On horizontal projection of rafter*

Beam: 16 plf

***Note:** Snow loads are normally given as a load on the horizontal projection of a roof.*

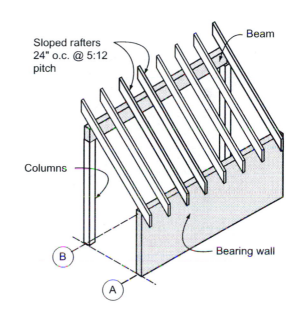

Figure 5.74   Isometric of a simple shed roof structure.

Solution:

Rafter analysis (Figures 5.76 and 5.77):

Roofing:	5 psf
Sheathing:	3 psf

$$(8\ psf) \times \left(\frac{24\ in.}{12\,{}^{in.}\!/_{ft.}} \right) = 16\,{}^{lb.}\!/_{ft.}$$

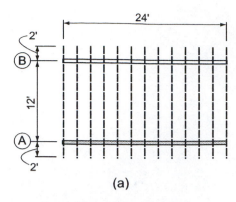

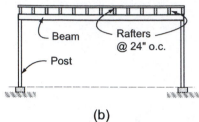

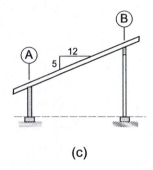

(a)

(b)

(c)

Figure 5.75   (a) Framing plan. (b) Front elevation. (c) Side elevation.

Rafter: $4\,{}^{lb.}/_{ft.}$

Total load along the rafter length:

$$16\,{}^{lb.}/_{ft.} + 4\,{}^{lb.}/_{ft.} = 20\,{}^{lb.}/_{ft.}$$

Adjusting the load to an equivalent on the horizontal projection of the rafter:

$$\omega_{DL\,(adj.)} = \frac{13}{12}\left(20\,{}^{lb.}/_{ft.}\right) = 21.7\,{}^{lb.}/_{ft.}$$

$$\text{Snow (SL)} = \left(40\,{}^{lb.}/_{ft.^2}\right) \times \left(\frac{24\text{ in.}}{12\,{}^{in.}/_{ft.}}\right) = 80\,{}^{lb.}/_{ft.}$$

$$\omega_{equiv.} = 21.7\,{}^{lb.}/_{ft.} + 80\,{}^{lb.}/_{ft.} = 101.7\,{}^{lb.}/_{ft.}$$

Note: The rafter is horizontally projected for ease of computation.
Using FBD sketches and computations, determine the load condition on (1) beam, (2) column, and (3) wall (typical stud).

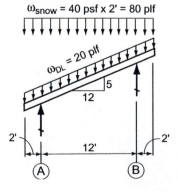

Figure 5.76   FBD of the inclined rafter.

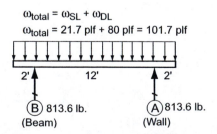

Figure 5.77   FBD of the equivalent horizontally projected rafter.

Load Tracing

1. **Beam analysis** (Figure 5.78):

   The reaction from a typical roof joist on top of the beam and stud wall is 813.6 lb. However, since the roof joists occur every two feet, the equivalent load ω is equal to:

   $$\omega = \frac{813.6 \text{ lb.}}{2 \text{ ft.}} = 406.8 \,^{\text{lb.}}/_{\text{ft.}} + \underset{\text{(beam wt.)}}{16 \,^{\text{lb.}}/_{\text{ft.}}} = 422.8 \,^{\text{lb.}}/_{\text{ft.}}$$

2. **Column analysis** (Figures 5.78 and 5.79):

   The column load is computed as:

   $$P = \frac{\omega L}{2} = \frac{\left(422.8 \,^{\text{lb.}}/_{\text{ft.}}\right) \times (24 \text{ ft.})}{2} = 5074 \text{ lb.}$$

   Note: This equation simply divides the total load on the rafter in half, since the rafter is symmetrically loaded.

3. **Stud wall** (Figures 5.80 and 5.81):

   Tributary wall length per stud is:

   $$16 \text{ in.} = \frac{16 \text{ in.}}{12 \,^{\text{in.}}/_{\text{ft.}}} = 1.33 \text{ ft.}$$

   $$P = 406.8 \,^{\text{lb.}}/_{\text{ft.}} \times 1.33 \text{ ft.} = 541 \text{ lb.}$$

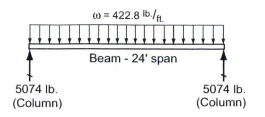

Figure 5.78   *FBD of the beam.*

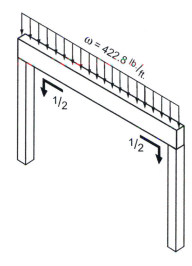

Figure 5.79   *Beam/column load trace.*

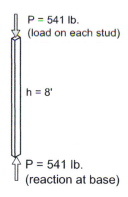

Figure 5.80   *Tributary load in each stud.*

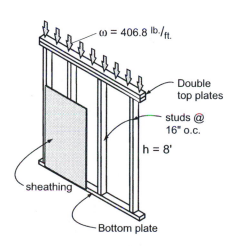

Figure 5.81   *Load on the bearing wall.*

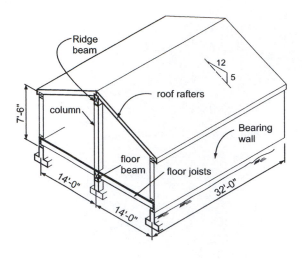

Figure 5.82   Isometric/section through the building.

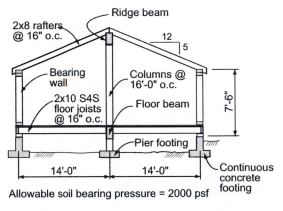

Figure 5.83   Section through building.

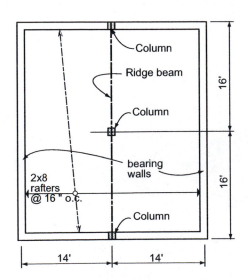

Figure 5.84   Framing plan—roof.

Example Problem 5.7

A simple light-framed wood building, shown in Figures 5.82 to 5.84, is subjected to the load conditions as specified. Using FBDs and equations of equilibrium, trace the loads through the building for the following elements:

1. Determine the equivalent (horizontally projected) load on the rafters.
2. Determine the load per foot on the bearing wall.
3. Determine the load on the ridge beam.
4. Determine the column loads.
5. Determine the minimum width of the continuous foundation.
6. Determine the size of the interior footings.

Load conditions:

Soil-bearing pressure	= 2000 psf
Flooring	= 2 psf
Subfloor	= 5 psf
Joists	= 4 psf
LL (occupancy)	= 40 psf
Snow	= 25 psf
Walls	= 7 psf

Along rafter length:

Roofing	= 5 psf
Sheathing	= 3 psf
Rafters	= 2 psf
Ceiling	= 2 psf

The ridge beam spans 16' from column support to column support.

Solution:

1. **Rafters:** (Figures 5.85 and 5.86)
   Snow = 25 psf: rafters spaced 16" o.c.

   $$\omega_{SL} = \left(25\,^{lb.}/_{ft.^2}\right)\left(\frac{16}{12}\,ft.\right) = 33.3\,^{lb.}/_{ft.}$$

   (Horizontal projection)

   Roof DL = 12 psf

   $$\omega_{DL} = \left(12\,^{lb.}/_{ft.^2}\right)\left(\frac{16}{12}\,ft.\right) = 16\,^{lb.}/_{ft.}$$

   (Along rafter length)

   $$\omega_{DL} = \left(\frac{13}{12}\right)\left(16\,^{lb.}/_{ft.}\right) = 17.3\,^{lb.}/_{ft.}$$

   (Equivalent load horizontally projected)

   $$\omega_{Total} = \omega_{SL} + \omega_{DL}$$

   $$\omega_{Total} = 33.3\,^{lb.}/_{ft.} + 17.3\,^{lb.}/_{ft.} = 50.6\,^{lb.}/_{ft.}$$

   The reaction at each rafter support may be determined using equilibrium equations. When uniform loads on a simply supported member are present, a simple formula may be used where:

   $$R = \frac{\omega L}{2} = \frac{\left(50.6\,^{lb.}/_{ft.}\right)(14\,ft.)}{2} = 354\,lb.$$

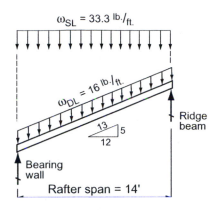

Figure 5.85   FBD of an inclined rafter.

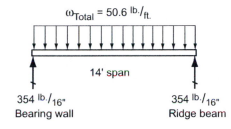

Figure 5.86   FBD of an equivalent horizontal rafter.

2. **Bearing wall:** (Figure 5.87)
   The reaction of the rafter onto the bearing wall is 354 lb. every 16". A conversion should be done to express the load at the top of the wall in pounds per lineal foot.

   $$\omega = \left(354\,^{lb.}/_{16\,in.}\right)\left(\frac{12}{16}\,ft.\right) = 266\,^{lb.}/_{ft.}$$

   A strip of wall 1 ft. wide and 7 ft.-6 in. tall weighs:

   $$\omega_{wall} = \left(7\,^{lb.}/_{ft.^2}\right)(7.5\,ft.) = 52.5\,^{lb.}/_{ft.}$$

   $$\omega = 266\,^{lb.}/_{ft.} + 52.5\,^{lb.}/_{ft.} = 318.5\,^{lb.}/_{ft.}$$

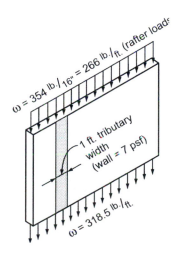

Figure 5.87   Loads on the bearing wall.

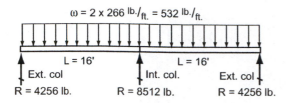

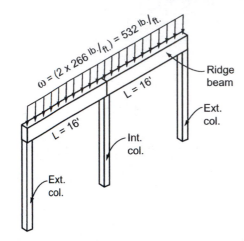

Figure 5.88   FBD of the ridge beam.

Figure 5.89   Ridge beam and column loads.

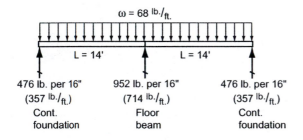

Figure 5.90   FBD of the floor joists.

3. **Ridge beam** (Figures 5.88 and 5.89):

    Rafter reactions are equal to 354 lb. per 16", or 266 $^{lb.}/_{ft.}$. Since the ridge beam is required to support rafters from both sides:

    $$\omega = 2\left(266\,^{lb.}/_{ft.}\right) = 532\,^{lb.}/_{ft.}$$

    Note: The ridge beams are treated as two simple span beams, each 16 ft. in length.

    Exterior columns supporting the ridge beam carry:

    $$P_{ext.} = \left(532\,^{lb.}/_{ft.}\right)(8\text{ ft.}) = 4256\text{ lb.}$$

    Note: The 8′ represents the tributary beam length that is supported by the exterior columns.

    Interior columns support a tributary beam length of 16 ft., therefore:

    $$P_{int.} = \left(532\,^{lb.}/_{ft.}\right)(16\text{ ft.}) = 8512\text{ lb.}$$

4. **Floor joists** (Figure 5.90): Joists are spaced 16" o.c., which also represents the tributary width of load assigned to each joist.

    Loads: DL + LL = 11 psf + 40 psf = 51 psf

    (Floor joist loads and reactions)

    $$\omega = 51\,^{lb.}/_{ft.} \times \left(\tfrac{16}{12}\right)\text{ft.} = 68\,^{lb.}/_{ft.}$$

    Foundation reactions may be obtained by:

    Foundation

    $$= \left(68\,^{lb.}/_{ft.}\right)(7\text{ ft. \{trib. length\}}) = 476\,^{lb.}/_{16\text{ in.}}$$

    The central floor beam supports a floor joist reaction equal to:

    $$\text{Beam} = \left(68\,^{lb.}/_{ft.}\right)(14\text{ ft. \{trib. length\}}) = 952\,^{lb.}/_{16\text{ in.}}$$

    Conversion of the floor joist reactions into load per foot results in:

    $$\text{Foundation: } \omega = \left(476\,^{lb.}/_{16\text{ in.}}\right)\left(\frac{12}{16}\text{ft.}\right) = 357\text{ lb.}$$

    $$\text{Beam: } \omega = \left(952\,^{lb.}/_{16\text{ in.}}\right)\left(\frac{12}{16}\text{ft.}\right) = 714\text{ lb.}$$

5. **Continuous foundation** (Figure 5.91):

The stem wall measures 8 in. thick and 2 ft. tall. The footing base is 8 in. thick and x wide.

Loads from the roof, wall, and floor are combined as a total load on top of the foundation stem:

$$\omega_{total} = 318.5\,^{lb.}/_{ft.} + 357\,^{lb.}/_{ft.} = 675.5\,^{lb.}/_{ft.}$$

The foundation wall stem adds additional load on the footing equal to:

$$\text{Stem weight} = \left(\frac{8}{12}\,\text{ft.}\right)(2\,\text{ft.})(150\,^{lb.}/_{ft.^3}) = 200\,^{lb.}/_{ft.}$$

Since the footing width x is unknown, the weight of the footing base must be computed in terms of pounds per square foot.

$$\text{Footing weight} = \left(\frac{8}{12}\,\text{ft.}\right)(150\,^{lb.}/_{ft.^3}) = 100\,^{lb.}/_{ft.^2}$$

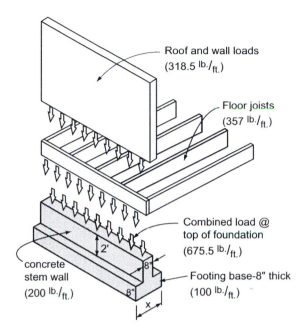

Figure 5.91   Wall/floor/foundation load path.

In determining the footing width, examine a unit length (1') of foundation as a representation of the entire length.

q = allowable soil bearing pressure = 2000 psf

$q_{net} = q$ − footing weight = 2000 psf − 100 psf = 1900 psf

This value of q_{net} represents the resistance of the soil available to safely support the loads of the roof, walls, floor, and foundation stem.

$$\omega_{total} = 675.5\,^{lb.}/_{ft.} + 200\,^{lb.}/_{ft.} = 875.5\,^{lb.}/_{ft.}$$

The minimum required resistance area of the footing per unit length is:

$$A = (1\,\text{ft.})(x)$$

$$q_{net} = \frac{\omega}{x}; \quad x = \frac{\omega}{q_{net}}$$

$$x = \frac{\left(875.5\,^{lb.}/_{ft.}\right)}{\left(1900\,^{lb.}/_{ft.^2}\right)} = 0.46\,\text{ft.} \approx 6\,\text{in.}$$

Note that a footing base of 6 in. would be less than the stem wall's thickness. A minimum footing width for a one-story, light-framed assembly should be 12 inches.

If a 12-in. base width is provided, the actual pressure on the soil will be:

Actual pressure

$$= \frac{\omega}{x = 1 \text{ ft.}} = \frac{875.5 \,{}^{\text{lb.}}\!/_{\text{ft.}}}{1 \text{ ft.}} = 875.5 \,{}^{\text{lb.}}\!/_{\text{ft.}^2} < q_{net}$$

∴ footing width is OK

6. **Interior spread footings** (Figures 5.92 and 5.93):

A determination of individual post loads is necessary before footing sizes can be computed.

Assume the spread footings have a thickness of 8 in. and the $q = 2000$ psf:

$$\omega_{footing} = \left(\frac{8}{12}\text{ft.}\right)\!\left(150\,{}^{\text{lb.}}\!/_{\text{ft.}^3}\right) = 100\,{}^{\text{lb.}}\!/_{\text{ft.}^2}$$

$q_{net} = q -$ footing wt. $= 2000$ psf $- 100$ psf $- 1900$ psf

Critical center column:

$$A = x^2 = \frac{P}{q_{net}} = \frac{14{,}224 \text{ lb.}}{1900\,{}^{\text{lb.}}\!/_{\text{ft.}^2}} = 7.5 \text{ ft.}^2$$

$$\therefore x = 2.74 \text{ ft.} = 2 \text{ ft. } 9 \text{ in. square}$$

Other footing:

$$x^2 = \frac{5712 \text{ lb.}}{1900\,{}^{\text{lb.}}\!/_{\text{ft.}^2}} = 3.00 \text{ ft.}^2;$$

$$\therefore x = 1.73 \text{ ft.} \approx 1 \text{ ft. } 9 \text{ in.}$$

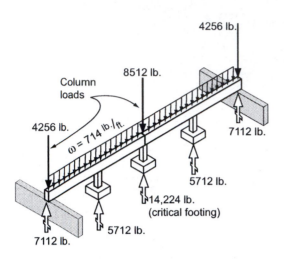

Figure 5.92   Floor beam and post loads.

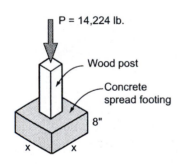

Figure 5.93   Load on the critical spread footing.

Supplementary Problems

In each of the load-tracing problems that follow, construct a series of FBDs and show the propagation of loads through the various structural elements.

5.1.1. Determine the column loads assuming:

DL (decking, flooring, etc.)	= 10 psf
LL (occupancy)	= <u>40 psf</u>
Total	50 psf

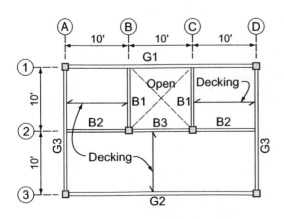

Problem 5.1.1   Typical-floor framing.

5.1.2. Loads:

Roof: DL = 10 psf

LL = 25 psf

(snow horizontally projected)

Ceiling: DL = 5 psf

LL = 10 psf

Bearing walls: DL = 10 psf (2nd and 3rd floors)

Floors: DL = 20 psf (2nd and 3rd floors)

LL = 40 psf (2nd and 3rd floors)

1. Determine the equivalent (horizontally projected) load on the rafters spaced at 2'0" on centers.
2. Determine the load per foot on the bearing walls.
3. Determine the loading and beam reactions for each of the steel wide-flange beams.

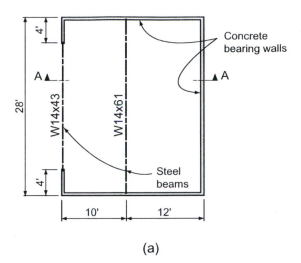

(a)

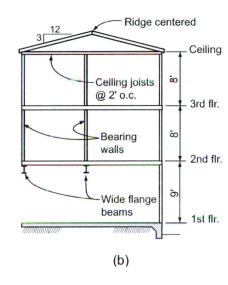

(b)

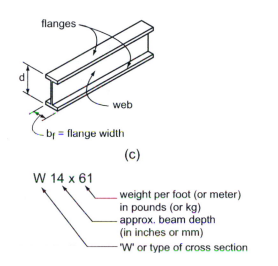

(c)

Problem 5.1.2   (a) First floor plan. (b) Section A-A. (c) Steel wide-flange section.

5.1.3. Trace the loads through the following elements in this structure.

1. Rafters.
2. Stud walls.
3. Roof beam.
4. Columns (interior and exterior).
5. Floor joist.
6. Floor beam.
7. Load on top of continuous footings.
8. Critical interior footing load.

Roof loads:

Snow = 20 psf

Roofing = 5 psf

Sheathing = 3 psf

Ceiling = 2 psf

Bearing walls = 8 psf; rafters = 3 psf; floor joists = 3 psf

Allowable soil bearing pressure = 3000 psf.

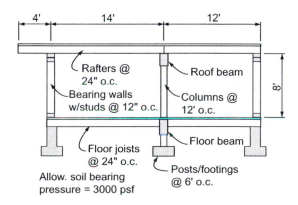

Problem 5.1.3   Section through a simple wood structure.

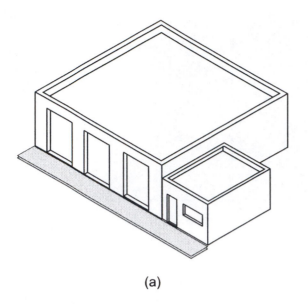

(a)

5.1.4. Draw FBDs and show load conditions for B-1, G-1, and the interior column.

Loads:

Snow load	= 25 psf
Roofing and joists (deck)	= 10 psf
Insul., mech., elec.	= 5 psf
Beams B-1	= 15 lb./ft.
Girders G-1	= 50 lb./ft.

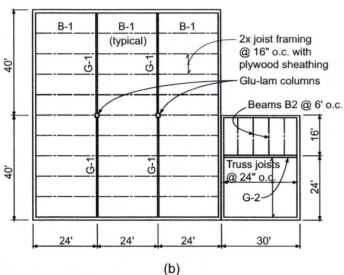

(b)

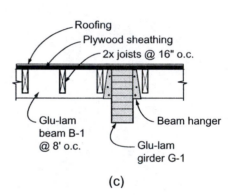

(c)

Problem 5.1.4   (a) Warehouse/office building.
(b) Roof framing plan. (c) Roof framing—typical section.

5.1.5. Using the building shown in Problem 5.1.4, draw the FBDs for beam B-2 and girder G-2. Beam B-2 is spaced at 6'-0" o.c.

Loads:

Snow = 25 psf
Roofing and decking = 10 psf
Truss joists = 3 psf
Insulation, mech., and elect. = 5 psf
Beam B-2 = 12 pounds per lineal foot
Girder G-2 = 40 plf.

5.1.6.

Roof loads:

Snow	= 20 psf
Shakes	= 5 psf
Plywood	= 2 psf
Insulation	= 5 psf
Joists	= 4 lb./ft.
Ridge beam	= 40 lb./ft.

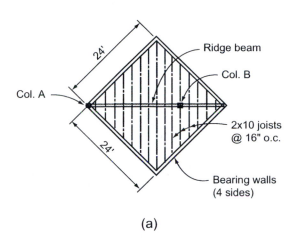

(a)

1. Show a sketch of the load and its magnitude acting on the 34 ft.-long ridge beam.
2. What is the force in columns *A* and *B*, which support the ridge beam?

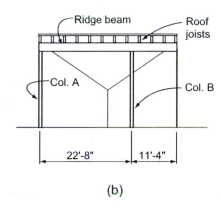

(b)

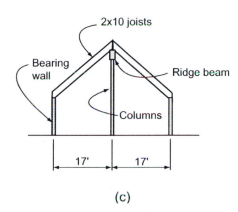

(c)

Problem 5.1.6   (a) Roof framing plan. (b) Interior elevation. (c) Section through building.

5.1.7. Show graphically (FBDs for each element) the load trace (load condition) for:

1. Rafter(s).
2. Roof beam.
3. Exterior stud wall(s).
4. Interior columns.
5. Floor joist(s).
6. Floor beam.
7. Floor post.
8. Exterior foundation width(s) (adequacy?).
9. Size of critical pier footing.

Load conditions:

Snow (horiz. proj.)	= 30 psf
Finish floor	= 2 psf
Subfloor	= 3 psf
Joists	= 3 psf
Insulation	= 2 psf
Occupancy (LL)	= 40 psf
$\gamma_{concrete}$	= 150 #/ft.^{3}
(density of concrete)	

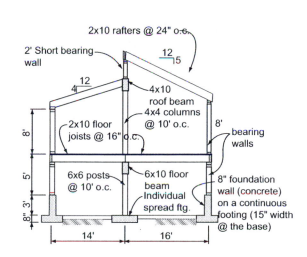

Problem 5.1.7   Building section (the building is 50' long).

Along rafter length:

Roofing	= 8 psf
Sheathing	= 2 psf
Rafters	= 3 psf
Ceiling	= 3 psf
Insulation	= 2 psf
Soil bearing pressure	= 2000 psf

5.1.8. For the hipped roof building shown, evaluate the load conditions on:

1. A typical jack rafter.
2. A hip rafter.
3. A ceiling joist.
4. Beams B-1, B-2, B-3.
5. An interior column.

Roof live loads:

Snow = 25 psf

Roof dead loads:

Roofing = 6 psf

Plywood roof sheathing = 1.5 psf

Joist framing = 4 lb./ft.

Ceiling loads:

Dead load	= 7 psf
Live load	= 20 psf

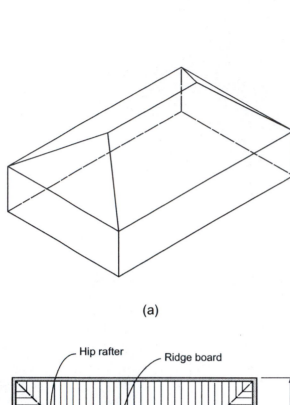

(a)

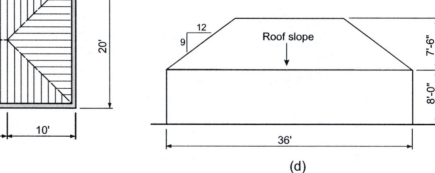

(b)

(d)

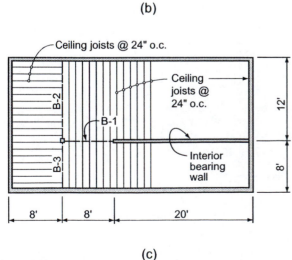

(c)

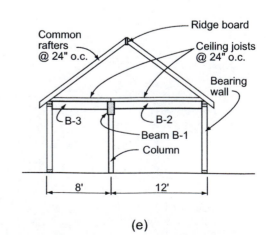

(e)

Problem 5.1.8   (a) Hipped roof structure. (b) Roof framing plan. (c) Ceiling framing plan. (d) Building elevation. (e) Building section.

6 Stress, Strain, and Deformation

6.1 STRESS—AXIAL, SHEAR, AND BEARING

Statics is essentially *force analysis* in which equations of equilibrium are used to determine the internal forces produced in elements and connections of a structural framework resisting externally applied forces. *Strength of materials* refers to the mechanical changes occurring in the material of the building element resisting the internal forces. *Structural design* involves the selection of the correct material, size, and shape of the structural member to safely support the loads imposed on the framework. Design of the structural parts or components must be adequate enough to withstand the forces safely and without excessive deformations. Good structural design incorporates the issues of safety, usability/comfort, and an economy of material and means.

The primary objective of a course of study in strength (mechanics) of materials is the development of the relationship between the loads applied to a nonrigid body and the resulting internal forces and deformations induced in the body. The chapters dealing with strength of materials are intended to give the student information and an elementary understanding of the mechanical properties of some of the more common building materials. There will be no attempt to replicate the theory and comprehensiveness of strength of materials courses typically found in engineering curricula.

In his book *Dialogues Concerning Two New Sciences* (1638), Galileo Galilei made reference to the strength of beams and the properties of structural materials. He became one of the early scholars who fostered the development of strength of materials as an area of study.

Knowledge of the fundamentals of statics and strength of materials provides the foundation for the structural design of elements and connections.

Definition of Stress

Stress, like pressure, is a term used to describe the intensity of a force—the quantity of force that acts on a unit of area. *Unit stress* refers to the magnitude of the stress per unit of

Galileo Galilei (1564–1642)

Galileo, pushed by his mathematician father to study medicine, was purposely kept from the study of mathemetics. Fate took a turn, however, and Galileo accidentally attended a lecture on geometry. He pursued the subject further, which eventually led him to the works of Archimedes. Galileo pleaded and, reluctantly, his father conceded and permitted him to pursue the study of mathematics and physics. Galileo's fundamental contribution to science was his emphasis on direct observation and experimentation rather than on blind faith in the authority of ancient scientists. His literary talent enabled him to describe his theories and present his quantitative method in an exquisite manner. Galileo is regarded as the founder of modern physical science, and his discoveries and the publication of his book Mechanics *served as the basis a century later for the three laws of motion propounded by Sir Isaac Newton.*

Galileo is perhaps best known for his views on free-falling bodies. Legend has it that he simultaneously dropped two cannon balls, one ten times heavier than the other, from the Leaning Tower of Pisa, both being seen and heard to touch the ground at the same time. This experiment has not been substantiated, but other experiments actually performed by Galileo were sufficient to cast doubt on Aristotelian physics.

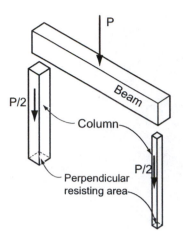

Figure 6.1   Two columns with the same load, different stress.

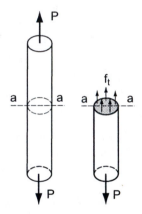

Figure 6.2   Normal tensile stress through section a-a.

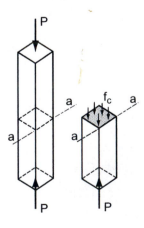

Figure 6.3   Normal compressive stress across section a-a.

area. *Force,* in structural design, has little significance until information is known about the resisting material, cross-sectional properties, and size of the element resisting the force (see Figure 6.1).

A stress can be classified according to the internal reaction that produces it. Axial (load passing through the center of the member cross-section) tensile or compressive forces, as shown in Figures 6.2 and 6.3, produce tensile or compressive stress, respectively. This type of stress is classified as a *normal stress* (the first of two basic stresses) since the stressed surface is normal (perpendicular) to the load direction. The stressed areas (section cut a-a) in Figures 6.2 and 6.3 are perpendicular to the axial load direction.

Unit stresses on *axially loaded* members are assumed to be uniform throughout the resisting area and assumes that the minimum and maximum unit stresses are relatively equal to the average unit stress.

The unit *normal stress*, the average value of the axial stress, may be represented mathematically as:

$$\underset{\text{(tension)}}{f_t} \quad \text{or} \quad \underset{\text{(compression)}}{f_c} \quad = \sigma = \frac{P}{A_\perp}$$

$$= \frac{\text{axial force}}{\text{perpendicular resisting area}}$$

where:

$f = \sigma$ (sigma) = the symbol(s) representing unit stress; units are expressed as:
psi $-\left(^{\text{lb.}}/_{\text{in.}^2}\right)$, ksi $-\left(^{\text{k}}/_{\text{in.}^2}\right)$, kilopascal or

kPa $-\left(^{\text{kN}}/_{\text{m}^2}\right)$, megapascal or MPa $-\left(^{\text{MN}}/_{\text{m}^2}\right)$.

Note: $1\left(^{\text{N}}/_{\text{m}^2}\right) = 1$ pascal $= 1$ Pa

P = applied load or force; units are in pounds (lb.), kips (k), Newtons (N), kilonewtons (kN), and meganewtons (MN).

A = resisting area; units are in.^{2}, ft.^{2}, mm^2, or m^2.

Example Problem 6.1 (Figure 6.4)

A short timber post supports an axial load of 10,000 lb. from a floor beam. The post is a Douglas-fir, 6×6 S4S member with actual cross-sectional dimensions of 5.5" × 5.5".

Determine the axial compressive stress that develops in the post.

Solution:

The short post resists the axial compressive load across its cross-section that is perpendicular to the load direction. Dividing the post load by the cross-sectional area of the post results in a stress equal to:

$$f_c = \frac{P}{A} = \frac{10{,}000 \text{ lb.}}{(5.5" \times 5.5")} = \frac{10{,}000 \text{ lb.}}{30.25 \text{ in.}^2} = 330.6 \text{ lb.}/\text{in.}^2$$

The stress value is compared against the capability of the post material (the allowable capacity—generally obtained from standard design manuals) to determine its adequacy in supporting the applied load.

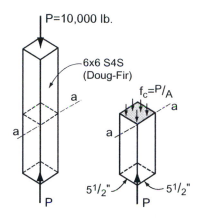

Figure 6.4   Short timber post in compression.

Example Problem 6.2 (Figure 6.5)

A steel rod with a diameter of 15 mm supports a weight equal to 6 kN. Determine the average unit tensile stress developed in the rod.

Solution:

This problem assumes an axial tensile load stresses the circular cross-section of the rod uniformly. The tension stress is computed as:

$$f_t = \frac{P}{A}$$

$$P = 6 \text{ kN}$$

$$A = \pi(r^2) = \frac{\pi D^2}{4} = \frac{(3.14)(0.015 \text{ m})^2}{4} = 0.000177 \text{ m}^2$$

$$f_t = \frac{6 \text{ kN}}{1.77 \times 10^{-5} \text{ m}^2} = 33{,}900 \text{ kN}/\text{m}^2$$

$$= 33{,}900 \text{ kPa} = 33.9 \text{ MPa}$$

Again, the actual tensile stress computed must be compared against the established safe limits of the steel to determine if the specified rod size is adequate to support the load.

Shear stress, the second of two basic stresses, develops on an area parallel to the direction of the applied load (Figure 6.6). Other types of stresses (bearing, torsional, or bending) are similar to or derivative of the two basic stresses, normal and shear.

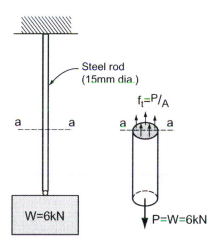

Figure 6.5   Steel rod loaded in tension.

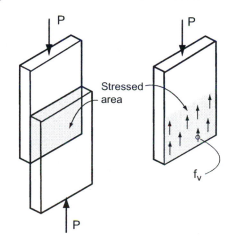

Figure 6.6   Shear stress between two glued blocks.

Unit shear stress is expressed mathematically as:

$$f_v = \tau \,(\text{tau}) = \frac{P}{A} = \frac{\text{applied force or load}}{\text{parallel resisting area}}$$

where:

$f_v = \tau =$ average unit shear stress (psi, ksi, Pa, kPa, MPa)

$P =$ applied load (lb., k, N, kN, MN)

$A =$ resisting cross-sectional area parallel to the load direction (in.^{2}, mm^2, or m^2)

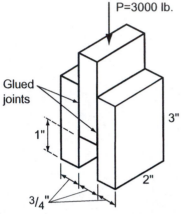

Figure 6.7   *Plywood shear blocks.*

Example Problem 6.3 (Figure 6.7)

A plywood fabricator uses the arrangement shown to determine the shear strength of a glued joint. If $P = 3000$ lb. at failure, what is the unit shearing stress that develops between the plywood pieces?

Solution:

Note that in this test arrangement, there are two surfaces in shear that resist the applied load.

$$f_v = \frac{P}{A} = \frac{3000 \text{ lb.}}{\underset{(\text{surfaces})}{2} \times (2'' \times 2'')}$$

$$= \frac{3000 \text{ lb.}}{8 \text{ in.}^2} = 375\,{}^{\text{lb.}}\!/_{\text{in.}^2}$$

Shear Stress in Bolted Connections

Another common example of shear stresses is found in structural bolted connections as shown in Figure 6.8. When a steel plate is lapped and bolted to another plate (called a *lap splice*), as shown in Figure 6.8, only one plane of shear develops in the bolt and thus the bolt is stressed in single shear.

$$f_v = \frac{P}{A} \, \ldots \quad (\text{single shear})$$

where:

$f_v =$ average shear stress through the bolt cross-section (psi, ksi, kPa, Mpa)

$A =$ bolt cross-sectional area multiplied by the number of bolts in the connection (in.^{2}, mm^2, and m^2)

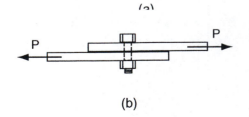

(b)

Figure 6.8   *(a) Two steel plates bolted together using one bolt. (b) Elevation showing the bolt in shear.*

If a middle plate is "sandwiched" between two thinner, outer plates and bolted together, a condition of double shear results in the bolts (Figure 6.9). Each bolt develops two shear planes and thus increases the resisting capacity of the bolt by two.

$$f_v = \frac{P}{2A} \quad \dots \quad \text{(double shear)}$$

where:

> f_v = average shear stress through the bolt cross-section (psi, ksi, kPa, Mpa)
>
> $2A$ = $2 \times$ (bolt cross-sectional area) multiplied by the number of bolts in the connection (in.², mm², and m²)

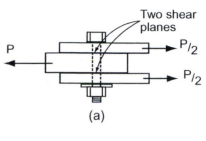

(a)

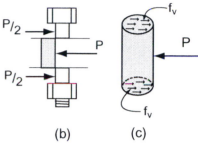

(b)   (c)

Figure 6.9   (a) Bolted connection in double shear. (b) FBD of the bolt. (c) FBD of the middle section.

Example Problem 6.4 (Figure 6.10)

Determine the shear stress developed in the bolts shown in the lap splice if the applied load is $P = 8$ k. The maximum shear stress that the bolts are capable of safely resisting is equal to 14.5 ksi. Are the bolts adequate?

Solution:

$$f_v = \frac{P}{A} \quad \dots \quad \text{(single shear)}$$

$$P = 8 \text{ k}$$

$$A = \underset{\text{(no. bolts)}}{2} \times \frac{\pi(D)^2}{4} = 2 \times \frac{3.14(0.625)^2}{4}$$

$$= 0.614 \text{ in.}^2$$

$$\therefore f_v = \frac{8 \text{ k}}{0.614 \text{ in.}^2} = 13.03 \,^k/_{\text{in.}^2} < 14.5 \,^k/_{\text{in.}^2} \dots \text{ OK}$$

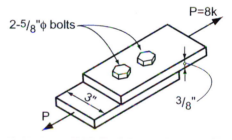

Figure 6.10   Bolted connection—single shear.

The general equation for direct stress can be written in three different ways, depending on the condition being evaluated. This simple equation has the capability of solving an incredibly wide variety of problems.

1. $f = {}^P/_A$ (Basic equation; used for analysis purposes in which the load, member size, and material are known. Given the completed design, is it adequate? That is, does it perform the function economically and without excessive deformation? What is the margin of safety allowed in each member? We call this *structural analysis*.)

2. $P = f \times A$ (Used in evaluating or checking the capacity of a member when the material and member size are known. Given a completed structure, what is its actual load-carrying capacity? The structure may have been designed for some purpose other than the one for which it is now to be used. Is the structure or its members adequate for the proposed new use? This is a *rating* problem.)

3. $A = {}^{P}/_{f}$ (Design version of the stress equation; member size can be determined if the load and material's allowable stress capability are known. Given a certain function to perform (the supporting of a roof system over a sports arena, the floor beams for a multistory office building), of what materials should the structure be constructed, and what should be the sizes and proportions of the various elements? This constitutes *structural design*, where there is often no single solution to a given problem, as there is in statics.)

Since the complete scope of these problems is obviously too comprehensive for coverage in a single text, this text will be restricted to the study of individual members and simple structural frameworks. Subsequent, more advanced structures books will consider the entire structure and will provide essential background for more thorough analysis and design.

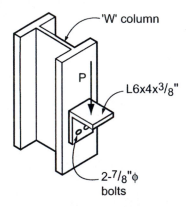

'W' column

P

L6x4x3/8"

2-7/8"φ
bolts

Figure 6.11   Steel column with a seat angle.

Example Problem 6.5 (Figure 6.11)

A steel seat angle supports a beam load P using 2- $^{7}/_{8}$" φ high-strength steel bolts (A325 with an allowable shear stress capacity F_v = 17 ksi). Determine the maximum permissible load P.

Solution:

In this example, the bolt material and the bolt size are known. The evaluation of the capacity of the connection (P) makes this a *rating* type problem. The bolts are in single shear.

Note: When the allowable stress is used, a capital letter is used for the stress designation.

$$P = F_v \times A$$
$$F_v = 17 \text{ ksi}$$
$$A = 2 \times \frac{\pi D^2}{4} = 2 \times \left(\frac{\pi \times 0.875^2}{4} \right) = 1.20 \text{ in.}^2$$
$$P_{\text{allowable}} = \left(17\,^{k}/_{\text{in.}^2} \right) \times (1.20 \text{ in.}^2) = 20.4 \text{ k}$$

Example Problem 6.6 (Figures 6.12 and 6.13)

A 10' × 20' hotel marquee hangs from two rods inclined at an angle of 30° (Figure 6.12). The dead load and snow load on the marquee add up to 100 psf. *Design* the two rods out of a structural steel that has an *allowable* tensile stress equal to $F_t = 22,000$ psi.

Solution:

The total load area of the marquee is equal to:

$$A = 10' \times 20' = 200 \text{ ft.}^2$$

Total load on the marquee: $200 \text{ ft.}^2 \times 100 \, ^{\text{lb.}}/_{\text{ft.}^2} = 20,000$ lb.

Since the framing is symmetrical, each rod carries an equal amount of the load.

$$T_{AB_y} = T_{AB} \times \sin 30° = 0.5 \, T_{AB}$$

$$\sum M_c = +(10 \text{ k})(5') - 0.5 \, T_{AB}(10') = 0$$

$$\therefore T_{AB} = 10 \text{ k}$$

$$f_t = \frac{P}{A}; \quad \text{rewriting the equation into a design form:}$$

$$A_{\text{req'd}} = \frac{P}{F_t} = \frac{10 \text{ k}}{22 \, ^k/_{\text{in.}^2}} = 0.46 \text{ in.}^2$$

The closest available rod diameter (to the nearest $^1/_{16}$") is:

$$\text{Rod diameter} = {}^{13}/_{16}" \, \phi, \, A = 0.519 \text{ in.}^2$$

Checking the actual tension stress against the allowable stress:

$$f_{t_{\text{actual}}} = \frac{10 \text{ k}}{0.519 \text{ in.}^2} = 19.27 \, ^k/_{\text{in.}^2} < 22_{\text{allowable}} \, ^k/_{\text{in.}^2}$$

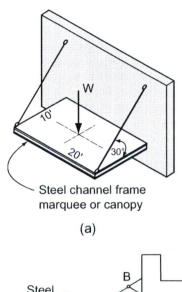

Steel channel frame
marquee or canopy

(a)

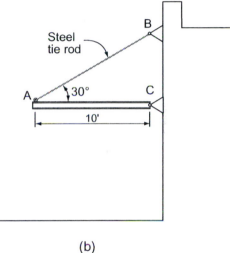

(b)

Figure 6.12

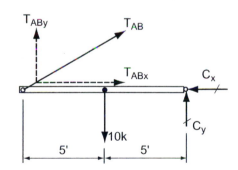

Figure 6.13   FBD of the marquee.

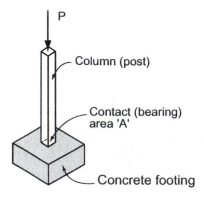

Figure 6.14 Bearing stress-post/footing/ground.

Bearing Stress

A type of normal stress, called bearing stress (see Figure 6.14), represents the intensity of force between a body and another body (i.e., the contact between beam–column, column–footing, footing–ground). The stressed surface is perpendicular to the direction of the applied load (the same as normal stress) and the average bearing stress is defined in terms of a force per unit area:

$$f_p = \frac{P}{A}$$

where:

f_p = unit bearing stress (psi, ksi, psf, ksf) or (pascal = N/m², kPa, and MPa)
P = applied load (lb. or k) or (N, kN, MN)
A = bearing contact area (in.², ft.²) or (mm², m²)

Both the column and footing may be assumed to be separate structural members and the bearing surface is the contact area between them. There also exists a bearing surface between the footing and the ground.

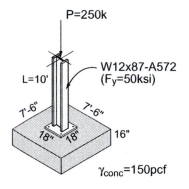

Figure 6.15 Column-base plate-footing.

Example Problem 6.7 (Figure 6.15)

A W12×87 steel column, 10' tall, is subjected to a load of 250 kips. The cross-sectional area of the column measures $A = 25.6$ in.² Determine the following:

    a. the compression stress developed in the column
    b. the bearing stress between the steel base plate and the concrete footing
    c. the bearing pressure (stress) between the base of the concrete footing and the soil

Solution:

    a. The normal compressive stress developed in the column is determined using the basic stress equation:

$$f_c = \frac{P}{A} = \frac{250 \text{ k}}{25.6 \text{ in.}^2} = 9.77 \text{ }^{k}/_{in.^2}$$

This value must be compared against the allowable compressive stress of the column to determine its suitability for use as a column.

    b. The bearing stress between the column and footing is:

$$f_{brg} = \frac{P}{A} = \frac{250 \text{ k}}{\underset{\text{(area of plate)}}{(18'' \times 18'')}} = 0.77 \text{ ksi}$$

If the column weight is included:

$$f_{brg} = \frac{250\,k + \left(0.087\,^{k}/_{ft.} \times 10\,ft.\right)}{(18'' \times 18'')} = 0.77\,ksi$$

which is essentially the same.

c. The bearing stress that develops between the footing and the ground is determined by including the pressure caused by the footing weight in addition to column load. Concrete has a density equal to: $\gamma_{concrete} = 150\,^{lb.}/_{ft.^3}$

$$f_{brg} = \frac{P_{load}}{A_{ftg}} + \text{footing weight}$$

$$\text{footing weight} = \left(^{16}/_{12}ft.\right) \times \left(.150\,^{k}/_{ft.^3}\right) = 0.2\,ksf$$
$$= 200\,psf = 0.2\,ksf$$

$$f_{brg} = \frac{251\,k}{(7.5' \times 7.5')} + 0.2\,k/ft^2$$
$$= 4.46 + 0.2 = 4.66\,ksf$$

This is a relatively high bearing stress and a very good soil will be needed for safe resistance to avoid excessive settling of the footing.

Torsional Stress

A twisting action in a *spandrel beam* (edge beam—Figure 6.16) or shaft (Figure 6.17) is resisted by the material principally by internal shearing forces. This twist is known as *torsion*.

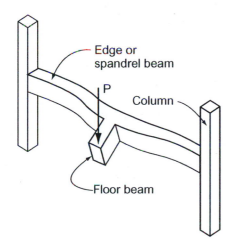

Figure 6.16   Torsion on a spandrel beam.

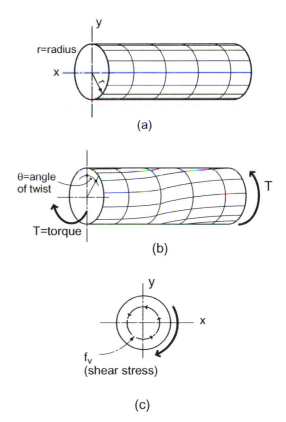

Figure 6.17   (a) Unloaded. (b) Torque applied. (c) Circular cross-section in torsion.

The twisting forces that produce a twisting moment are commonly called *torque*. Members in torsion are subjected to twisting action along their longitudinal axes caused by a moment couple or eccentric load. Most building members subjected to torsional effects are also experiencing either bending, shear, tensile, and/or compressive stresses; therefore, it is relatively uncommon to design specifically for torsion. However, designs involving machinery and motors with shafts are extremely sensitive to the stresses resulting from torsion.

C. A. Coulomb, a French engineer of the 18th century, was the first to explain torsion in a solid or hollow circular shaft. He experimentally developed a relationship between the applied torque (T) and the resulting deformation (angle of twist) of circular rods. From the distortion of the rod shown in Figure 6.17b, it is clear that shearing stresses f_v must exist. In an elastic material (such as steel), the stresses increase in magnitude proportionately to the distance from the center of the circular cross-section and flow circularly around the area. Coulomb derived by means of equilibrium concepts the relationship:

$$T = \frac{\pi r^3 f_v}{2}$$

where:

$\quad\quad T$ = externally applied torsional moment (torque)
$\quad \pi r^2$ = cross-sectional area of the rod (shaft)
$\quad\quad r$ = radius of the rod
$\quad\quad f_v$ = internal shear stress on the transverse plane of the rod

Hollow, circular cross-sections (pipes) offer the greatest torque resistance per unit volume of material, since material located near the center is at a low stress level and thus less effective.

Non-circular cross-sectioned members such as rectangular or I-shaped beams develop a completely different distribution of shear stress when subjected to torsion.

A full century after Coulomb, Barre de Saint-Venant developed a theory to explain the difference between circular torsion and non-circular torsion. Circumferential lines of a circular rod remain in their original cross-sectional plane under torsional forces (Figure 6.17). No numerical problems dealing with torsion will be covered in this book.

Further example problems involving direct, shear, and bearing stress are illustrated next.

Example Problem 6.8 (Figures 6.18 to 6.20)

A pipe storage rack is used for storing pipe in a shop. The support rack beam is fastened to the main floor beam using steel straps $\frac{1}{2}$" × 2" in dimension. Round bolts are used to fasten the strap to the floor beam in single shear. (a) If the weight of the pipes impose a maximum tension load of 10,000 pounds in each strap, determine the tension stress developed in the steel strap. (b) Also, what diameter bolt is necessary to fasten the strap to the floor beam if the allowable shear stress for the bolts equals $F_v = 15,000\,{}^{lb.}/_{in.^2}$?

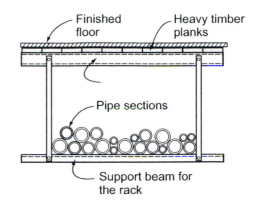

Figure 6.18   Pipe storage rack.

Solution:

a. The tensile stress developed in the steel strap (Figure 6.19) can be determined using the direct stress formula.

$$f_t = \frac{P}{A} = \frac{10,000 \text{ lb.}}{\left(\frac{1}{2}\text{"} \times 2\text{"}\right)} = 10,000\,{}^{lb.}/_{in.^2}$$

In mild steel (A36), the maximum permissible tensile stress (allowable) is equal to

$$F_t {}_{\text{(allowable)}} = 22,000 \text{ psi}$$

Therefore, the strap size is adequate to support the tensile load safely.

b. To determine the size bolt necessary to carry the load safely in single shear, the design form of the equation must be used.

$$f_v = \frac{P}{A}; \quad A = \frac{P}{F_v} = \frac{10,000 \text{ lb.}}{15,000\,{}^{lb.}/_{in.^2}} = 0.67 \text{ in.}^2$$

$$A = \frac{\pi D^2}{4}; \quad D^2 = \frac{4 \times A}{\pi} = \frac{4 \times 0.67 \text{ in.}^2}{3.14}$$

$$= 0.854 \text{ in.}^2$$

$$D = 0.92 \text{ in.}; \quad \text{Use: } 1\text{"}\phi \text{ bolt.}$$

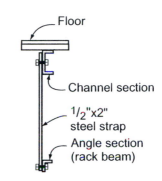

Figure 6.19   Section.

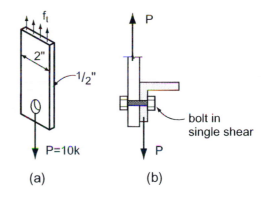

Figure 6.20   Bolt in single shear.

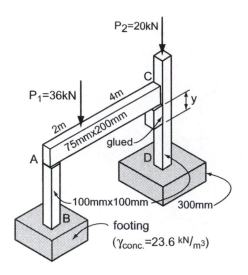

Figure 6.21   *Beam and column arrangement.*

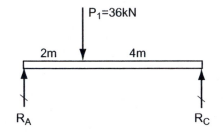

Figure 6.22   *FBD of beam AC.*

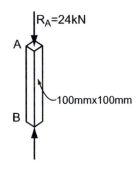

Figure 6.23   *FBD of column AB.*

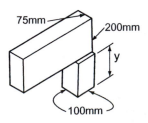

Figure 6.24   *Beam-shear block.*

Example Problem 6.9 (Figures 6.21 to 6.26)

A 75 mm × 200 mm "rough cut" beam is supported by columns at both ends. Column *AB* supports the beam in bearing while column *CD* utilizes a shear block at *C*. Both columns bear on concrete footings on the ground.

a. What is the compressive stress developed in column *AB*?
b. What is the bearing stress that develops at *C* between the beam and shear block made from a 100 mm × 100 mm block cut from a post?
c. What is the required depth *y* necessary to resist the shear force developed at the glued joint between the shear block and post? Assume that the glue is capable of safely resisting 500 kPa (72.5 psi) in shear.
d. Determine the size of square footing required to take the maximum column load if the allowable soil pressure $q = 73\,^{kN}/_{m^2} = 73$ kPa (1525 psf).

Solution:

a. Construct a FBD of the beam *AC*. Using the equations of equilibrium, determine the support reactions.

$$R_A = 24 \text{ kN (5.4 k)}$$

$$R_C = 12 \text{ kN (2.70 k)}$$

The force in column *AB* is the reaction force developed on the beam at *A*.

The normal compressive stress is computed using the basic stress equation:

$$f_{comp.} = \frac{P}{A} = \frac{R_A}{A}$$

$$\text{Area} = A = 100 \text{ mm} \times 100 \text{ mm}$$

$$= 10,000 \text{ mm}^2 = 0.01 \text{ m}^2$$

$$\therefore f_c = \frac{24 \text{ kN}}{0.01 \text{ m}^2} = 2400\,^{kN}/_{m^2} = 2400 \text{ kPa}$$

$$= 2.4 \text{ MPa (348 psi)}$$

b. The bearing area is the contact area between the beam and shear block.

$$A_{brg} = 75 \text{ mm} \times 100 \text{ mm}$$

$$= 7,500 \text{ mm}^2 = 0.0075 \text{ m}^2$$

$$f_{brg} = \frac{P}{A} = \frac{R_c}{A} = \frac{12 \text{ kN}}{7.5 \times 10^{-3} \text{ m}^2} = 1600 \text{ }^{kN}/_{m^2}$$

$$= 1.6 \text{ MPa (232 psi)}$$

c. The interface between the shear block and the post represents the shear area parallel to the load direction. The minimum distance y required for the glue to develop the necessary resistance to the shear load is found by rewriting the basic stress equation into its design form.

$$f_v = \frac{R_c}{A}; \qquad A_{req'd} = \frac{R_c}{F_v}$$

$$F_v = 500 \text{ kPa}; \quad R_c = 12 \text{ kN}$$

$$A_{req'd} = \frac{12 \text{ kN}}{500 \text{ kPa}} = \frac{12 \text{ kN}}{500 \text{ }^{kN}/_{m^2}} = 0.024 \text{ m}^2$$

Since the shear block is 100-mm wide,

$$A = 0.10 \text{ m} \times y;$$

$$y = \frac{A}{0.10 \text{ m}} = \frac{0.024 \text{ m}^2}{0.10 \text{ m}}$$

$$= 0.24 \text{ m} = 240 \text{ mm (9.5 in.)}$$

d. Load through column CD = 12 kN + 20 kN = 32 kN

The load from column CD must be transmitted through the footing and down to the ground. Again, the stress equation is written in its design form where the minimum footing area is computed.

$$f_{brg} = \frac{P}{A}; \qquad A = x^2 = \frac{P}{q_{net}}$$

q_{net} is the bearing pressure remaining after accounting for the weight of the footing.

$$q_{net} = q - \text{pressure exerted by the} \\ \text{column weight}$$

$$\therefore q_{net} = 73 \text{ kPa} - (\gamma_{conc.} \times t)$$

$$= 73 \text{ kPa} - \left(23.6 \text{ }^{kN}/_{m^3} \times 0.3 \text{ m}\right)$$

$$q_{net} = \underset{(1525 \text{ psf})}{73 \text{ kPa}} - \underset{(148 \text{ psf})}{7.1 \text{ kPa}} = \underset{(1377 \text{ psf})}{65.9 \text{ kPa}}$$

$$x^2 = \frac{P}{q_{net}} = \frac{32 \text{ kN}}{65.9 \text{ }^{kN}/_{m^2}} = 0.486 \text{ m}^2 \text{ (5.23 ft.}^2\text{)}$$

$$\therefore x = 0.70 \text{ m (2.3 ft.)}$$

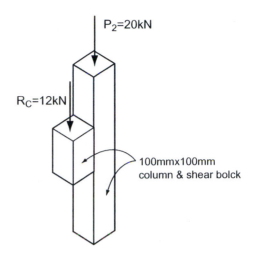

Figure 6.25   Beam and column load.

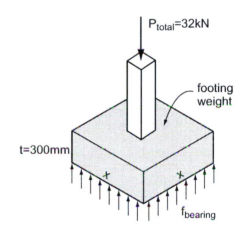

Figure 6.26   Footing detail.

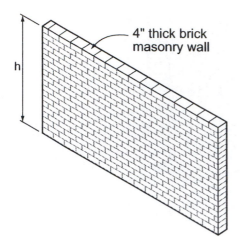

Figure 6.27   Masonry wall.

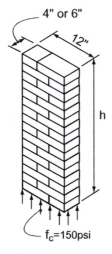

Figure 6.28   1' ft. tributary width of the masonry wall.

Example Problem 6.10 (Figures 6.27 and 6.28)

Masonry brickwork has a density of 120 lb./ft.^{3} Determine the maximum height of a brick wall if the allowable compressive stress is limited to 150 lb./in.^{2} and the brick is: (a) 4" wide and (b) 6" wide.

Solution:

Note that the length of the wall is not given since it is unnecessary when determining the stress in the wall. Instead, a 1 ft. tributary strip will be used as a generaliztion of what the entire wall is experiencing.

a. Assume a strip of wall 1 ft. wide, 4" thick, and h high.

$$f_c = \frac{P}{A}; \quad P = A \times F_c = (4" \times 12") \times 150^{\text{lb.}}/_{\text{in.}^2}$$

$$= 7200 \text{ lb.}$$

But, $P = \gamma_{\text{brick}} \times A \times h$:

$$h = \frac{P}{\gamma_{\text{brick}} \times A} = \frac{7200 \text{ lb.}}{\left(120^{\text{lb.}}/_{\text{ft.}^3}\right) \times \left(\dfrac{48 \text{ in.}^2}{144^{\text{in.}^2}/_{\text{ft.}^2}}\right)}$$

$$= 180 \text{ ft.}$$

b. Again assuming a 1 ft. (12") wide strip and 6" thick:

$$P = A \times F_c = (6" \times 12") \times 150^{\text{lb.}}/_{\text{in.}^2}$$

$$= 10{,}800 \text{ lb.}$$

But:

$$P = \gamma_{\text{brick}} \times A \times h$$

$$A = 72 \text{ in.}^2 = \left(\frac{72 \text{ in.}^2}{144^{\text{in.}^2}/_{\text{ft.}^2}}\right) = 0.5 \text{ ft.}^2$$

$$\therefore h = \frac{10{,}800 \text{ lb.}}{\left(120^{\text{lb.}}/_{\text{ft.}^3}\right) \times 0.5 \text{ ft.}^2} = 180 \text{ ft.}$$

Note: The answer in both cases resulted in the same theoretical height h = 180 ft. *If the wall thickness were increased to 8", the answer would still remain the same. Stress is a load per unit area and as the cross-sectional area of the wall changes, a proportional change in the weight occurs, keeping the unit stress the same. If a column of wall measuring 1" × 1" × h' tall is used, the same height of 180 ft. would produce a weight that results in a stress of 150 psi.*

Example Problem 6.11 (Figures 6.29 and 6.30)

A rigid steel beam 20 feet long supports a 5 kip load as shown in Figure 6.29. Assume that the beam weighs 1 kip, concentrated at the midlength of the beam.

a. Determine the tensile stress developed in the steel rod that has a diameter of 0.75 inches.
b. What is the minimum diameter rod that can safely resist the load if the maximum (allowable) tension stress is equal to $F_t = 22$ ksi?

Solution:

a. Begin by constructing a FBD of the beam (Figure 6.30) and solve for the tensile force generated by the rod BC by summing moments about the support pin at A.

$$\sum M_A = +BC\,(10.61') - 5\,\text{k}\,(14.14')$$

$$- 1\,\text{k}(7.07') = 0$$

$$\therefore BC = \frac{5\,\text{k}(14.14') - 1\,\text{k}(7.07')}{(10.61')} = 6.0\,\text{k}$$

From the direct normal stress equation, we get:

$$f_{BC} = \frac{P}{A} = \frac{BC}{A}$$

The cross-sectional area of the 3/4" ϕ rod is equal to:

$$A = \frac{\pi D^2}{4} = \frac{(3.14)(0.75\ \text{in.})^2}{4} = 0.442\ \text{in.}^2$$

$$\therefore f_{BC} = \frac{6.0\,\text{k}}{0.442\ \text{in.}^2} = 13.6\,{}^{\text{k}}/_{\text{in.}^2}$$

b. Rewrite the basic stress equation into its design form and solve the required cross-sectional area when the tensile stress reaches a maximum of 22 ksi.

$$f_t = \frac{P}{A}; \qquad A_{\text{req'd}} = \frac{P}{F_t} = \frac{6.0\,\text{k}}{22\,{}^{\text{k}}/_{\text{in.}^2}} = 0.273\ \text{in.}^2$$

$$A = \frac{\pi D^2}{4}; \qquad D^2 = \frac{4A}{\pi} = \frac{4(0.273\ \text{in.}^2)}{(3.14)}$$

$$= 0.424\ \text{in.}^2$$

$$\therefore D = 0.59" \ \phi$$

Use: 5/8" ϕ ($D = 0.625$ in.)

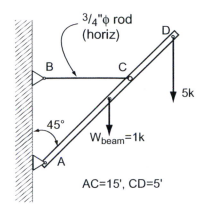

Figure 6.29 Inclined beam with vertical loads.

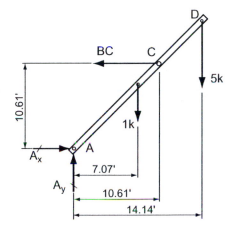

Figure 6.30 FBD of member ACD.

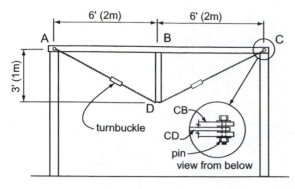

Figure 6.31   Trussed beam with turnbuckles.

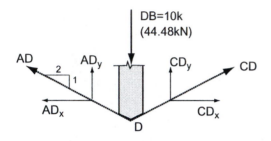

Figure 6.32   FBD of joint D.

Example Problem 6.12 (Figures 6.31 and 6.32)

Turnbuckles on a trussed beam, as shown in Figure 6.31, are tightened until the compression block DB exerts a force of 10 k (44.48 kN) on the beam at B. Member DB is a hollow shaft of inner diameter 1.0 inch (25.4 mm) and outer diameter of 2.0 inches (50.8 mm). Rods AD and CD each have cross-sectional areas of 1.0 in.^{2} (645 mm^2). Pin C has a diameter of 0.75 in. (19.0 mm). Determine, using U.S. customary and SI metric units:

a. the axial stress in BD
b. the axial stress in CD
c. the shearing stress in pin C.

Solution:

The procedure begins with the construction of the FBD of the concurrent joint at D (Figure 6.32). The strut force in member BD and the rods DA and DC are all two-force members with known lines of action. Member force BD has a known magnitude equal to 10 k (44.48 kN). Sum forces in the vertical direction and solve for the tensile forces developed in rods AD and CD.

$$\sum F_y = -10\,k + AD_y + CD_y = 0$$

$$AD_y = CD_y$$

$$\therefore 2CD_y = 10\,k; \qquad CD_y = 5\,k$$

$$CD = 5\sqrt{5} = 11.18\,k$$

Solving this using SI units:

$$\sum F_y = -44.48\,kN + AD_y + CD_y = 0$$

$$AD_y = CD_y$$

$$\therefore 2CD_y = 44.48\,kN; \qquad CD_y = 22.24\,kN$$

$$CD = 22.24\sqrt{5} = 49.73\,kN$$

a. The axial stress in member BD:

$$f_c = \frac{P}{A} = \frac{10\,k}{\left(\dfrac{\pi(2")^2}{4}\right) - \left(\dfrac{\pi(1")^2}{4}\right)}$$

$$= \frac{10\,k}{(3.14\,in.^2) - (0.785\,in.^2)} = 4.25\,ksi$$

In SI units:

$$f_c = \frac{P}{A} = \frac{44.48 \text{ kN}}{\left(\dfrac{\pi(50.8 \text{ mm})^2}{4}\right) - \left(\dfrac{\pi(25.4 \text{ mm})^2}{4}\right)}$$

$$= \frac{44.48 \text{ kN}}{1520 \text{ mm}^2} = \frac{44.48 \text{ kN}}{1.52 \times 10^{-3} \text{ m}^2}$$

$$f_c = 29,260 \, {}^{\text{kN}}/_{\text{m}^2} = 29.3 \, {}^{\text{MN}}/_{\text{m}^2} = 29.3 \text{ MPa}$$

b. Both rods, *CD* and *AD*, carry the same tensile force of 11.2 k (49.73 kN). The tensile stress is computed as:

$$f_t = \frac{P}{A} = \frac{CD}{A} = \frac{11.18 \text{ k}}{1.0 \text{ in.}^2} = 11.18 \text{ ksi}$$

In SI units:

$$f_t = \frac{CD}{A} = \frac{49.73 \text{ kN}}{645 \text{ mm}^2} = \frac{49.73 \text{ kN}}{.000645 \text{ m}^2}$$

$$= 77,100 \, {}^{\text{kN}}/_{\text{m}^2} = 77.1 \text{ MPa}$$

c. The pin at *C* is subjected to double shear. Therefore, the shear stress is equal to:

$$f_v = \frac{P}{2A} = \frac{11.18 \text{ k}}{2 \times \left(\dfrac{\pi \times .75^2}{4}\right)}$$

$$= \frac{11.18 \text{ k}}{2 \times 0.442 \text{ in.}^2} = 12.65 \text{ ksi}$$

In SI units:

$$f_v = \frac{P}{2A} = \frac{49.73 \text{ kN}}{2 \times \left(\dfrac{\pi \times 19.0 \text{ mm}^2}{4}\right)}$$

$$= \frac{49.73 \text{ kN}}{2 \times 283 \text{ mm}^2} = 0.088 \, {}^{\text{kN}}/_{\text{mm}^2}$$

$$f_v = 88,000 \, {}^{\text{kN}}/_{\text{m}^2} = 88 \, {}^{\text{MN}}/_{\text{m}^2} = 88 \text{ MPa}$$

6.2 AXIAL STRAIN—HOOKE'S LAW, STRESS-STRAIN DIAGRAM, AND MECHANICAL PROPERTIES OF MATERIALS

Deformation and Strain

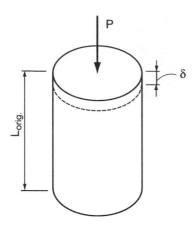

Figure 6.33   Concrete cylinder in compression.

Concrete columns, steel tension rods, and masonry arches all deform when subjected to applied loads. Physical changes, in the form of altered size or shape, occur in the structural elements resisting internal stresses. Tension stresses tend to elongate the structural members and the change in length is called *deformation*, designated by the symbol δ (delta). Deformation, either *elongation* or *contraction* (from compressive loads), generally represents the total change in dimension experienced by the member. The *unit change*, or deformation per unit length, is called *strain* (ε). Deformation (δ) or strain (ε) may be the result of internal stresses or *temperature change*. This deformation behavior is typical of most materials because solids deform to some extent under applied loads and no truly "rigid bodies" exist in structural design.

In Figure 6.33, a short concrete cylinder is subjected to a compressive load that results in a dimensional change in the direction of the force. Similarly, in Figure 6.34, a steel flat bar supports a tensile load that produces an elongation δ The shortening or elongation is measured in inches or millimeters.

Strain, or the deformation per unit length, is expressed as:

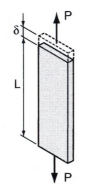

Figure 6.34   Steel bar in tension.

$$\varepsilon = \frac{\delta}{L}$$

where:

ε = unit strain $\left(^{in.}/_{in.}\right)$ or $\left(^{mm}/_{mm}\right)$

δ = total deformation, elongation, or contraction (in. or mm)

L = original length of the specimen in the load direction (in. or mm)

Note that unit strain ε is a dimensionless ratio since δ and L have the same dimensional units.

Members subjected to a shear stress undergo a deformation that results in a change in shape.

Rather than an elongation or shortening, shearing stress causes an angular deformation of the body. The square shown in Figure 6.35 becomes a parallelogram when acted upon by shear stresses. Shearing strain, represented by γ (gamma), is expressed as:

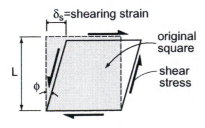

Figure 6.35   Shear deformation.

$$\gamma = \frac{\delta_s}{L} = \tan \phi \cong \phi$$

When the angle ϕ is small, $\tan \phi \cong \phi$ where ϕ is the angle expressed in radians.

Example Problem 6.13 (Figure 6.36)

A steel column is compressed 0.055". The original column length is 10 ft. What is the resulting unit strain?

Solution:

The equation for unit strain is:

$$\varepsilon = \frac{\delta}{L} = \frac{0.055 \text{ in.}}{10 \text{ ft.} \times 12^{\text{in.}}/_{\text{ft.}}} = 0.000458 \, ^{\text{in.}}/_{\text{in.}}$$

$$= 4.58 \times 10^{-4} \, ^{\text{in.}}/_{\text{in.}}$$

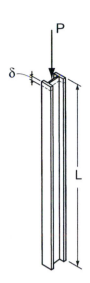

Figure 6.36   Column in compression.

Example Problem 6.14 (Figure 6.37)

A concrete cylinder is tested in a compression machine that registers a strain of 0.003. The test cylinder measures 203-mm tall and 102 mm in diameter. What is the deformation that results just as the concrete fails in compression?

Solution:

$$\varepsilon = \frac{\delta}{L}; \qquad \delta = \varepsilon L$$

$$\therefore \delta = \varepsilon L = \left(0.003 \, ^{\text{mm}}/_{\text{mm}}\right) \times (203 \text{ mm}) = 0.61 \text{ mm}$$

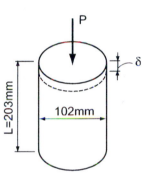

Figure 6.37   Concrete test cylinder.

Example Problem 6.15

The maximum permissible strain in an elevator cable is $0.00045 \left(^{\text{in.}}/_{\text{in.}} \text{ or } ^{\text{mm}}/_{\text{mm}}\right)$. Determine the elongation in the cable that measures 200 ft. (61 m).

Solution:

$$\varepsilon = \frac{\delta}{L};$$

$$\delta = \varepsilon L = \left(0.00045 \, ^{\text{in.}}/_{\text{in.}}\right) \times (200 \text{ ft.} \times 12^{\text{in.}}/_{\text{ft.}})$$

$$= 1.08 \text{ in.}$$

Solving in SI units:

$$\delta = \varepsilon L = \left(0.00045 \, ^{\text{mm}}/_{\text{mm}}\right) \times (61{,}000 \text{ mm})$$

$$= 27.5 \text{ mm}$$

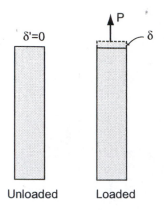

Figure 6.38   Steel bar under tensile load.

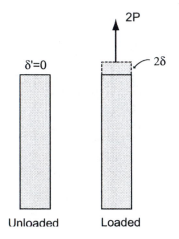

Figure 6.39   Load doubled on test specimen.

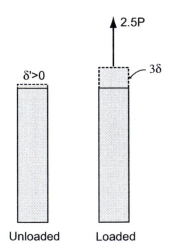

Figure 6.40   Permanent deformation in the test specimen.

Hooke's Law and the Stress-Strain Diagram

A wide variety of materials are presently used in architectural structures: stone, brick, concrete, steel, timber, aluminum, plastics, and so on. All have essential properties that make them applicable for a given purpose in a structure. The criteria for selection, at a very basic level, is the material's ability to withstand forces without excessive deformations or actual failures.

One major consideration that must be accounted for in any structural design is deformation. Deformation in structures cannot increase indefinitely, and it should disappear after the applied load is removed. *Elasticity* is a material property in which deformations disappear with the disappearance of the load (Figures 6.38 to 6.40).

All structural materials are elastic to some extent. As loads are applied and deformations result, the deformations will vanish as the load is removed, as long as a certain limit is not exceeded. This limit is the *elastic limit,* within which no permanent deformations result from the application and removal of the load. If this limit of loading is exceeded, a permanent deformation results. The behavior of the material is then termed as *plastic* or *inelastic.*

In some materials, when the elastic limit is exceeded, the molecular bonds within the material are unable to reform, thus causing cracks or separation of the material. Such materials are termed *brittle.* Cast iron, high-carbon steel, and ceramics are considered brittle; low-carbon steel, aluminum, copper, and gold exhibit properties of *ductility* (which is a measure of *plasticity*).

Materials that have molecular bonds reforming after exceeding the elastic limit will result in permanent deformations; however, the material still remains in one piece without any significant loss in strength. This type of material behavior is termed *ductile.*

Ductile materials give warning of impending failure, whereas brittle materials do not.

One of the most important discoveries in the science of mechanics of materials was undoubtedly that pertaining to the elastic character of materials. This discovery, in 1678 by Robert Hooke, an English scientist, mathematically relates stress to strain. The relationship, known as Hooke's law, states that in elastic materials stress and strain are proportional. Hooke observed this stress-strain relationship by experimentally loading various materials in tension and measuring the subsequent deformations. Since Robert Hooke's initial experiment, techniques and testing equipment have improved; however, the relationship between stress and strain and the determination of elastic and plastic properties of materials still use his basic concept.

Today, universal testing machines are employed to apply precise loads at precise rates to standardized tensile and compressive specimens (Figure 6.41). The tensile test is the most common test applied to materials. A variety of devices for measuring and recording strain or deformation can be attached to the test specimen to obtain data for plotting *stress-strain diagrams* (or load-deformation curves).

The *stress-strain curves* obtained from tension or compression tests conducted on various materials reveal several characteristic patterns (Figure 6.42). Ductile rolled steels, such as ordinary, low-carbon structural steel, stretch considerably after following a straight-line variation of stress and strain. For steels alloyed with increasing amounts of carbon and other strengthening materials, such as chromium, nickel, silicon, and manganese, the tendency to produce such an intermediate stretching point becomes increasingly remote. The stress-strain curves for heavily alloyed steels are generally straight to a point a short distance from the rupture point. In contrast to such straight-line stress-strain curves are those obtained for materials such as cast iron, brass, concrete, wood, and so forth, which are often curved throughout most of their length.

The stress-strain curve for low-carbon (<0.30% carbon) steel (Figure 6.43) will form the basis for the ensuing remarks concerning several familiar strength values. This diagram plots strain along the abscissa and stress along the ordinate. The stress is defined as the load in pounds, kips, kN, or MN divided by the original cross-sectional area of the specimen.

As the test proceeds, with larger loads being applied at a specified rate, the actual cross-sectional area of the specimen decreases. At high stresses, this reduction in area becomes appreciable.

The significant points on the stress-strain curve shown in Figure 6.43 are defined as follows:

1. **Proportional limit.** The proportional limit is that stress beyond which the ratio of stress to strain no longer remains constant. It is the greatest

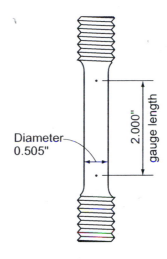

Figure 6.41   *Typical test specimen (metal).*

Photo 6.2.1   *Steel test specimen before and after loading.*

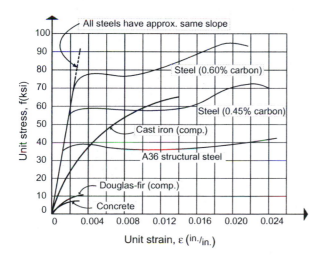

Figure 6.42   *Stress-strain diagram for various materials*

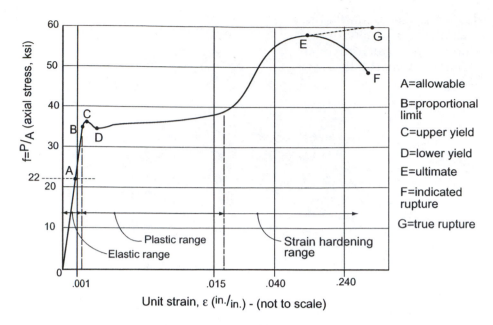

Figure 6.43   Stress-strain diagram for mild (A36) steel.

stress that a material is capable of developing without deviation from Hooke's law of stress-strain proportionality.

2. **Elastic limit.** Located close to the proportional limit, yet of entirely different meaning, is the *elastic limit*. The elastic limit is that maximum unit stress that can be developed in a material without causing a permanent set (permanent deformation). A specimen stressed to a point below its elastic limit will assume its original dimensions when the load is released. If the stress should exceed its elastic limit, the specimen will deform plastically and will no longer attain its original dimensions when unloaded (see Figure 6.40). It is then said to have incurred a *permanent set*.

3. **Yield point.** When the load on the test specimen is increased beyond the elastic limit, a stress level is reached where the material continues to elongate without an increase of load. The *yield point* is defined as the stress at which a marked increase in strain occurs without a concurrent increase in applied stress. After the initial *yielding* (upper yield point) is reached, the force resisting deformation decreases due to the yielding of the material. The value of stress after initial yielding is known as the lower yield point and is usually taken to be the true material characteristic to be used as the basis for the determination of allowable stress (for design purposes). Many materials do not exhibit well-defined *yield points*, and the yield strength is defined as the stress at

which the material exhibits a specified limiting permanent set. When a test specimen is stressed beyond its elastic limit and then has its load released, such a load cycle does not necessarily damage a material even if the imposed stress exceeds the elastic limit. *Ductility* may be lowered, but the *hardness* (ability of a material to resist indentation) and *elastic stress limit* of the material will generally increase.

4. **Ultimate strength.** The *ultimate strength* of a material is defined as the stress obtained by dividing the maximum load reached before the specimen breaks by the original cross-sectional area. The ultimate strength (often called the *tensile strength*) of the material is sometimes used as a basis for establishing the allowable design stresses for a material.

5. **Rupture strength (breaking strength, fracture strength).** In a ductile material, rupture does not usually occur at the ultimate unit stress. After the ultimate unit stress has been reached, the material will generally neck down (decrease in cross-section) and its rapidly increasing elongation will be accompanied by a decrease in load. This decrease becomes more rapid as the rupture point is approached. The *rupture strength*, obtained by dividing the load at rupture by the original area (indicated rupture), has little or no value in design. A more correct evaluation of the variation of stress following the attainment of the ultimate unit stress is obtained by dividing the loads by the simultaneously occurring decreasing areas (*true rupture strength*).

6. **Elongation.** *Elongation* is a measure of the ability of a material to undergo deformation without rupture. *Percentage elongation* is a measure of the ductility of a material. *Ductility* is a desirable and necessary property, and a member must possess it to prevent failure due to local overstressing.

7. **Reduction of area.** As the load on the material undergoing testing is increased, the original cross-sectional area decreases until it is at a minimum at the instant of fracture. It is customary to express this reduction in area as the ratio (as a percentage) of the change in area to the original specimen cross-sectional area. The failed specimen exhibits a local decrease in diameter known as *necking down* in the region where failure occurs. The percent reduction in area can be used as a measure of ductility. Brittle materials exhibit almost no reduction in area, while ductile materials exhibit a high percent reduction in area.

Additonal Mechanical Properties of Materials

8. **Allowable stress.** A *working stress* or *allowable stress* (see Tables 6.2a and 6.2b) is defined as the maximum stress permitted in a design computation. It is the stress derived from the results of many tests and the accumulated experience obtained from many years of firsthand observation in the performance of members in actual service.

9. **Factor of safety.** The *factor of safety* may be defined as the ratio of a failure-producing load to the estimated actual load. The ratio may be calculated as the ultimate stress (or yield-point stress) to the allowable working stress.

10. **Compression tests.** The compression test is used primarily to test brittle materials such as cast iron and concrete (Figure 6.44). The universal testing machine is used for this test, and data is taken in a manner similar to that discussed for the tension test. The results of the compression test generally define an elastic range, a proportional limit, and a yield strength. In the compression test, the cross-sectional area of the specimen increases as loads increase, which produces a continuously rising stress-strain curve.

11. **Creep.** The deformation that most structural materials undergo when stressed to their allowable limit at room temperature is a completed deformation and will not increase no matter how long the stress is applied. At some higher temperature, however, these same materials will reveal a continuing deformation or creep that, if permitted to continue, will ultimately lead to excessive displacements or rupture (Figure 6.45).

12. **Cyclic stress (fatigue).** Members subjected to repeated conditions of loading or unloading, or to repeated stress reversals, will fail at a stress considerably lower than the ultimate stress obtained in a simple tension test. Failures that occur as a result of this type of repeated loading are known as *fatigue* failures.

A theory of fatigue failure assumes a sudden change in the shape of the loaded member and inconsistencies in the material cause localized stresses that are far above the average stress in the material to develop. These localized stresses exceed the material's yield strength and cause permanent deformations to occur locally. Repeated permanent deformations in a small area eventually cause hairline cracks to develop. This cracking process continues until the average stress on the resisting area reaches the ultimate strength of the material.

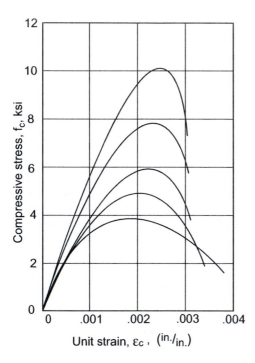

Figure 6.44   Stress-strain curves for normal-density concrete.

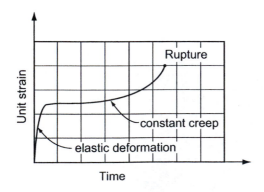

Figure 6.45   Stress with respect to time (creep).

13. **Poisson's ratio.** The cross-sectional reduction (necking down) during a steel tensile test has a definite relationship to the increase in length (elongation) experienced by the specimen. When a material is loaded in one direction, it will undergo strains perpendicular to the direction of the load as well as parallel to it. The ratio of the lateral or perpendicular strain to the longitudinal or axial strain is called *Poisson's ratio*. Poisson's ratio varies from 0.2 to 0.4 for most metals. Most steels have values in the range of 0.283 to 0.292. The symbol µ (mu) is used for Poisson's ratio, which is given by the equation:

$$\mu = \frac{\varepsilon_{lateral}}{\varepsilon_{longitudinal}}$$

14. **Toughness.** The area under the stress-strain curve (Figure 6.46) is a measure of the work required to cause *fracture* to occur. This ability of a material to absorb energy up to fracture is also used by designers as a characteristic property of a material and is called *toughness*. Toughness may be important in applications where stresses in the plastic range of the material may be approached, but where the resulting permanent set is not critical—and sometimes even desirable. The stress-strain diagrams indicate that low-carbon (mild) steels are much "tougher" than high-carbon (higher strength) steels. This concept is sometimes in opposition to the instinct of the engineer to specify the use of a "stronger" steel when a structure fails or seems in danger of failing. This could be a mistake in larger structures, since even in mild steel much of the strength is not really being used. Failure of a structure may be controlled by the *brittleness* of the material, not by its *strength*.

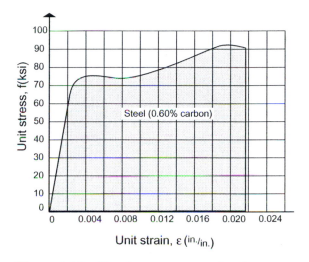

Figure 6.46   Toughness—area under the stress-strain diagram.

15. **Ductility.** *Ductility* is a property of a material enabling it to undergo considerable *plastic deformation* under tensile load before actual rupture. A ductile material is one that can be drawn into a long thin wire by a tensile force without failure.

16. **Malleability.** Most materials that are very ductile are also quite malleable. *Malleability* is a property of a material enabling it to undergo considerable plastic deformation under compressive load before actual rupture. A hammering or rolling operation would require a malleable material due to the extensive compressive deformation that accompanies the process.

17. **Brittleness.** A brittle material is neither ductile nor malleable and will fail suddenly without warning because of the absence of any plastic

deformation prior to failure. A brittle material exhibits no yield point or necking down process and has a rupture strength approximately equal to its ultimate strength.

18. **Strength.** This is a property determined by the greatest stress that the material can withstand prior to failure. Depending on how *failure* is defined, *strength* may be the proportional limit, the yield point, or the ultimate strength. No single value is adequate to define strength since the behavior of a material under load differs with the kind of stress present and the nature of the loading.

19. **Resilience.** *Resilience* is a property of a material enabling it to endure high impact loads without inducing a stress in excess of the elastic limit. It implies that the energy absorbed during the blow is stored and recovered when the body is unloaded. The measure of resilience is furnished by the area under the elastic portion of the stress-strain curve from the origin through the elastic limit (note that this is similar to the toughness diagram that used the rupture point).

20. **Stiffness.** The property of a material that enables it to withstand high stress without great strain is referred to as material stiffness. It is a resistance to any sort of deformation. Stiffness of a material is a function of its modulus of elasticity (Young's modulus) *E*. Materials with high *E* values will exhibit a greater stiffness than materials with lower *E* values.

Sir Thomas Young (1779–1824)

Young was a prodigy in his infancy who purportedly could read at the age of 2 and had read the entire Bible twice before the age of 4. At Cambridge, his incredible abilities earned him the nickname "Phenomenon Young." He matured into an adult prodigy, was knowledgeable in 12 languages, and could play a variety of musical instruments. As a physician, he was interested in sense perception. From the eye and then to light itself, it fell upon Young to demonstrate the wave nature of light. Turning more and more to physics, he introduced the concept of energy in its modern form in 1807. In the same year, he suggested the ratio formed by dividing a unit stress by its corresponding value of strain (Young's modulus) as a means of evaluating the stiffness of various materials. He was also an accomplished Egyptologist, who was instrumental in deciphering the Rosetta Stone, the key to Egyptian hieroglyphics.

Modulus of Elasticity (Young's Modulus)

In 1678, Sir Robert Hooke observed that when rolled materials were subjected to equal increments of stress, they resulted in equal increments of strain (in other words, stress is proportional to strain). The ratio formed by dividing a unit stress by its corresponding value of strain was suggested by Thomas Young in 1807 as a means of evaluating the *relative stiffness* of various materials. This ratio is called *Young's modulus* or the *modulus of elasticity* (E) and is the slope of the straight-line portion of the stress-strain diagram:

$$E = \frac{f}{\varepsilon}$$

where:

E = modulus of elasticity (ksi or psi), (MPa, Pa)

f = stress (ksi or psi), (MPa or Pa)

e = strain (in./in.), (mm/mm)

This ratio of stress to strain remains constant for all steels and many other structural materials within their useful range. Generally, a high modulus of elasticity is desirable since E is often referred to as a *stiffness factor*. Materials exhibiting high E values are more resistant to deformation and, in the case of beams, suffer much less deflection under load. Note in Figure 6.47 that of the three materials shown, the steel specimen has a much steeper slope in the elastic range than aluminum or wood, and therefore will be much more resistant to deformation. The Young's modulus equation may also be written in a very useful expanded form whenever the stress and deformation are caused by *axial loads*.

$$f = \frac{P}{A}; \quad \varepsilon = \frac{\delta}{L}; \quad E = \frac{f}{\varepsilon} = \frac{P/A}{\delta/L} = \frac{PL}{A\delta}$$

$$\therefore \ \delta = \frac{PL}{AE} \quad \text{(Elastic equation)}$$

where:

  δ = deformation (in., mm)
  P = applied axial load (lb., k; or N, kN)
  L = length of the member (in., mm, or m)
  A = cross-sectional area of member (in.² or mm²)
  E = modulus of elasticity of material (psi, ksi; Pa, MPa)

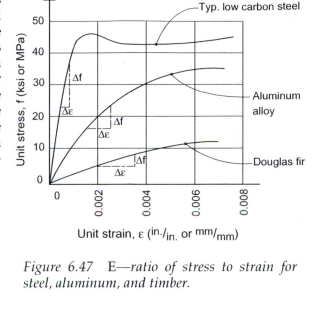

Figure 6.47   E—ratio of stress to strain for steel, aluminum, and timber.

Example Problem 6.16 (Figure 6.48)

A vertical 6×8 S4S timber column of Douglas fir is 10'-6" tall and is subjected to a compressive load of 30,000 pounds. Calculate the unit stress and the total shortening that occurs under this load.

Note: *A 6×8 S4S has a dressed (actual) dimension of $5\frac{1}{2}$"* $\times 7\frac{1}{2}$".

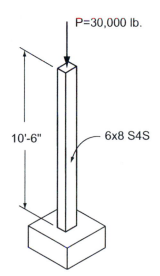

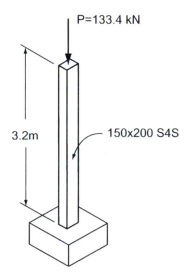

Figure 6.48   (a) Axially loaded timber column (U.S. conventional). (b) Axially loaded timber column (SI metric).

Solution:

Compute the area of the 6×8 S4S section.

$$A = 5.5'' \times 7.5'' = 41.25 \text{ in.}^2$$

Then write the unit stress equation:

$$f_c = \frac{P}{A} = \frac{30,000 \text{ lb.}}{41.25 \text{ in.}^2} = 727 \text{ psi}$$

To determine the elastic shortening of the column, the deformation equation is written as:

$$\delta = \frac{PL}{AE} = \frac{fL}{E}$$

The modulus of elasticity of the timber column is obtained from Table 6.1a or Table 6.2a.

$$\therefore \delta = \frac{fL}{E} = \frac{\left(727^{\text{lb.}}/_{\text{in.}^2}\right)(126'')}{\left(1.6 \times 10^6 \,{}^{\text{lb.}}/_{\text{in.}^2}\right)}$$

$$= 0.057'' \left(\text{less than } {}^1/_{16}''\right)$$

Table 6.1a   Average strength values for selected engineering materials (U.S. customary).

Materials	Yield Stress or Proportional Limit (ksi)			Ultimate Strength (ksi)			Modulus of Elasticity (ksi)	
	Tension	Compression	Shear	Tension	Compression	Shear	Tens. or Comp.	Shear
Steel:								
A36 Structural	36	36	22	58		40	29,000	12,000
A572 Grade 50* (also A992)	50	50	30	65		45	30,000	12,000
A572 Grade 65	65	65	32.5	80		65.5	30,000	12,000
Iron:								
Malleable	25	25	12	50		40	24,000	11,000
Wrought	30	30	18	58		38	25,000	10,000
Aluminum Alloy:								
6061-T6 rolled/extruded	35	35	20	38		24	10,000	3,800
6063-T6 extruded tubes	25	25	14	30		19	10,000	3,800
Other Metals:								
Brass: 70%, Cu: 30% Zinc	25		15	55		48	14,000	6,000
Bronze, cast heat treated	55		37	75		56	12,000	5,000
Timber, air dry:								
Yellow pine		6.2			8.4	1.0	1,600	
Douglas fir		5.4			6.8	0.8	1,600	
Spruce		4.0			5.0	0.75	1,200	
Concrete:								
Concrete: 1:2:4 mix, 28 days					3.0		3,000	

* With special requirements per AISC Tech. Bulletin #3.

Table 6.1b   Average strength values for selected engineering materials (SI Metric).

Materials	Yield Stress or Proportional Limit (MPa)			Ultimate Strength (MPa)			Modulus of Elasticity (MPa)	
	Tension	Compression	Shear	Tension	Compression	Shear	Tens. or Comp.	Shear
Steel:								
A36 Structural	250	250	150	400		275	200,000	83,000
A572 Grade 50* (also A992)	345	345	210	450		310	207,000	83,000
A572 Grade 65	450	450	225	550		450	207,000	83,000
Iron:								
Malleable	170	170	83	345		275	166,000	76,000
Wrought	210	210	125	400		260	172,000	70,000
Aluminum Alloy:								
6061-T6 rolled/extruded	240	240	140	260		165	70,000	26,000
6063-T6 extruded tubes	170	170	97	207		130	70,000	26,000
Other Metals:								
Brass; 70% Cu, 30% Zinc	170		104	380		330	97,000	41,000
Bronze, cast heat treated	380		255	520		390	83,000	34,500
Timber, air dry:								
Yellow pine		43			58	7.0	11,000	
Douglas fir		37			47	5.5	11,000	
Spruce		28			34	5.2	8,300	
Concrete:								
Concrete: 1:2:4 mix, 28 days					21		21,000	

* With special requirements per AISC Tech. Bulletin #3.

This same problem will be solved using SI metric units.

A 6×8 S4S is referenced as a 150×200 nominal size.

The actual dimensions are 140 mm × 191 mm.

$$A = (140 \text{ mm} \times 191 \text{ mm}) = 26,700 \text{ mm}^2$$
$$= 26.7 \times 10^3 \text{ mm}^2 = 0.0267 \text{ m}^2$$

The length of the column $L = 10.5' = 126" = 3200 \text{ mm} = 3.2 \text{ m}$

$$P = 30,000 \text{ lb.} = 133,400 \text{ N} = 133.4 \text{ kN}$$

$$\therefore f_c = \frac{P}{A} = \frac{133.4 \text{ kN}}{0.0267 \text{ m}^2}$$

$$= 5000 \text{ }^{kN}/_{m^2} = 5 \text{ }^{MN}/_{m^2} = 5 \text{ MPa}$$

To determine the column shortening:

$$\delta = \frac{PL}{AE} = \frac{fL}{E} = \frac{(5 \text{ MPa})(3.2 \text{ m})}{(11,000 \text{ MPa})}$$

$$= 0.001450 \text{ m} = 1.45 \text{ mm}$$

Table 6.2a Allowable stresses for selected engineering materials (U.S. Customary).

Materials	Unit Weight (density) γ (pcf)	Modulus of Elasticity E (ksi)	Allowable Tension Stress F_t	Allowable Axial Compress. F_c	Allowable Compress. Bearing	Bending Stress F_b	Allowable Shear Stress F_v
Metals							
A36 Steel $F_y = 36$ ksi	490	29,000	22 ksi	22 ksi		22 ksi	14.5 ksi
A572 Grade 50A992	490	30,000	30 ksi	30 ksi		30 ksi	20 ksi
A572 Steel $F_y = 65$ ksi	490	30,000	39 ksi	39 ksi		39 ksi	26 ksi
Aluminum	165	10,000	16 ksi	16 ksi		16 ksi	10 ksi
Iron (cast)	450	15,000	5 ksi	20 ksi		5 ksi	7.5 ksi
Brittle Materials							
Concrete	145	3,000	100 psi	1,000 psi			100 psi
Stone masonry	165	1,000	10 psi	100 psi			10 psi
Brick masonry	120	1,500	20 psi	300 psi			30 psi
Wood					$F_c\perp$		
Doug-fir larch north*							
• Joist & rafters (No. 2)	35	1,700	650 psi	1,050 psi	625 psi	1,450 psi	95 psi
• Beams & posts (No. 1)	35	1,600	700 psi	1,000 psi	625 psi	1,300 psi	85 psi
Southern pine*							
• Joists & rafters (No. 2)	35	1,600	625 psi	1,000 psi	565 psi	1,400 psi	90 psi
• Beams & posts (D-No. 1)	35	1,600	1,050 psi	975 psi	440 psi	1,550 psi	110 psi
Hem-fir*							
• Joists & rafters (No. 2)	30	1,400	800 psi	1,050 psi	405 psi	1,150 psi	75 psi
• Beams & posts (No. 1)	30	1,300	600 psi	850 psi	405 psi	1,000 psi	70 psi
Wood Products							
Glu-lam beams	35	1,800	1,100 psi	1,650 psi	650 psi	2,400 psi	165 psi
Micro-lam beams	37	1,800	1,850 psi	2,460 psi	750 psi	2,600 psi	285 psi
Parallam beams	45	2,000	2,000 ksi	2,900 psi	650 psi	2,900 psi	290 psi

Example Problem 6.17 (Figures 6.29 and 6.30)

Using the information from Example Problem 6.11, determine the elongation that develops in the $\frac{3}{4}$"ϕ diameter steel rod.

Solution:

From the solution found in Example Problem 6.11, the stress in the rod is: $f_{BC} = 13.6$ ksi.

The elongation is determined by using the elastic equation:

$$\delta = \frac{PL}{AE} = \frac{fL}{E} = \frac{\left(13.6\,^{k}/_{in.^2}\right)\left(10.61' \times 12\,^{in.}/_{ft.}\right)}{29 \times 10^{3}\,^{k}/_{in.^2}}$$

$$= 0.060''$$

Table 6.2b   Allowable stresses for selected engineering materials (SI Metric).

Materials	Unit Weight (density) $\gamma \left({}^{kg}/_{m^3} \right)$	Modulus of Elasticity E (MPa)	Allowable Tension Stress F_t (MPa)	Allowable Axial Compress. F_c (MPa)	Allowable Compress. Bearing F_{brg} (MPa)	Bending Stress F_b (MPa)	Allowable Shear Stress F_v (MPa)
Metals							
A36 Steel F_y = 36 ksi	7,850	200,000	150	150		150	100
A572 Grade 50/A992	7,850	207,000	207	207		207	138
A572 Steel F_y = 65 ksi	7,850	207,000	270	270		270	180
Aluminum	2,640	70,000	110	110		110	70
Iron (cast)	7,210	105,000	35	140		35	52
Brittle Materials							
Concrete	2,400	21,000	4.8	6.9			0.7
Stone masonry	2,600	7,000	0.07	0.7			0.1
Brick masonry	1,900	10,500	0.14	2.1			0.2
Wood					$F_c \perp$		
Doug-fir larch north*							
• Joist & rafters (No. 2)	560	11,700	4.48	7.24	4.31	10.00	0.66
• Beams & posts (No. 1)	560	11,000	4.83	6.90	4.31	8.97	0.59
Southern pine*							
• Joists & rafters (No. 2)	560	11,000	4.31	6.90	3.90	9.66	0.62
• Beams & posts (D-No. 1)	560	11,000	7.24	6.72	3.03	10.69	0.76
Hem-Fir*							
• Joists & rafters (No. 2)	480	9,700	5.52	7.24	2.79	7.93	0.52
• Beams & posts (No. 1)	480	9,000	4.14	5.86	2.79	6.90	0.48
Wood Products							
Glu-lam beams	560	12,400	7.59	11.38	4.48	16.55	1.14
Micro-lam beams	590	12,400	12.76	16.97	5.17	17.93	1.97
Parallam beams	720	13,800	13.79	20.00	4.48	20.00	2.00

Example Problem 6.18 (Figure 6.49)

The steel towers of a large suspension bridge are planned to equal a total of 250 ft. (unstressed). They will be designed to have an average compressive stress of 50% of the yield stress of the high-strength steel (F_y = 100 ksi and E = 30,000 ksi or F_y = 689.5 MPa and E = 207×10^3 MPa). Determine the total loss of height in the tower.

Solution:

Since the compressive stress is limited to 50% of the yield stress, the steel remains within the elastic range and the deformation equation can be expressed as:

$$\delta = \frac{PL}{AE} = \frac{fL}{E} = \frac{\left(0.5 \times 100^{k}/_{in.^2}\right)\left(250' \times 12^{in.}/_{ft.}\right)}{30 \times 10^{3\,k}/_{in.^2}}$$

$$= 5.0''$$

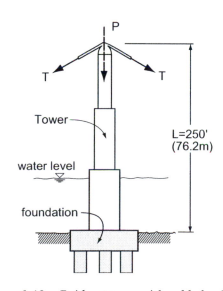

Figure 6.49   Bridge tower with cable loads.

Solving the same problem in SI units:

$$\delta = \frac{fL}{E} = \frac{(0.5 \times 689.5 \text{ MPa}) \times (76.2 \text{ m})}{(207 \times 10^3 \text{ MPa})}$$

$$= 0.127 \text{ m} = 127 \text{ mm}$$

Example Problem 6.19 (Figure 6.50)

A skywalk in the atrium area of a major hotel uses a system of $1\frac{1}{4}''$ ϕ steel rods to support the dead and occupancy loads. Assuming $F_t = 22$ ksi and $E = 29 \times 10^3$ ksi, determine the following:

a. the allowable load capacity in each rod;
b. if the rods supporting the third-floor walkway are 30 ft. in length, how much elongation would occur if the rod was loaded to its maximum allowable stress limit;
c. the ultimate load capacity of the rod at failure assuming A36 steel (see Table 6.2a).

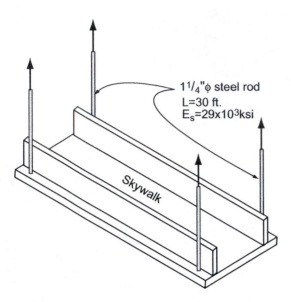

1¼"ϕ steel rod
L=30 ft.
E_s=29x10³ksi

Skywalk

Figure 6.50   Partial framing of a skywalk.

Solution:

a. The allowable load capacity is determined by using the basic stress equation rewritten as:

$$f = \frac{P}{A}; \quad A = \frac{\pi(1.25^2) \text{ in.}^2}{4} = 1.23 \text{ in.}^2$$

$$P_{\text{allow.}} = F_t \times A = \left(22\,^k\!/_{\text{in.}^2}\right) \times (1.23 \text{ in.}^2)$$

$$= 27 \text{ k (or 120 kN)}$$

b. The elongation is computed using the elastic equation.

$$\delta = \frac{PL}{AE} = \frac{fL}{E} = \frac{\left(22\,^k\!/_{\text{in.}^2}\right)\left(30' \times 12\,^{\text{in.}}\!/_{\text{ft.}}\right)}{29 \times 10^3\,^k\!/_{\text{in.}^2}}$$

$$= 0.273'' \text{ (or 6.9 mm)}$$

c. From Table 6.2a, the ultimate stress for A36 steel is:

$$F_{\text{ult.}} = 58 \text{ ksi}; \quad \therefore P_{\text{ult.}} = F_u \times A$$

$$= \left(58\,^k\!/_{\text{in.}^2}\right) \times (1.23 \text{ in.}^2) = 71.3 \text{ k}$$

Example Problem 6.20 (Figures 6.51 and 6.52)

A log post supports a floor load P. Assuming that the post has a diameter $D_1 = 12$ in. (0.305 m), (a) determine the allowable load P if the compression stress in the post is limited to $F_{allow.} = 800\,^{lb.}/_{in.^2}$ (5.52 MPa). Once P has been determined, (b) find the deformation (shortening) that results. Also, (c) determine the size D_2 of the circular footing necessary if the allowable soil pressure is: $q = 6000$ psf (287 kPa).

$$E_{timber} = 1.5 \times 10^3 \text{ ksi (10.34 MPa)};$$

$$\gamma_{concrete} = 150\,^{lb.}/_{ft.^3}\left(23.6\,^{kN}/_{m^3}\right)$$

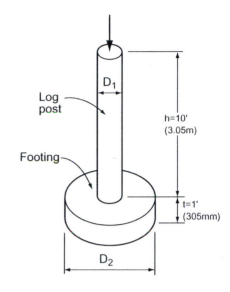

Figure 6.51   Log post with footing.

Solution:

a. $f = \dfrac{P}{A}$; $P_{allow.} = F_{allow.} \times A$

$$A = \frac{\pi D_1^2}{4} = \frac{\pi(12'')^2}{4} = 113 \text{ in.}^2;$$

$$A = \frac{\pi(0.305 \text{ m})^2}{4} = 0.0730 \text{ m}^2$$

$$P_{allow.} = \left(800\,^{lb.}/_{in.^2}\right) \times (113 \text{ in.}^2) = 90,400 \text{ lb.}$$

or $P_{allow.} = \left(5520\,^{kN}/_{m^2}\right) \times (0.073 \text{ m}^2) = 403 \text{ kN}$

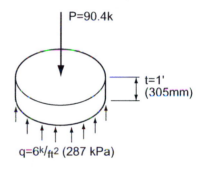

Figure 6.52   Loads on the footing.

b. $\delta = \dfrac{PL}{AE} = \dfrac{(90.4k)(120'')}{(113 \text{ in.}^2)\left(1.5 \times 10^3\,^{k}/_{in.^2}\right)} = 0.064''$

$$\delta = \frac{(403 \text{ kN})(3.05 \text{ m})}{(0.0730 \text{ m}^2)\left(10.34 \times 10^6\,^{kN}/_{m^2}\right)}$$

$$= 0.001600 \text{ m} = 1.6 \text{ mm}$$

c. $q_{net} = q - $ footing weight

ftg. wt. $= (\gamma_{conc.}) \times$ (ftg. thickness)

$$= \left(150\,^{lb.}/_{ft.^3}\right) \times (1') = 150\,^{lb.}/_{ft.^2}$$

or   ftg. wt. $= \left(23.6\,^{kN}/_{m^3}\right) \times (0.305 \text{ m})$

$$= 7.20\,^{kN}/_{m^2}$$

$\therefore q_{net} = 6000 \text{ psf} - 150 \text{ psf} = 5850 \text{ psf}$

or $q_{net} = 287 \text{ kPa} - 7.2 \text{ kPa} = 279.8 \text{ kPa}$

$$q_{net} = f = \frac{P}{A};\quad A = \frac{P}{q_{net}} = \frac{90.4 \text{ k}}{5.85\,^{k}/_{ft.^2}}$$

$$= 15.45 \text{ ft.}^2$$

$$\text{or} \qquad A = \frac{P}{q_{\text{net}}} = \frac{403 \text{ kN}}{279.8 \,^{\text{kN}}\!/_{\text{m}^2}} = 1.44 \text{ m}^2$$

$$A = \frac{\pi D_2^2}{4} = 15.45 \text{ ft.}^2;$$

$$D_2^2 = 19.68 \text{ ft.}^2; \qquad \therefore D_2 = 4.44'$$

$$\text{or} \qquad A = \frac{\pi D_2^2}{4} = 1.44 \text{ m}^2; \quad D_2 = 1.35 \text{ m}$$

6.3   STRESS CONCENTRATION

In our initial discussion on normal stress (see Figure 6.53), we made the assumption that if a load is applied centrically (through the axis of the member), the stress developed on the normal plane could be assumed to be uniform (Figure 6.54). For most cases of normal stress, this is a practical assumption to make for a static load condition.

If, however, the geometry of the member is changed so as to include discontinuities or changing cross-sections, stress can no longer be assumed to be uniform across the surface. Stress trajectories, also called *isostatic lines*, connect points of equal principal stress and represent stress paths through a member. This concept provides a visual picture of the stress distribution of a member or structure under various loading conditions. Stress trajectories are normally drawn at equal increments of stress to denote a uniform stress. A crowding of the stress trajectory lines indicates stress concentration, or high stress, just as contour lines on a topographic map indicate steep grades (Figure 6.55).

A French mathematician named *Barre de Saint-Venant* (1797–1886) observed that localized distortions occurred in areas of discontinuity and that stress concentrations developed, causing an uneven distribution of stress across the stressed surface. However, these localized effects disappeared at some distance from such locations. This is known as *Saint-Venant's principle*.

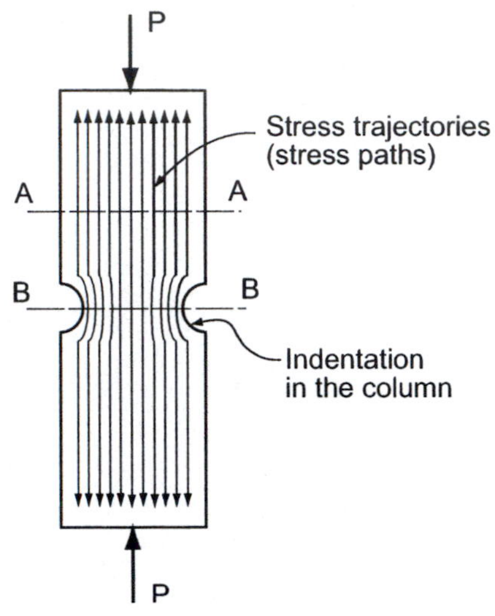

Load concentrations, reentrant corners, notches, openings, and other discontinuities will cause stress concentrations. These, however, do not necessarily produce structural failures, even if the maximum stress exceeds the allowable working stress. For example, in structural steel, extreme stress conditions may be relieved because steel has a tendency to yield (give), thus causing a redistribution of some of the stress across more of the cross-section. This redistribution of stress enables the greater part of the structural member to be within the permissible stress range.

In the past, photoelasticity (the shining of a polarized light on a transparent material) was often used to produce stress patterns for various structural members under loading. Today, computer modeling and analysis software are capable of generating colorful stress contour mapping, visually

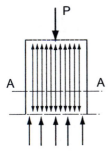

Figure 6.53   Normal stress.

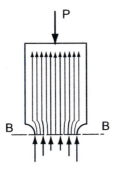

Figure 6.54   Uniform stress distribution.

Figure 6.55   Non-uniform stress distribution.

representing the stress intensity for both individual members and the structure as a whole.

Often, structural elements will have discontinuities (holes in a beam for mechanical ducts, window openings in walls) that interrupt the stress paths (called *stress trajectories*). The stress at the discontinuity may be considerably greater than the average stress due to centric loading; thus, there exists a stress concentration at the discontinuity (see Figure 6.56).

Stress concentrations are usually not significantly critical in the case of static loading of ductile material, because the material will yield inelastically in the high-stress areas, and redistribution of stress results. Equilibrium is established and no harm is done. However, in cases of dynamic or impact loading, or static loading of brittle material, stress concentrations become very critical and cannot be ignored. Stress redistribution does not result to such an extent that equilibrium is maintained.

6.4   THERMAL EFFECTS

Most structural materials expand in volume when subjected to heat and contract when cooled. Whenever a design prevents the change in length of a member subjected to temperature variation, internal stresses develop. Sometimes these *thermal stresses* may be sufficiently high to exceed the elastic limit and cause serious damage. Free, *unrestrained members* experience no stress changes with temperature changes, but dimensional change results. For example, it is common practice to provide expansion joints between sidewalk pavements to allow movement during hot summer days. Prevention of expansion on a hot day would undoubtedly result in severe buckling of the pavement.

The dimensional change due to temperature changes is usually described in terms of the change in a linear dimension. The change in length of a structural member, ΔL, is directly proportional to both the temperature change (ΔT) and the original length of the member L. Thermal sensitivity, called the *coefficient of linear expansion* (α), has been determined for all engineering materials (see Table 6.3). Careful measurements have shown that the ratio of strain ε to temperature change ΔT is a constant:

$$\alpha = \frac{\text{strain}}{\text{temp. change}} = \frac{\varepsilon}{\Delta T} = \frac{\delta/L}{\Delta T}$$

Solving this equation for the deformation:

where:

$$\delta = \alpha L \Delta T$$

where:

α = coefficient of thermal expansion or contraction
L = original length of the member (inches or mm)
ΔT = change in temperature (°F or °C)
δ = total change in length (in. or mm)

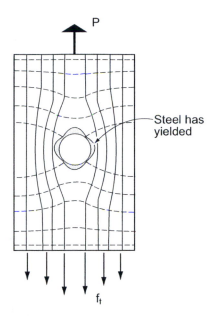

Figure 6.56   Stress trajectories around a hole.

Table 6.3   Linear coefficients of thermal expansion or contraction.

Material	Coefficient α (in./in./°F)	Coefficient α (mm/mm/°C)
Wood	3.0×10^{-6}	5.4×10^{-6}
Glass	4.4×10^{-6}	8.0×10^{-6}
Concrete	5.5×10^{-6}	9.9×10^{-6}
Cast iron	5.9×10^{-6}	10.6×10^{-6}
Steel	6.5×10^{-6}	11.7×10^{-6}
Wrought iron	6.7×10^{-6}	12.0×10^{-6}
Copper	9.3×10^{-6}	16.8×10^{-6}
Bronze	10.1×10^{-6}	18.1×10^{-6}
Brass	10.4×10^{-6}	18.8×10^{-6}
Aluminum	12.8×10^{-6}	23.1×10^{-6}

Of perhaps even greater importance in engineering design are the stresses developed by restraining the free expansion and contraction of members subjected to temperature variations. To calculate these temperature stresses, it is useful to determine first the free expansion or contraction of the member involved and, second, the force and unit stress developed in forcing the member to attain its original length. The problem from this point on is exactly the same as those solved in the earlier portions of this chapter dealing with axial stresses, strains, and deformations. The amount of stress developed by restoring a bar to its original length L is:

$$f = \varepsilon E = \frac{\delta}{L}E = \frac{\alpha L \Delta T E}{L} = \alpha \Delta T E$$

$$\therefore f = \alpha \Delta T E$$

Example Problem 6.21 (Figure 6.57)

A 60' length of steel rail is laid on a day when the temperature is 40°F. In order to prevent the rail from developing any internal stresses due to a thermal increase of 60°F, what is the amount of deformation that needs to be accommodated with respect to the slotted connection at the rail end(s)? $E_{st} = 29 \times 10^3$ ksi.

Solution:

Steel has a coefficient of expansion $\alpha = 6.5 \times 10^{-6}/°F$ (see Table 6.3).

Using the deformation equation due to thermal change:

$$\delta = \alpha L \Delta T = (6.5 \times 10^{-6}/°F)\left(60' \times 12^{\text{in.}}/_{\text{ft.}}\right)(60°F)$$

$$= 0.28''$$

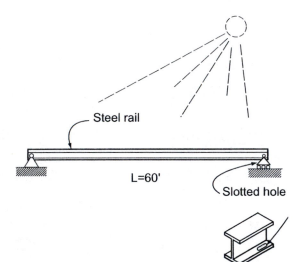

Figure 6.57   Steel rail subjected to thermal change.

This amount of deformation (0.28'') for a 60'-long rail section may not seem large but if there are no provisions made to allow movement during thermal changes, large internal stress may result. If the rail section in this example has a cross-sectional area of $A = 10.5$ in.^{2}, determine the amount of internal compressive stress that can result if the rail is restrained from moving.

$$f = \alpha \Delta T E = (6.5 \times 10^{-6}/°F)(60°F)\left(29 \times 10^{3}{}^{k}/_{\text{in.}^2}\right)$$

$$= 11.31 \text{ ksi}$$

(a very large internal stress which can potentially cause the rail to buckle)

Example Problem 6.22

A steel surveying tape measures exactly 100' (30.5 m) at 68°F (20°C) when subjected to a pull of 10 lb. (44.48 N). If the tape cross-section is 1/32" × 5/8" (0.794 mm × 15.875 mm), what correction should be made in the reading of the tape if it is subjected to a temperature of 100°F (37.8°C) and a pull of 25 pounds (111.2 N)? $E_{st} = 30 \times 10^6$ psi (207×10^3 MPa)?

Solution:

This solution involves two deformation equations, temperature and load. Both conditons cause elongation to occur.

$$\delta_{\text{temp.}} = \alpha \Delta T L$$

$$= (6.5 \times 10^{-6}/°F)(100°F - 68°F)(100' \times 12 \, ^{\text{in.}}/_{\text{ft.}})$$

$$= 0.250''$$

or, in SI units,

$$\delta_{\text{temp.}}$$

$$= (11.7 \times 10^{-6}/°C)(37.8°C - 20°C)(30{,}500 \text{ mm})$$

$$= 6.35 \text{ mm}$$

$$\delta_{\text{load}} = \frac{\Delta PL}{AE} = \frac{(25 \text{ lb.} - 10 \text{ lb.})(1200'')}{\left(\frac{1}{32}'' \times \frac{5}{8}''\right)\left(30 \times 10^6 \, ^{\text{lb.}}/_{\text{in.}^2}\right)}$$

$$= 0.031''$$

In SI units,

$$\delta_{\text{load}} = \frac{\Delta PL}{AE}$$

$$= \frac{(111.2 \text{ N} - 44.5 \text{ N})(30{,}500 \text{ mm})}{(0.794 \text{ mm} \times 15.875 \text{ mm})\left(207 \times 10^3 \, ^{\text{N}}/_{\text{mm}^2}\right)}$$

$$= 0.780 \text{ mm}$$

Note: $1 \text{ MPa} = 1 \times 10^6 \, ^{\text{N}}/_{\text{m}^2}$

$$= 1 \times 10^6 \, ^{\text{N}}/_{\text{m}^2} \times 10^{-6} \, ^{\text{mm}^2}/_{\text{m}^2} = 1 \, ^{\text{N}}/_{\text{mm}^2}$$

$$\therefore E_{st} = 207{,}000 \text{ MPa} = 207 \times 10^3 \, ^{\text{N}}/_{\text{mm}^2}$$

$$\delta_{\text{total}} = \delta_{\text{temp}} + \delta_{\text{load}} = 0.250'' + 0.031'' = 0.281''$$

In SI units;

$$\delta_{\text{total}} = \delta_{\text{temp}} + \delta_{\text{load}} = 6.35 \text{ mm} + 0.78 \text{ mm} = 7.13 \text{ mm}$$

What is the average axial tension stress developed in the steel tape at a pull of 25 pounds? Is the stress within the elastic range?

$$f = \frac{P}{A} = \frac{25 \text{ lb.}}{\left(\frac{1}{32}'' \times \frac{5}{8}''\right)} = \frac{25 \text{ lb.}}{0.0195 \text{ in.}^2} = 1282 \text{ psi}$$

(the tensile stress is well below the elastic limit)

$$f = \frac{P}{A} = \frac{111.2 \text{ N}}{(0.794 \text{ mm} \times 15.875 \text{ mm})} = 8.83 \, ^{\text{N}}/_{\text{mm}^2}$$

$$= 8.83 \times 10^6 \, ^{\text{N}}/_{\text{m}^2} = 8.83 \text{ MPa}$$

Example Problem 6.23

A thin steel hoop of 559-mm inside diameter is to be shrunk over a 559.5-mm diameter cylinder. (a) To what temperature must the hoop be raised so that it will just slip over the cylinder, assuming the room temperature to be 20.5°C? (b) What average stress will be induced in the ring when it cools to room temperature assuming that the cylinder does not contract during cooling?

Solution:

a. The inside diameter of the hoop must expand to 559.5(+) mm to just fit over the cylinder. Therefore, the hoop needs to expand 1.57 mm in length during the temperature increase.

$$\delta = \pi(d_1 - d_2) = \pi(559.5 \text{ mm} - 559 \text{ mm}) = 1.57 \text{ mm}$$

But, $\delta = \alpha \Delta T L = 1.57 \text{ mm}$

$$\Delta T = \frac{\delta}{\alpha L} = \frac{1.57 \text{ mm}}{(11.7 \times 10^{-6}/°C)(\pi \times 559.5 \text{ mm})}$$

$$= 76.4°C$$

The final temperature is equal to:

$$T_{final} = T_{room} + \Delta T = 20.5°C + 76.4°C = 96.9°C$$

Boiling water at 100°C is probably sufficient to accomplish the task.

b. When the hoop cools to room temperature, $\Delta T = 76.4°C$, the restriction of deformation is equal to

$$\delta = \pi(559.5 - 559.0) = 1.57 \text{ mm}$$

$$\delta = \frac{PL}{AE} = \frac{fL}{E};$$

$$f = \frac{\delta E}{L} = \frac{(1.57 \text{ mm})(207 \times 10^3 \text{ N/mm}^2)}{(\pi \times 559.5 \text{ mm})}$$

$$= 185.0 \text{ N/mm}^2 = 185 \text{ MPa}$$

The stress is very high but it is still within the elastic range (the yield point of steel $F_y = 345$ MPa is very close to the elastic limit).

Example Problem 6.24 (Figures 6.58 and 6.59)

A W8×67 steel beam, 20 ft. in length, is rigidly attached at one end of a concrete wall. If a gap of 0.010 in. exists at the opposite end when the temperature is 45°F, what results when the temperature rises to 95°F?

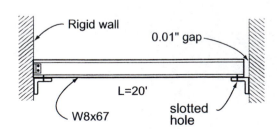

Figure 6.58   Steel beam with a gap.

Solution:

If the beam were totally free to expand;

$$\delta_{total} = \alpha \Delta T L$$

$$= (6.5 \times 10^{-6}/°F)(95°F - 45°F)(20' \times 12^{in.}/_{ft.})$$

$$= 0.078"$$

Since, $\delta_{total} > \delta_{gap} = 0.01"$; internal stresses develop

$$\delta_{constrained} = \delta_{total} - \delta_{gap} = 0.078" - 0.01" = 0.068"$$

$$\delta_{constrained} = 0.068" = \frac{PL}{AE} = \frac{fL}{E}$$

$$\therefore f = \frac{\delta_{con.}E}{L} = \frac{(0.068")(29 \times 10^{6\,lb.}/_{in.^2})}{240"} = 8220 \text{ psi}$$

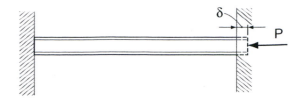

Figure 6.59   Beam constrained between two supports.

Example Problem 6.25

A bottle made of aluminum is sealed with a steel cap. Unfortunately, during the capping process the top is screwed on too tightly. In order to loosen the cap, would you recommend placing the capped bottle in your refrigerator or under hot water? Why?

Solution:

The inside diameter of the cap and the outside diameter of the bottle are essentially the same. Therefore, the solution involves the comparison of the coefficients of thermal expansion between the two materials.

Bottle: $\alpha_{alum.} = 12.8 \times 10^{-6}/°F$

Cap: $\alpha_{steel} = 6.5 \times 10^{-6}/°F$

The rate of expansion of the aluminum is approximately twice as fast as that of steel. Placing the bottle in the refrigerator would cause the aluminum (bottle) to constrict faster than steel (cap), thereby loosening the cap. Hot water would cause the cap to get even tighter.

Supplementary Problems

Section 6.1—Axial, Shear, and Bearing Stress

6.1.1   A short 8×8 S4S timber post temporarily carries a small load of $P = 1800$ lb. without the use of a footing. Will the load cause a potentially large settlement (due to exceeding the allowable bearing pressure) or is this arrangement satisfactory? Show computations to support your conclusion.

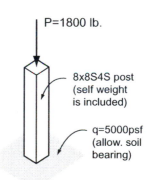

P=1800 lb.

8x8S4S post
(self weight
is included)

q=5000psf
(allow. soil
bearing)

Problem 6.1.1

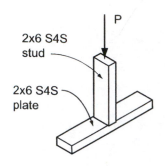

Problem 6.1.2

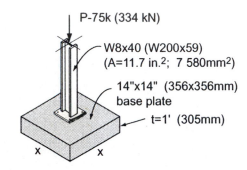

Problem 6.1.3

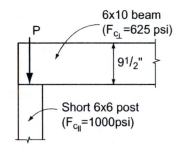

Problem 6.1.4

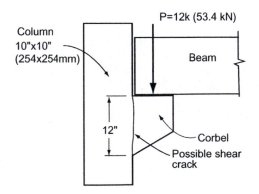

Problem 6.1.5

6.1.2 Determine the maximum load capacity of a stud wall arrangement if the allowable capacities of the wood are:

$$F_{c_{||}} = 1000 \text{ psi (compression parallel to the wood grain)}$$

$$F_{c_{\perp}} = 400 \text{ psi (compression perpendicular to the wood grain)}$$

Note: The actual dimensions of a 2×6 S4S member measures $1\,\frac{1}{2}'' \times 5\,\frac{1}{2}''$.

6.1.3 A steel column carries a building load of 75 k (334 kN) to a 14" × 14" (356 mm × 356 mm) base plate that is bolted to a concrete footing pad that measures 1 foot (305 mm) in thickness. The column has a cross-sectional area $A = 11.7$ in.^{2} ($A = 7580 \text{ mm}^2 = 7.58 \times 10^{-3} \text{ m}^2$). Determine the following:

    a. the average compressive stress developed in the W8 × 40 (W200 × 59) column
    b. the bearing stress between the steel base plate and the concrete footing
    c. the footing size, assuming that the allowable soil bearing pressure is $q = 4$ ksf (191 kPa) and the density of concrete is 150 pcf $\left(2400\,\frac{\text{kg}}{\text{m}^3} = 23.6\,\frac{\text{kN}}{\text{m}^3}\right)$

6.1.4   a. Determine the maximum reaction load P based on the beam's bearing capacity and/or the post's compression allowable. See Table 6.2a, assuming Douglas fir-larch species; 6×6 S4S (5.5" × 5.5"); 6×10 S4S (5.5" × 9.5")
    b. If the beam is simply supported, uniformly loaded, and spanning 14', what is the maximum load per foot (ω) that the beam can safely support based on the reaction P found in part (a)?

6.1.5 In the precast concrete beam-column connection shown, a corbel (concrete bracket) is used to transmit the load from the beam to the column. The column measures 10 inches square (254 mm × 254 mm) with the corbel being 12" (305-mm) deep. Assuming that the concrete might fail in shear as is shown in the diagram, determine if the corbel can "theoretically" support a load of $P = 12$ k (53.4 kN) without steel reinforcement if the concrete has a shear stress capacity of 120 psi (827 kPa).

6.1.6   A 4×6 S4S lower cord member in a timber roof truss must be spliced using two, 2×6 S4S members on either side with 2-$\frac{1}{2}$" ϕ bolts on each half. Determine the maximum capacity of the connection based on:

   a. shear stress in the bolts ($F_v = 10$ ksi)
   b. bearing stress between the bolts and the wood members. ($F_c = 1000\,^{\text{lb.}}/_{\text{in.}^2}$ in the wood members).

Note: Analyze the connection on either side of the section cut/splice. The bearing area of the bolts on the wood member is computed as:

   (no. of bolts) x (bolt diameter) x (critical length of bolt)

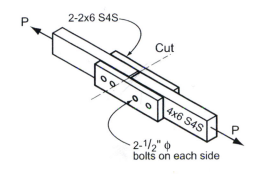

Problem 6.1.6

6.1.7   Three 2×4 S4S blocks are glued together as shown. Assuming the glue has a shear capacity of 80 psi, determine the minimum length L required.

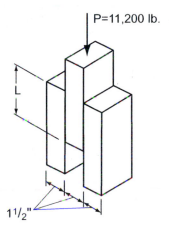

Problem 6.1.7

6.1.8   A hoist, consisting of a 6" × 8" wooden post *AB* and a 1 $\frac{1}{8}$" ϕ diameter steel rod *BC*, supports a load *P* = 9500 lb. as shown. Determine (a) the axial stresses in *AB* and *BC* and (b) the average shear stress in the 1-$\frac{1}{2}$" diameter pin at *C*. The pin at *C* is in double shear.

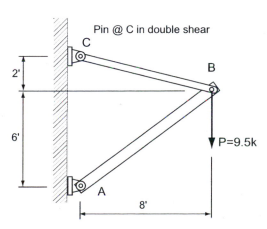

Problem 6.1.8

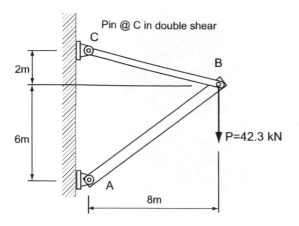

Problem 6.1.9

6.1.9 A hoist, consisting of a 152 mm × 203 mm wooden post *AB* and a 28.5-mm diameter steel rod *BC*, supports a load $P = 42.3$ kN as shown. Determine (a) the axial stresses in *AB* and *BC* and (b) the average shear stress in the 38-mm diameter pin at *C*. The pin at *C* is in double shear.

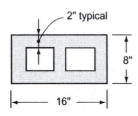

Problem 6.1.10 (a)

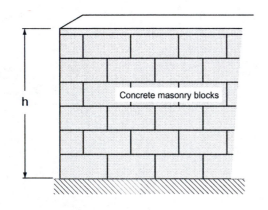

Problem 6.1.10 (b)

6.1.10 An unreinforced masonry wall (cells are not filled with mortar) has the cross-sectional dimensions as shown. What is the maximum permissible height *h* of the wall if the allowable compressive stress for the hollow core masonry block is $F_c = 250$ psi? Assume that the blocks have a density of $\gamma_{cmu} = 100 \, {}^{lb.}/_{ft.^3}$.

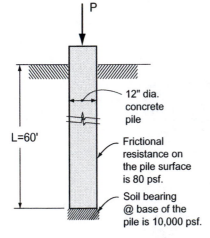

6.1.11 A precast concrete pile supports a building load *P* as shown. Assuming that the capacity of the pile is based on its surface friction plus the bearing at the base of the pile, determine *P*.

Problem 6.1.11

Section 6.2—Axial Strain and Elasticity

6.2.1 A concrete test cylinder is loaded with P = 100 k (445 kN) and a resulting shortening of 0.036" (0.0914 mm). Determine the unit strain developed in the concrete. The height of the cylinder is 12" (305 mm) and the diameter is 6" (152 mm).

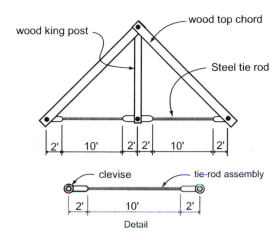

Problem 6.2.1

6.2.2 A truss tie-rod has the dimensions as shown. Upon loading, it is found that an elongation of 0.400 in. occurred in each tie-rod assembly. If the unit strain on the rod portion equaled 0.0026, what was the unit strain on the two end clevises?

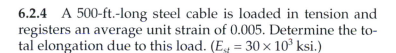

Problem 6.2.2

6.2.3 The midpoint C of a cable drops to C' when a weight W is suspended from it. Find the strain in the cable.

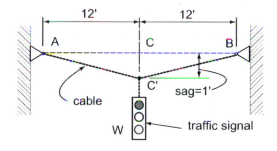

6.2.4 A 500-ft.-long steel cable is loaded in tension and registers an average unit strain of 0.005. Determine the total elongation due to this load. (E_{st} = 30 × 10³ ksi.)

Problem 6.2.3

6.2.5 Determine the modulus of elasticity of a rod that has a cross-sectional area A = 1 in.², L = 8'-0", P = 25 k, and an elongation of δ = 0.10" when loaded.

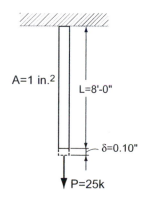

Problem 6.2.5

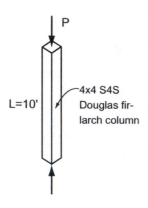

Problem 6.2.6

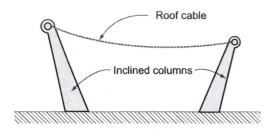

Problem 6.2.7

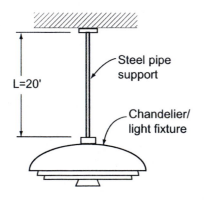

Problem 6.2.8

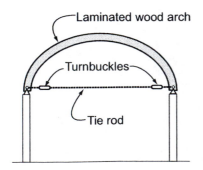

Problem 6.2.9

6.2.6 Determine the allowable capacity of the column shown, assuming that the allowable compressive stress ($F_c = 600$ psi) accounts for the buckling potential. Refer to the standard wood tables in the Appendix for information on the cross-sectional properties.

6.2.7 A 600'-long (183 m) roof cable cannot be permitted to stretch more than 3' (0.91 m) when loaded or the roof geometry will be affected too drastically. If $E_{st} = 30 \times 10^3$ ksi (207×10^3 MPa) and the load is 1500 k (6.67 MN), determine the required cable diameter needed to avoid excessive elongation or overstress. The allowable tensile stress $F_t = 100$ ksi (689.5 MPa).

6.2.8 A large chandelier weighing 1500 lb. is suspended from the roof of a theater lobby. The steel pipe from which it hangs is 20 feet long. Determine the size of pipe necessary to carry the chandelier safely. Use A36 steel. What is the resulting elongation of the pipe? Use the section properties table in the Appendix to determine the appropriate pipe size.

6.2.9 The ends of the laminated-wood roof arch shown are tied together with a horizontal steel rod 90'-10" long, which must withstand a total load of 60 k. Two turnbuckles are used. All threaded rod ends are upset (see Figure 6.41).

    a. Determine the required diameter D of the rod if the maximum allowable stress is 20 ksi.
    b. If the unstressed length of the rod is 90'-10" and there are 4 threads per inch on the upset ends, how many turns of one turnbuckle will bring it back to its original length after it has elongated under the full allowable tensile stress? $E_{st} = 29 \times 10^3$ ksi.

6.2.10 A W14×74 steel column, 14' tall, is subjected to a load of 400 kips. Determine the following:

   a. the average compressive stress developed in the column,

   b. the bearing stress between the steel base plate and the concrete footing (include the weight of the column),

   c. the bearing pressure (stress) between the base of the concrete footing and the soil (including the column weight),

   d. the elastic shortening experienced by the column.

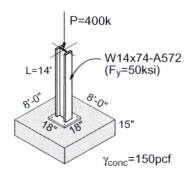

Column-base plate-footing

Problem 6.2.10

6.2.11 A bungee jumper is using a 1" φ (25.4 mm) cord that has a lot of "give." If the jumper has an estimated equivalent "dynamic" weight of 400 lb. (1.78 kN), determine the following:

   a. The maximum elongation that would occur.

   b. Is the stress developed in the cord within the elastic limit?

   c. Does the jumper "bonk" or does he survive?

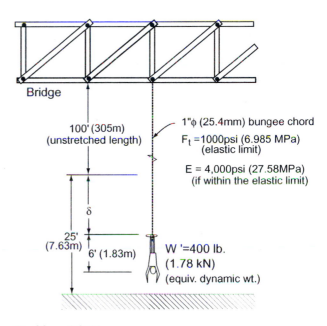

Problem 6.2.11

Section 6.4—Thermal Stresses

6.4.1 A W18×35 wide-flange beam (see steel tables in the Appendix) is used as a support beam for a bridge. If a temperature change (rise) of 40°F occurs, determine the deformation that results.

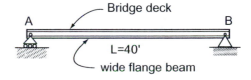

Problem 6.4.1

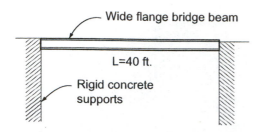

Wide flange bridge beam

L=40 ft.

Rigid concrete supports

Problem 6.4.2

6.4.2 In the previous problem, what would happen if both ends of the beam were solidly imbedded in concrete foundations on both sides? Assume it is initially unstressed axially.

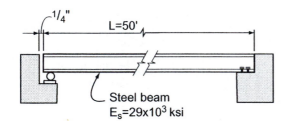

Immovable concrete piers

L=200' (61m)

Problem 6.4.3

6.4.3 The straight bottom chord of a steel bridge is 200 ft. (61 m) long.

  a. Calculate the change in its length caused by a change in temperature of 70°F (38.9°C), which is the possible difference between day and night temperatures.
  b. What average unit stress will be produced in this chord if both of its ends are fully restrained against linear expansion?
  c. If the lower chord is fully restrained between two virtually "immovable" concrete piers and its average cross-sectional area contains 20 in.² (12.9 × 10³ mm² = 12.9 × 10³ m²) of steel, what total force P would it exert against its fastenings at the piers?

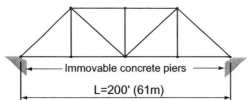

−¹/₄"

L=50'

Steel beam
$E_s = 29 \times 10^3$ ksi

Problem 6.4.4

6.4.4 Compute the thermal stress induced due to a temperature change of 100°F, assuming that the beam is able to expand freely for ¹/₄".

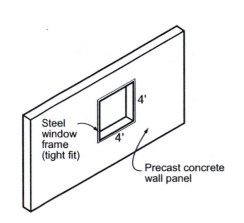

Steel window frame (tight fit)

4'

4'

Precast concrete wall panel

Problem 6.4.5

6.4.5 A square, steel window frame fits very tightly into a concrete wall opening at the time of installation. Assuming a thermal rise of 85°F occurs in the steel window frame and a 10°F change results in the thermally massive concrete wall, calculate the amount of stress that develops in the frame (length or height). If there was sufficient tolerance allowed for the frame to elongate or shorten freely due to thermal changes, what would the resulting deformation be? See Tables 6.2a and 6.3 for pertinent data on steel.

6.4.6 A long concrete bearing wall has vertical expansion joints placed every 40′ (12.2 m). Determine the required width of the gap in a joint if it is wide open at 20°F (−6.7°C) and just barely closed at 80°F (26.7°C). Assume $\alpha = 6.0 \times 10^{-6}/°F$ ($10.8 \times 10^{-6}/°C$).

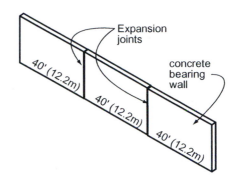

Problem 6.4.6

6.4.7 A steel wide-flange beam (W8×31) is used to brace two shoring walls as shown. If the walls move 0.01 inches outward when the beam is subjected to a 100°F temperature change, determine the stress in the beam.

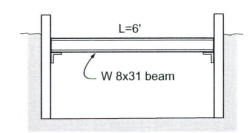

Problem 6.4.7

6.4.8 An aluminum curtain wall panel 12′ (3.66 m) high is attached to large concrete columns (top and bottom) when the temperature is 65°F (18.3°C). No provision is made for differential thermal movement vertically. Because of insulation between them, the sun heats up the wall panel to 120°F (48.9°C) but the column only to 80°F (26.7°C). Determine the consequent compressive stress in the curtain wall.

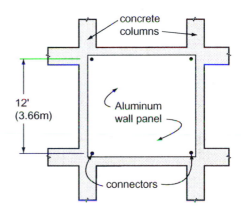

Problem 6.4.8

6.4.9 A 350-lb. weight is suspended from a $^3/_8$″ ϕ diameter steel rod, 20 feet long. Determine the following:

    a. The elongation of the rod due to the weight.
    b. The change in temperature that would cause the rod to return to its unloaded position.

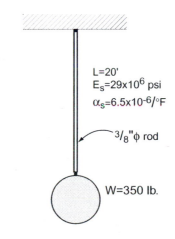

Problem 6.4.9

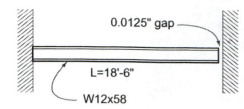

0.0125" gap

L=18'-6"

W12x58

Problem 6.4.10

6.4.10 A W12×58 steel beam is rigidly attached at one end of a concrete wall. If a gap of 0.0125" exists at the opposite end when the temperature is 35°F, what results when the temperature rises to 100°F? If the gap closes fully and some of the thermal deformation is restrained, determine the resulting stress in the beam.

6.4.11 A question of magnitude from Galileo's *Dialogues Concerning Two New Sciences*:

> "... It would be impossible to build up the bony structures of men, horses, or other animals so as to hold together and perform their normal functions if these animals were to be increased enormously in height; for this increase in height can be accomplished only by employing a material(s) which is harder and stronger than usual, or by enlarging the size of the bones, thus changing their shape until the form and appearance of the animals suggest a monstrosity. ... If the size of the body be diminished, the strength of that body is not diminished in the same proportion; indeed, the smaller the body the greater is its relative strength. Thus a small dog could probably carry on its back two or three dogs of his own size; but I believe that a horse could not carry even one of his own size."

Discuss Galileo's observation in relation to the concepts discussed thus far in strength of materials.

7

Centroids and Cross-Sectional Properties of Structural Members

Introduction

Beam design requires the knowledge of material strengths (allowable stresses), critical shear and moment values, and information about their cross-section. The shape and proportion of a beam cross-section is critical in keeping bending and shear stresses within allowable limits and moderating the amount of deflection that will result from the loads. Why does a 2×8 joist standing on edge deflect less when loaded at midspan than the same 2×8 used as a plank? Columns with improperly configured cross-sections may be highly susceptible to buckling under relatively moderate loads. Are circular pipe columns then better at supporting axial loads than columns with a cruciform cross-section?

In subsequent chapters, it will be necessary to calculate two cross-sectional properties crucial to the design of beams and columns. The two properties are the *centroid* and the *moment of inertia*.

7.1 CENTER OF GRAVITY AND CENTROIDS

Center of gravity (CG) or *center of mass* refers to masses or weights and can be thought of as a single point at which the weight could be held and be in balance in all directions. If the weight or object were homogeneous, the center of gravity and centroid would coincide. In the case of a hammer with a wooden handle, its center of gravity would be close to the heavy steel head. The centroid, which is found by ignoring weight and considering only volume, would be closer to the middle of the handle. Due to varying densities, the center of gravity and centroid do not coincide. Centroid usually refers to the centers of lines, areas, and volumes. The centroid of cross-sectional areas (of beams and columns) will be used later as the *reference origin* for computing other section properties.

The method of locating the center of gravity of a mass or an area is based on the method of determining the resultants

Photo 7.1.1   Where's the center of gravity?

239

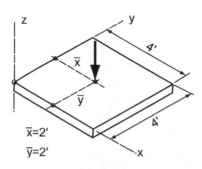

Figure 7.1   Centroid of the entire plate.

of parallel force systems. If an area or mass is divided into a large number of small, equal areas, each of which is represented by a vector (weight) acting at its centroid, the resultant vector of the entire area would act through the center of gravity of the total mass area.

The center of gravity of a mass or of an area is the theoretical point at which the entire mass or area can be considered to be concentrated (see Figure 7.1).

To develop the equations necessary for calculating the centroidal axes of an area, consider a simple, square plate of uniform thickness (Figure 7.1).

$$\bar{x} = \text{centroidal } x \text{ distance} = 2' \text{ from the reference } y \text{ axis}$$

$$\bar{y} = \text{centroidal } y \text{ distance} = 2' \text{ from the reference } x \text{ axis}$$

It may be obvious that the centroid is located at $\bar{x} = 2'$ and $\bar{y} = 2'$, but a methodology may be necessary to handle odd-shaped or complicated areas.

The first step, using the method described below, is to divide the area into smaller increments called *components* (see Figure 7.2).

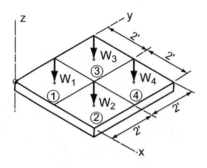

Figure 7.2   Plate divided into four equal components.

Each component has its own weight and centroid, with all weights directed perpendicular to the surface area (x-y plane). The magnitude of the resultant is equal to the algebraic sum of the component weights.

$$W = \sum \Delta W$$

where:

   W = total weight of the plate
   ΔW = component weight

Centroids are obtained by taking moments about the x and y axes, respectively. The principle involved may be stated as:

   The moment of an area about an axis equals the algebraic sum of the moments of its component areas about the same axis.

Using the diagrams in Figures 7.3 and 7.4, write the moment equations:

$$\sum M_y: \bar{x}W = W_1x_2 + W_2x_1 + W_3x_2 + W_4x_1; \text{ and}$$

$$\sum M_x: \bar{y}W = W_1y_1 + W_2y_1 + W_3y_2 + W_4y_2$$

Therefore:

$$\bar{x} = \frac{\sum(x\Delta W)}{W}$$

$$\bar{y} = \frac{\sum(y\Delta W)}{W}$$

Assuming that the plate is of uniform thickness t:

$$W = \gamma\, tA$$

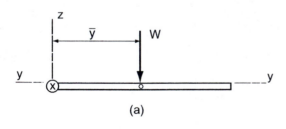

(a)

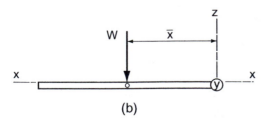

(b)

Figure 7.3   Elevations of the whole plate.

where:

W = total weight of the plate

γ = density of the plate material

t = plate thickness

A = surface area of the plate

Correspondingly, for the component parts (areas) of the plate with uniform thickness t:

$$\Delta W = \gamma t \Delta A$$

where:

ΔW = weight of component plate area

ΔA = surface area of component

If we return to the previous moment equations and substitute the values $\gamma t A$ for W and $\gamma t \Delta A$ for ΔW, we find that if the plate is homogeneous and of constant thickness, γt cancels out of the equation(s).

The resulting moment equation would then be written as:

$$\sum M_y\colon \bar{x}W = x_2\Delta A_1 + x_1\Delta A_2 + x_2\Delta A_3 + x_1\Delta A_4$$

$$\sum M_x\colon \bar{y}W = y_1\Delta A_1 + y_1\Delta A_2 + y_2\Delta A_3 + y_2\Delta A_4$$

$$\bar{x} = \frac{\sum(x\Delta A)}{A}$$

$$\bar{y} = \frac{(y\Delta A)}{A}$$

where:

$$A = \sum \Delta A$$

The moment of an area (or the first moment of an area) is defined as the product of the area multiplied by the perpendicular distance from the moment axis to the centroid of the area.

The previous derivation shows how the centroid of an area can be located by dividing an area into elemental areas and summing the moments of the elemental areas about an axis. In finding the centroid of a more complex area (i.e., a *composite area*), a similar methodology is used.

Composite or more complex areas are first divided into simpler geometric shapes (as shown in Table 7.1) with known centroids. A reference origin is chosen (usually the lower left corner) to establish the reference x and y axes. Then moments of area are summed about the reference x and y axes, respectively. The centroid locates the new reference origin for subsequent computations of other cross-sectional properties (*moment of inertia* and *radius of gyration*).

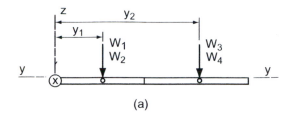

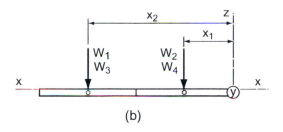

Figure 7.4   Elevations of the quartered plate.

Table 7.1   Centroid of simple geometric shapes.

Shape	Centroid location	$\bar{x}$	$\bar{y}$	Area
Rectangle		$b/2$	$h/2$	bh
Triangle		$b/3$	$h/3$	$bh/2$
Semicircle		0	$4r/3\pi$	$\pi r^2/2$
Quarter circle		$4r/3\pi$	$4r/3\pi$	$\pi r^2/4$
Parabolic segment		$5b/8$	$2h/5$	$2bh/3$
Complement of a parabolic segment		$3b/4$	$3h/10$	$bh/3$

Example Problem 7.1: Centroids (Figures 7.5 and 7.6)

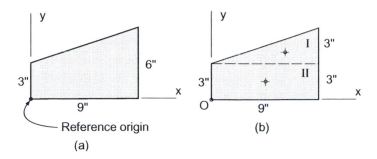

Figure 7.5 *Trapezoidal area.*

Determine the centroidal x and y distances for the composite area shown. Use the lower left corner of the trapezoid as the reference origin.

Solution:

Select a convenient reference origin. It is usually advantageous to select a reference origin such that x and y distances are measured in the positive x and y directions to avoid dealing with negative signs.

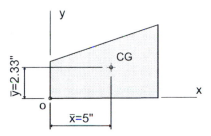

Figure 7.6 *Location of the centroid.*

Divide the trapezoid into simpler geometric shapes: a rectangle and a triangle.

As composite areas become more complex, it may be convenient to use a tabular format for centroidal axis calculations.

Component	Area (ΔA) (in.²)	x (in.)	$x\Delta A$ (in.³)	y (in.)	$y\Delta A$ (in.³)
(a)	$\dfrac{9"(3")}{2} = 13.5$ in.²	6"	81 in.³	4"	54 in.³
(b)	$9" (3") = 27$ in.²	4.5"	121.5 in.³	1.5"	40.5 in.³
	$A = \sum \Delta A = 40.5$ in.²		$\sum x\Delta A = 202.5$ in.³		$\sum y\Delta A = 94.5$ in.³

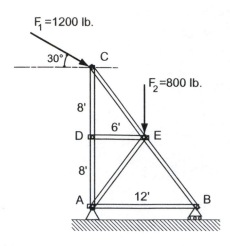

Problem 2.4.8

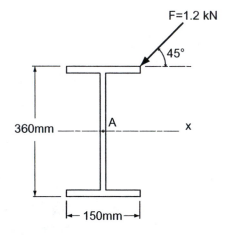

Problem 2.4.9

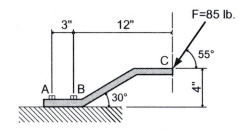

Problem 2.4.10

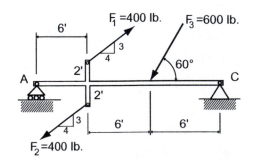

Problem 2.4.11

2.4.8 A vertical truss supports two applied forces F_1 and F_2. Determine the moment at supports A and B.

2.4.9 Determine the moment due to force F about the center point A of a wide-flange beam cross-section. The resulting moment is a twisting action on the beam section known as *torsion*.

Section 2.4—Moment of a Couple(s)

2.4.10 An 85-lb. force is applied to the bent plate as shown. Determine an equivalent force-couple system at A and at B.

2.4.11 Three forces are acting on the beam as shown. Determine the resultant moment about A from forces F_1, F_2, and F_3. What is the moment at support C?

Component	Area (ΔA) (mm²)	x (mm)	$x\Delta A$ (mm³)	y (mm)	$y\Delta A$ (mm³)
(a)	16×10^3	-20	-320×10^3	$+360$	5.76×10^6
(b)	19.2×10^3	$+40$	768×10^3	$+200$	3.84×10^6
(c)	19.2×10^3	$+120$	2.304×10^6	$+40$	0.77×10^6
	$A = \sum \Delta A = 54.4 \times 10^3$		$\sum x\Delta A = 2.75 \times 10^6$		$\sum y\Delta A = 10.37 \times 10^6$

Example Problem 7.3a (Figures 7.10 to 7.13)

A 6" thick concrete wall panel is precast to the dimensions as shown. Using the lower left corner as the reference origin, determine the center of gravity (centroid) of the panel.

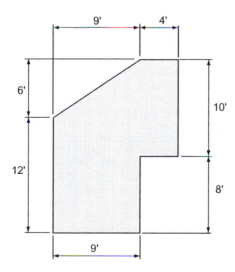

Figure 7.10   Concrete wall panel.

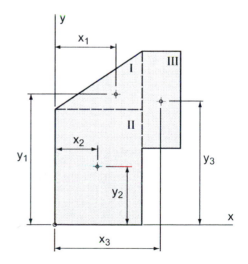

Figure 7.11   Subdivided into three components.

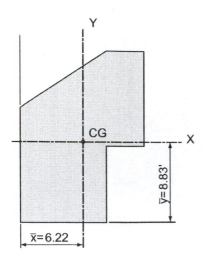

Figure 7.12   Centroid location.

Solution:

The concrete panel can be divided into three components, a triangle and two rectangles. By choosing the lower left corner as the reference origin, a negative sign can be avoided in the x and y distance measurements.

$$\bar{x} = \frac{\sum x \Delta A}{A} = \frac{1088 \text{ ft.}^3}{175 \text{ ft.}^2} = +6.22 \text{ ft.}$$

$$\bar{y} = \frac{\sum y \Delta A}{A} = \frac{1546 \text{ ft.}^3}{175 \text{ ft.}^2} = +8.83 \text{ ft.}$$

Component	Area (ΔA) (ft.²)	x (ft.)	$x\Delta A$ (ft.³)	y (ft.)	$y\Delta A$ (ft.³)
(a)	27	6	162	14	378
(b)	108	4.5	486	6	648
(c)	40	11	440	13	520
	$\sum \Delta A = 175$ ft.²		$\sum x\Delta A = 1088$ ft.³		$\sum y\Delta A = 1546$ ft.³

Example Problem 7.3b (Figure 7.13)

An alternate method that can be employed in solving this problem is referred to as the *negative area method*. In this example, fictitious components I and II are added to the wall panel to create a simple rectangular shape that measures 13' × 18'. Areas I and II will be treated as negative areas and subtracted from the gross rectangular area. Here, signs must be carefully accomodated in the algebraic summations done for the total area and *first moment of the areas*.

Note that the final results are identical to the centroidal distances found earlier.

$$\bar{x} = \frac{\sum x \Delta A}{A} = \frac{1088 \text{ ft.}^3}{175 \text{ ft.}^2} = +6.22 \text{ ft.}$$

$$\bar{y} = \frac{\sum y \Delta A}{A} = \frac{1546 \text{ ft.}^3}{175 \text{ ft.}^2} = +8.83 \text{ ft.}$$

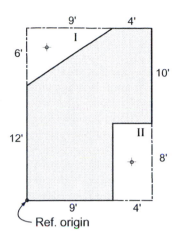

Figure 7.13   Negative area method.

Column	Area (ΔA) (ft.^{2})	x (ft.)	$x \Delta A$ (ft.^{3})	y (ft.)	$y \Delta A$ (ft.^{3})
(d)	+234	+6.5	+1521	+9	+2106
(e)	−27	+3	−81	+16	−432
(f)	−32	+11	−352	+4	−128
	$\sum \Delta A = 175$ ft.^{2}		$\sum x \Delta A = 1088$ ft.^{3}		$\sum y \Delta A = 1546$ ft.^{3}

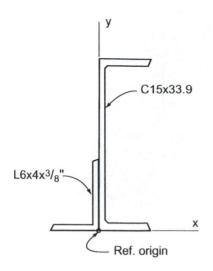

Figure 7.14 Composite cross-section.

Example Problem 7.4 (Figures 7.14 and 7.15)

Two standard, rolled steel sections are welded together to form a new cross-sectional shape. The angle and channel sections are placed back-to-back with the 6" leg of the angle being placed in the vertical orientation. Since dimensions for the rolled sections are often referenced from the straight back or leg, the reference origin in this example was selected to simplify the measurements. Standard tables found in the Appendix will be used to obtain the dimensional properties required for the solution.

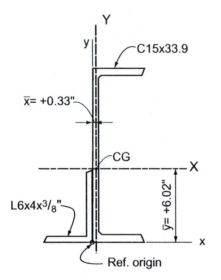

Figure 7.15 Location of the centroid.

Solution:

$$\bar{x} = \frac{\sum x \Delta A}{A} = \frac{4.44 \text{ in.}^3}{13.57 \text{ in.}^2} = 0.33 \text{ in.}$$

$$\bar{y} = \frac{\sum y \Delta A}{A} = \frac{81.7 \text{ in.}^3}{13.57 \text{ in.}^2} = 6.02 \text{ in.}$$

Component	A (in.^{2})	x (in.)	$x\Delta A$ (in.^{3})	y (in.)	$y\Delta A$ (in.^{3})
	3.61	−0.941	−3.40	+1.94	+7.00
	9.96	+0.787	+7.84	+7.5	+74.7
	$\sum \Delta A = 13.57 \text{ in.}^2$		$\sum x\Delta A = +4.44 \text{ in.}^3$		$\sum y\Delta A = +81.7 \text{ in.}^3$

Example Problem 7.5 (Figures 7.16 and 7.17)

A composite or built-up cross-section for a beam is fabricated using two $\frac{1}{2}'' \times 10''$ vertical plates with a C12 × 20.7 channel section welded to the top and a W12 × 16 section welded to the bottom as shown. Determine the location of the major x-axis using the center of the W12 × 16's web as the reference origin.

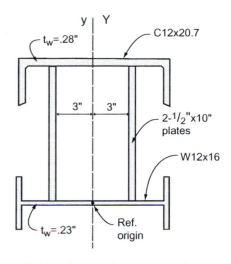

Figure 7.16   Composite cross-section.

Solution:

Since the cross-section is symmetrical about the y-axis, there is no need to determine the location of the major Y-axis; it's already established.

$$\bar{y} = \frac{\sum y \Delta A}{A} = \frac{110.23 \text{ in.}^3}{20.80 \text{ in.}^2} = 5.26''$$

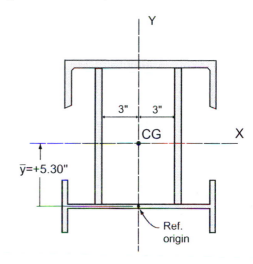

Figure 7.17   Centroid location.

Component	A (in.²)	y (in.)	$y\Delta A$ (in.³)
	6.09	$\frac{0.23}{2} + 10 + 0.28 - 0.70 = 9.70''$	59.07
	5 (ea.) = 10	$\frac{0.23}{2} + \frac{10}{2} = 5.12''$	51.2
	4.71	0	0
	$\sum \Delta A = 20.80 \text{ in.}^2$		$\sum y\Delta A = 110.3 \text{ in.}^3$

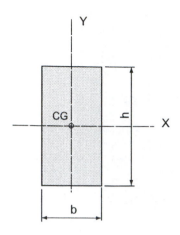

Figure 7.18   Rectangular beam cross-section.

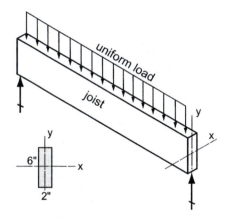

Figure 7.19   2" × 6" (full dimension) joist.

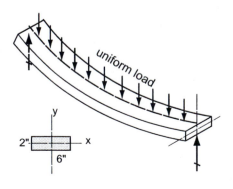

Figure 7.20   6" × 2" plank.

7.2   MOMENT OF INERTIA OF AN AREA

The *moment of inertia,* or *second-moment* as it is sometimes called, is a mathematical expression used in the study of the strength of beams and columns. It measures the effect of the cross-sectional shape of a beam on the beam's resistance to bending stress and deflection. *Elastic instability* or *buckling* of slender columns is also affected by the moment of inertia of its cross-section. The moment of inertia, or *I*-value, is a shape factor that quantifies the relative location of material in a cross-section in terms of effectiveness. A beam section with a large moment of inertia will have smaller stresses and deflections under a given load than one with a lesser *I*-value. A long, slender column will not be as susceptible to buckling laterally if the moment of inertia of its cross-section is sufficient. Moment of inertia is a measure of *cross-sectional stiffness,* whereas the modulus of elasticity *E* (studied in Chapter 6) is a measure of *material stiffness.*

The concept of moment of inertia is vital to understanding the behavior of most structures, and it is unfortunate that it has no accurate physical analogy or description. The dimensional unit for moment of inertia *I* is inches to the fourth power (in.^{4}) or millimeters to the fourth power (mm^4). Assuming the rectangular cross-section shown in Figure 7.18, a physical description can be given as follows:

Area = $b \times h$ (in.^{2}) and Perimeter = $2b + 2h$ (in.)

However, the moment of inertia for the cross-section is:

$$I = \frac{bh^3}{12} \text{ (in.}^4) \text{ (for a rectangular cross-section)}$$

The second moment of an area (area times a distance squared) is quite abstract and difficult to visualize as a physical property.

If we consider two prismatic beams made of the same material, but of different cross-sections, the beam whose cross-sectional area had the greater moment of inertia would have the greater resistance to bending. To have a greater moment of inertia does not necessarily imply, however, a greater cross-sectional area. Orientation of a cross-section with respect to its bending axis is crucial in obtaining a large moment of inertia.

A 2" by 6" rectangular cross-section is used as a joist in Figure 7.19 and as a plank in Figure 7.20. From experience, it is already known that the joist is much more resistant to bending/deflection than the plank. Like many structural elements, the rectangle has a strong axis (orientation) and a weak axis. It is far more efficient to load a cross-section so that bending occurs about its strong axis (*x*-axis).

It may help to understand the concept of moment of inertia if we draw an analogy based upon real inertia due to motion and mass. Imagine the two shapes shown in Figures 7.21 and 7.22 to be cut out of a $\frac{1}{2}$ in. steel plate and placed and spun about axis *x-x*. The two shapes have equal areas, but the one in Figure 7.21 has a much higher moment of inertia I_{xx} with respect to the axis of spin. It would be much harder to start it spinning and, once moving, much harder to stop. The same principle is involved when a figure skater spins on ice. With arms held close in, the skater will rotate rapidly; with arms outstretched (creating increased resistance to spin and/or more inertia), the skater slows down.

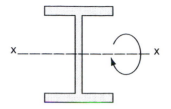

Figure 7.21   Wide-flange shape.

In our discussion of the method of finding the center of gravity of areas, each area was subdivided into small elemental areas (ΔA's) that were multiplied by their respective perpendicular distances (*x* or *y*) from the reference axis. The procedure is somewhat similar for determining the moment of inertia. The moment of inertia about the same axis would require, however, that each elemental area (ΔA) be multiplied by the square of the respective perpendicular distances (x^2 or y^2) from the reference axis.

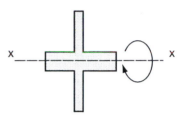

Figure 7.22   Cruciform shape.

A moment of inertia value can be computed for any shape with respect to any reference axis (Figure 7.23).

Suppose we wanted to find the moment of inertia of an irregular area, shown in Figure 7.23, about the *x*-axis. We would first consider the area to consist of many very small areas ΔA's. Considering the ΔA shown, its moment of inertia about the *x*-axis would be $y^2\,\Delta A$. But this product is only a small portion of the whole moment of inertia. Each ΔA making up part of the total area, when multiplied by the square of its corresponding moment arm *y* and summed together, will yield the moment of inertia of the entire area about the *x*-axis. The smaller the elements (ΔA's) selected, the more accurate the approximation will be. A calculus approach using an integration process would yield an exact answer because the ΔA's would be considered infinitely small.

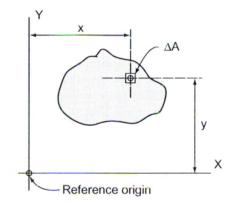

Figure 7.23   Moment of inertia of an irregular area.

> *The moment of inertia of an area about a given axis is defined as the sum of the products of all the elemental areas and the square of their respective distances to that axis.*

Therefore, the following two equations from Figure 7.23 are:

$$I_x = \sum y^2 \Delta A$$

$$I_y = \sum x^2 \Delta A$$

The moment of inertia *I* is always computed relative to a reference point or axis. *The most useful and often used reference axes are the centroidal x and y axes* (now you see the reason for finding the centroid of an area or cross-section). More importantly, it means that elements or areas that are relatively far away from the axis will contribute substantially more to an *I*-value than those that are close by.

Moments of inertia calculated by the integration method (exact) for some basic geometric shapes about their own respective x and y axes are shown in Table 7.2.

Table 7.2   Moments of inertia (exact) for selected simple geometric shapes.

Shape	Moment of Inertia
	$$I_x = \frac{bh^3}{12}$$ $$I_y = \frac{hb^3}{12}$$
	$$I_x = \frac{b_1 h_1^3 - b_2 h_2^3}{12}$$ $$I_y = \frac{h_1 b_1^3 - h_2 b_2^3}{12}$$
	$$I_x = \frac{bh^3}{36}$$ $$I_y = \frac{hb^3}{36}$$
	$$I_x = I_y = \frac{\pi r^4}{4} = \frac{\pi D^4}{64}$$
	$$I_x = I_y = \frac{\pi(D_1^4 - D_2^4)}{64}$$

Example Problem 7.6 (Figures 7.24 to 7.26)

Determine the I about the centroidal x-axis.

Solution:

This example will be solved using the *negative area method.*

$$I_{solid} = \frac{b_1 h_1^3}{12} = \frac{(6")(6")^3}{12} = 108 \text{ in.}^4$$

$$I_{voids} = 2 \times \frac{b_2 h_2^3}{12} = 2 \times \frac{(2")(4")^3}{12} = 21.3 \text{ in.}^4$$

$$\therefore I_x = 108 \text{ in.}^4 - 21.3 \text{ in.}^4 = 86.7 \text{ in.}^4$$

Note: I_x *is an exact value here, since the computations were based on an exact equation for rectangles (see Table 7.2).*

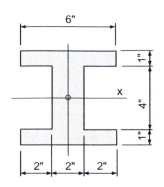

Figure 7.24   I cross-section.

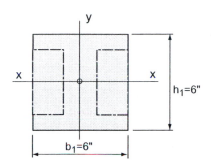

Figure 7.25   Solids.

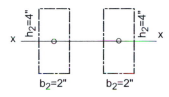

Figure 7.26   Voids.

7.3 MOMENT OF INERTIA OF COMPOSITE AREAS

In steel and concrete construction, the cross-sections usually employed for beams and columns are not like the simple geometric shapes shown in Table 7.2. Most structural shapes are a composite of two or more of the simple shapes combined in configurations to produce structural efficiency or construction expediency. We call these shapes *composite areas* (see Figure 7.27).

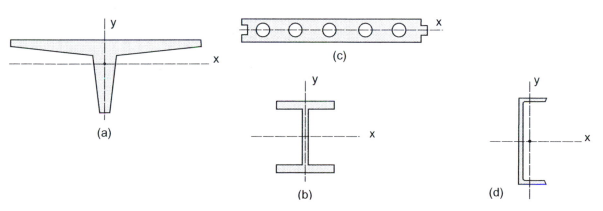

Figure 7.27   (a) Precast concrete T-section. (b) Steel wide-flange section. (c) Precast concrete plank. (d) Steel channel section.

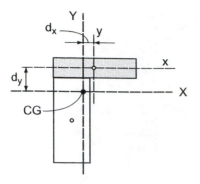

Figure 7.28   Transfer of the top flange area about the x-axis.

In structural design, the moment of inertia about the centroidal axis of the cross-section is an important section property. Since moments of inertia can be computed with respect to any reference axis, a standard reference was necessary in developing consistency when comparing the cross-sectional stiffness of one cross-section relative to another.

The *parallel axis theorem* provides a simple way to compute the moment of inertia of a shape about any axis parallel to the centroidal axis (see Figure 7.28). The principle of the parallel axis theorem may be stated as:

> *The moment of inertia of an area with respect to any axis not through its centroid is equal to the moment of inertia of that area with respect to its own parallel centroidal axis plus the product of the area and the square of the distance between the two axes.*

The parallel axis theorem expressed in equation form is:

$$I_x = \sum I_{xc} + \sum A d_y^2$$

where:

I_x = moment of inertia of the total cross-section about the major centroidal x-axis (in.^{4} or mm^4)

I_{xc} = moment of inertia of the component area about its own centroidal x-axis (in.^{4} or mm^4)

A = area of the component (in.^{2} or mm^2)

d_y = perpendicular distance between the major centroidal x-axis and the parallel axis that passes through the centroid of the component (in. or mm)

The transfer formula for the major y-axis is expressed as:

$$I_y = \sum I_{yc} + \sum A d_x^2$$

I_{yc} = moment of inertia of the component area about its own centroidal y-axis (in.^{4} or mm^4)

A = area of the component (in.^{2} or mm^2)

d_x = perpendicular distance between the major centroidal y-axis and the parallel y-axis that passes through the centroid of the component (in. or mm)

Example Problem 7.7 (Figures 7.29 and 7.30)

Determine the moment of inertia I_x about the centroidal x-axis.

Note: *This is identical to Example Problem 7.6.*

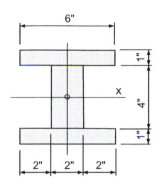

Figure 7.29 I cross-section.

Solution:

Instead of the negative area method, the *transfer formula* will be used, employing the parallel axis theorem.

The cross-section will be divided into three components as shown (two flanges and the web).

Each component of the composite area has its own centroid and is denoted as x_c, and the major centroidal axis of the entire cross-section is X. Note that the major centroidal axis X coincides with the component axis x_{c2} of component 2.

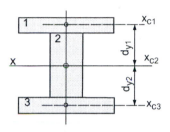

Figure 7.30 Component centroids to the major x axis.

As composite areas become more complex, it may be advisable to use a tabular format to minimize error and confusion.

The moment of inertia of the entire composite area is:

$$I_x = \sum I_{xc} + \sum Ad_y^2$$
$$= 11.7 \text{ in.}^4 + 75.0 \text{ in.}^4 = 86.7 \text{ in.}^4$$

The concept of the transfer formula involves the additional inertia required to merge or transfer an axis from one location to another. The term Ad_y^2 represents the additional inertia developed about the major axis X due to components 1 and 2.

Component	I_{xc} (in.^{4})	A (in.^{2})	d_y (in.)	Ad_y^2 (in.^{4})
(flange, 6" × 1", top)	$\dfrac{bh^3}{12} = \dfrac{(6")(1")^3}{12} = 0.5$	6	2.5	+37.5
(web, 2" × 4")	$\dfrac{(2")(4")^3}{12} = 10.67$	8	0	0
(flange, 6" × 1", bottom)	$= 0.5$	6	-2.5	$+37.5$
	$\sum I_{xc} = 11.7 \text{ in.}^4$			$\sum Ad_y^2 = 75.0 \text{ in.}^4$

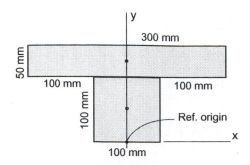

Figure 7.31   T cross-section.

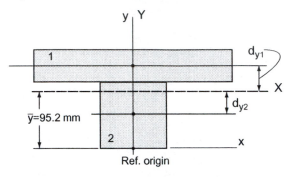

Figure 7.32   Centroidal distances to the major x axis.

Example Problem 7.8 (Figures 7.31 and 7.32)

Determine the moment of inertia about the centroidal x and y axes for the composite area shown.

Solution:

Usually the first step in determining the moment of inertia about the major or centroidal axis requires the determination of the centroid, particularly in unsymmetrical cross-sections.

$$\overline{x} = 0$$

$$\overline{y} = \frac{\sum y \Delta A}{A} = \frac{2.38 \times 10^6 \text{ mm}^3}{25 \times 10^3 \text{ mm}^2} = 95.2 \text{ mm}$$

Centroid Table

Component	ΔA (mm²)	x (mm)	xA (mm³)	y (mm)	$y\Delta A$ (mm³)
	15×10^3	0	0	125	1.875×10^6
	10×10^3	0	0	50	0.50×10^6
	$\sum \Delta A = 25 \times 10^3$		$\sum x\Delta A = 0$		$\sum y\Delta A = 2.38 \times 10^6$

$$d_{y1} = y_1 - \overline{y} = 125 \text{ mm} - 95.2 \text{ mm} = 29.8 \text{ mm}$$

$$d_{y2} = \overline{y} - y_2 = 95.2 \text{ mm} - 50 \text{ mm} = 45.2 \text{ mm}$$

$$I_x = \sum I_{xc} + \sum A d_y^2$$

$$= 11.46 \times 10^6 + 33.75 \times 10^6 = 45.2 \times 10^6 \text{ mm}^4$$

$$I_y = \sum I_{yc} + \sum A d_x^2$$

$$= 120.83 \times 10^6 + 0 = 120.8 \times 10^6 \text{ mm}^4$$

Moment of Inertia Table

Component	A (mm²)	I_{xc} (mm⁴)	d_y (mm)	Ad_y^2 (mm⁴)	I_{yc} (mm⁴)	d_x (mm)	Ad_x^2 (mm⁴)
(50, 300)	15×10^3	$\dfrac{(300)(50)^3}{12} =$ 3.125×10^6	29.8	13.32×10^6	$\dfrac{(50)(300)^3}{12} =$ 112.5×10^6	0	0
(100, 100)	10×10^3	$\dfrac{(100)(100)^3}{12} =$ 8.33×10^6	45.2	20.43×10^6	$\dfrac{(100)(100)^3}{12} =$ 8.33×10^6	0	0
		$\sum I_{xc} =$ 11.46×10^6 mm⁴		$\sum Ad_y^2 =$ 33.75×10^6 mm⁴	$\sum I_{yc} =$ 120.83×10^6 mm⁴		$\sum Ad_x^2 =$ 0

Example Problem 7.9 (Figures 7.33 and 7.34)

Determine the moment of inertia for the unsymmetrical beam cross-section about the centroidal x and y axes.

Solution:

Again, the first step is to locate the centroid of the cross-section.

$$\bar{x} = \frac{\sum x \Delta A}{A} = \frac{33.5 \text{ in.}^3}{11 \text{ in.}^2} = 3.05 \text{ in.}$$

$$\bar{y} = \frac{\sum y \Delta A}{A} = \frac{11.5 \text{ in.}^3}{11 \text{ in.}^2} = 1.05 \text{ in.}$$

$$d_{y_1} = y_1 - \bar{y} = 2" - 1.05" = 0.95"$$

$$d_{y_2} = \bar{y} - y_2 = 1.05" - 0.50" = 0.55"$$

$$d_{x_1} = \bar{x} - x_1 = 3.05" - 0.50" = 2.55"$$

$$d_{x_2} = x_2 - \bar{x} = 4.50 - 3.05" = 1.45"$$

$$I_x = \sum I_{xc} + \sum Ad_y^2$$

$$= 5.91 \text{ in.}^4 + 5.73 \text{ in.}^4 = 11.64 \text{ in.}^4$$

$$I_y = \sum I_{yc} + \sum Ad_x^2$$

$$= 28.91 \text{ in.}^4 + 40.73 \text{ in.}^4 = 69.64 \text{ in.}^4$$

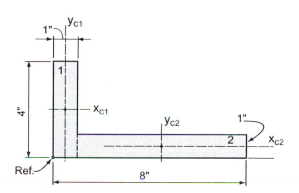

Figure 7.33   L-shaped cross-section.

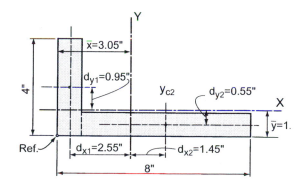

Figure 7.34   Centroidal distances to the major axes.

Centroid Table

Component	ΔA (in.^{2})	x (in.)	$x\Delta A$ (in.^{3})	y (in.)	$y\Delta A$ (in.^{3})
y_{c1} ... $4"$ ─x_{c1} ... $1"$	4	+0.5	+2	+2	+8
y_{c2} ... $1"$ ─ x_{c2} ... $7"$	7	+4.5	+31.5	+0.5	+3.5
	$\sum \Delta A = 11.0$		$\sum x\Delta A = 33.5$		$\sum y\Delta A = 11.5$

Moment of Inertia Table

Component	I_{xc} (in.^{4})	d_y (in.)	Ad_y^2 (in.^{4})	I_{yc} (in.^{4})	d_x (in.)	Ad_x^2 (in.^{4})
y_{c1} ... $4"$ ─x_{c1} ... $1"$	$\dfrac{(1)(4)^3}{12} = 5.33$	0.95	3.61	$\dfrac{(4)(1)^3}{12} = 0.33$	2.55	26.01
y_{c2} ... $1"$ ─ x_{c2} ... $7"$	$\dfrac{(7)(1)^3}{12} = 0.58$	0.55	2.12	$\dfrac{(1)(7)^3}{12} = 28.58$	1.45	14.72
	$\sum I_{xc} = 5.91$		$\sum Ad_y^2 = 5.73$	$\sum I_{yc} = 28.91$		$\sum Ad_x = 40.73$

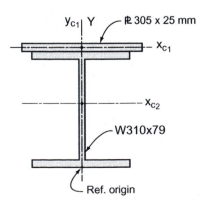

Figure 7.35 *Composite beam cross-section.*

Example Problem 7.10 (Figures 7.35 and 7.36)

Locate the centroidal x and y axes for the cross-section shown. Use the reference origin indicated and assume that the steel plate is centered over the flange of the wide-flange section. Compute the I_x and I_y about the major centroidal axes.

Solution:

$$\bar{y} = \frac{\sum y \Delta A}{A} = \frac{3.96 \times 10^6 \text{ mm}^3}{1.763 \times 10^4 \text{ mm}^2} = 225 \text{ mm}$$

$$\bar{x} = \frac{\sum x \Delta A}{A} = 0$$

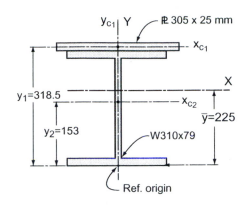

Figure 7.36   Centroidal distances from the reference origin.

Centroid Table

Component	ΔA (mm²)	y (mm)	yA (mm³)	x (mm)	$x\Delta A$ (mm³)
y_{c1} ⊢ x_{c1}	7,625	$\left(306 + \dfrac{25}{2}\right) = 318.5$	2.43×10^6	0	0
y_{c2} ⊢ x_{c2}	10,000	$\left(\dfrac{306}{2}\right) = 153$	1.53×10^6	0	0
	$\sum \Delta A = 17.63 \times 10^3$		$\sum y\Delta A = 3.96 \times 10^6$		$\sum x\Delta A = 0$

Moment of Inertia Table

Component	ΔA (mm²)	I_{xc} (mm⁴)	d_y (mm)	Ad_y^2 (mm⁴)	I_{yc} (mm⁴)	d_x (mm)	Ad_x^2 (mm⁴)
y_{c1} ⊢ x_{c1}	7,625	$\dfrac{(305)(25)^3}{12} = 0.397 \times 10^6$	$y_1 - \bar{y} = 93.5$	66.7×10^6	$\dfrac{(25)(305)^3}{12} = 59.1 \times 10^6$	0	0
y_{c2} ⊢ x_{c2}	10,000	177×10^6	$\bar{y} - y_2 = 72$	51.8×10^6	39.9×10^6	0	0
		$\sum I_{xc} = 177.4 \times 10^6$		$\sum Ad_y^2 = 118.5 \times 10^6$	$\sum I_{yc} = 99.0 \times 10^6$		$\sum Ad_x = 0$

$$I_x = \sum I_{xc} + \sum Ad_y^2$$

$$= 177.4 \times 10^6 + 118.5 \times 10^6 = 296 \times 10^6 \text{ mm}^4$$

$$I_y = \sum I_{yc} + \sum Ad_x^2$$

$$= 99.0 \times 10^6 + 0 = 99.0 \times 10^6 \text{ mm}^4$$

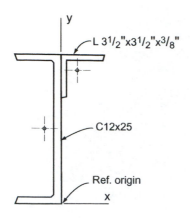

Figure 7.37   Composite beam cross-section

Example Problem 7.11 (Figures 7.37 and 7.38)

Determine the centroid of the composite cross-section shown, using a reference origin at the bottom right corner of the channel section. Then, compute the moment of inertia I_x and I_y about the major centroidal axes.

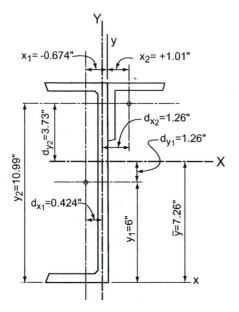

Figure 7.38   Measurements to the centroidal x and y axes.

Solution:

$$\bar{x} = \frac{\sum x \Delta A}{A} = \frac{-2.45 \text{ in.}^3}{9.83 \text{ in.}^2} = -0.25 \text{ in.}$$

$$\bar{y} = \frac{\sum y \Delta A}{A} = \frac{71.40 \text{ in.}^3}{9.83 \text{ in.}^2} = +7.26 \text{ in.}$$

Centroid Table

Component	ΔA (in.²)	x (in.)	$x \Delta A$ (in.³)	y (in.)	$y \Delta A$ (in.³)
⊐	7.35	−0.674	−4.95	6	+44.10
⌐	2.48	+1.01	+2.50	(12.00−1.01) = 10.99	+27.26
	$\sum \Delta A = 9.83$		$\sum x \Delta A = -2.45$		$\sum y \Delta A = 71.40$

Moment of Inertia Table

Component	ΔA (in.^{2})	I_{xc} (in.^{4})	d_y (in.)	Ad_y^2 (in.^{4})	I_{yc} (in.^{4})	d_x (in.)	Ad_x^2 (in.^{4})
⌐	7.35	144.0	$\bar{y} - y_1 =$ $(7.26 - 6.0) =$ 1.26	11.7	4.47	$x_1 - \bar{x} =$ $(0.624 - 0.25) =$ 0.424	1.32
⌐	2.48	2.87	$y_2 - \bar{y} =$ $(10.99 - 7.26) =$ 3.73	34.5	2.87	$x_2 + \bar{x} =$ $(1.01 + 0.25) =$ 1.26	3.94
		$\sum I_{xc} = 147.0$		$\sum Ad_y^2 = 46.2$	$\sum I_{yc} = 7.34$		$\sum Ad_x = 5.26$

$$I_x = \sum I_{xc} + \sum Ad_y^2 = 147.0 + 46.2 = 193.2 \text{ in.}^4$$
$$I_y = \sum I_{yc} + \sum Ad_x^2 = 7.3 + 5.3 = 12.6 \text{ in.}^4$$

7.4   RADIUS OF GYRATION

In the study of column behavior we will be using the term *radius of gyration* (r). The radius of gyration expresses the relationship between the area of a cross-section and a centroidal moment of inertia. It is a *shape factor* that measures a column's resistance to buckling about an axis. Assume that a W14 × 90 (W360 × 134) steel column is axially loaded until it fails in a buckling mode (Figure 7.39).

An examination of the buckled column will reveal that failure occurs about the y-axis. A measure of the column's ability to resist buckling is its radius of gyration (r) value. For the W14 × 90, the radius of gyration values about the x and y axes are:

$$\text{W14} \times 90: \quad r_x = 6.14'' \text{ (156 mm)}$$
$$\text{(W360} \times 134) \quad r_y = 3.70'' \text{ (94 mm)}$$

The larger the **r** value, the more resistance offered against buckling.

The radius of gyration of a cross-section (area) is defined as that distance from its moment of inertia axis at which the entire area could be considered as being concentrated (like a black hole in space) without changing its moment of inertia (see Figure 7.40).

If: $A_1 = A_2$ and $I_{x1} = I_{x2}$; then $I_x = Ar^2$ and

$$r_x = \sqrt{\frac{I_x}{A}} \quad \text{and} \quad r_y = \sqrt{\frac{I_y}{A}}$$

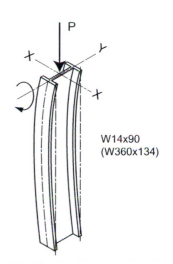

Figure 7.39   Column buckling about the weak axis.

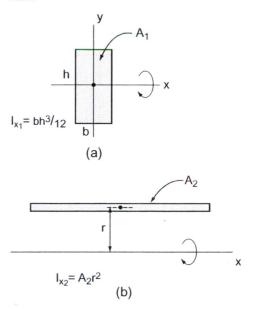

Figure 7.40   Radius of gyration for A_1 and A_2.

For all standard rolled shapes in steel, the radius of gyration values are given in the steel section properties table found in the Appendix.

Example Problem 7.12

Using the cross-section shown in Example Problem 7.11, we found that:

$$A = 9.83 \text{ in.}^2; \quad I_x = 193.2 \text{ in.}^4; \quad I_y = 12.6 \text{ in.}^4$$

The radii of gyration for the two centroidal axes are computed as:

$$r_x = \sqrt{\frac{I_x}{A}} = \sqrt{\frac{193.2 \text{ in.}}{9.83 \text{ in.}^2}} = 4.43 \text{ in.}$$

$$r_y = \sqrt{\frac{I_y}{A}} = \sqrt{\frac{12.6 \text{ in.}^4}{9.83 \text{ in.}^2}} = 1.13 \text{ in.}$$

If this section were used as a column, buckling would occur about the y-axis.

Example Problem 7.13 (Figure 7.41)

Two timber columns are used to support the end reaction of a roof beam. One column is a 6×6 S4S (150×150) and the other is a rectangular section with nominal dimensions of 4×10 S4S (100×250). If a column's capacity is limited by its *least (smallest) radius of gyration*, which section will support a larger axial load?

Solution:

The section properties of both sections can be obtained from standard timber tables found in the Appendix.

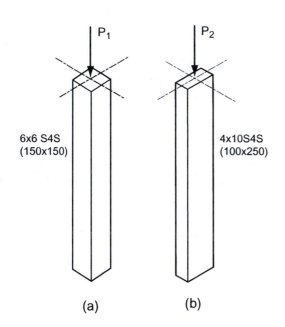

6×6 S4S (150×150)—nominal size

> $5.5" \times 5.5"$ (140 mm × 140 mm) - dressed size
>
> $A = 30.2 \text{ in.}^2$ ($19.6 \times 10^3 \text{ mm}^2$)
>
> $I_x = I_y = 76.3 \text{ in.}^4$ ($32 \times 10^6 \text{ mm}^4$)

4×10 S4S (100×250)—nominal size

> $3.5" \times 9.25"$ (89 mm × 235 mm) - dressed size
>
> $A = 32.4 \text{ in.}^2$ ($20.9 \times 10^3 \text{ mm}^2$)
>
> $I_x = 231 \text{ in.}^4$ ($96.3 \times 10^6 \text{ mm}^4$)
>
> $I_y = 33 \text{ in.}^4$ ($13.8 \times 10^6 \text{ mm}^4$)

Figure 7.41 Timber columns subjected to compressive loads.

P₁   6x6 S4S (150x150)   (a)   P₂   4x10S4S (100x250)   (b)

Computing the radii of gyration for each of the sections:

6×6 S4S:

$$r_x = r_y = \sqrt{\frac{I_{x,y}}{A}}$$

$$= \sqrt{\frac{76.3 \text{ in.}^4}{30.2 \text{ in.}^2}} = \sqrt{2.53 \text{ in.}^2} = 1.59 \text{ in. or}$$

$$r_x = r_y = \sqrt{\frac{I_{x,y}}{A}}$$

$$= \sqrt{\frac{32 \times 10^6 \text{ mm}^4}{19.6 \times 10^3 \text{ mm}^2}} = \sqrt{1633 \text{ mm}^2} = 40.4 \text{ mm}$$

4×10 S4S:

$$r_x = \sqrt{\frac{I_x}{A}} = \sqrt{\frac{231 \text{ in.}^4}{32.4 \text{ in.}^2}} = 2.67 \text{ in.};$$

$$r_x = \sqrt{\frac{96.3 \times 10^6 \text{ mm}^4}{20.9 \times 10^3 \text{ mm}^2}} = 67.9 \text{ mm}$$

$$r_y = \sqrt{\frac{I_y}{A}} = \sqrt{\frac{33 \text{ in.}^4}{32.4 \text{ in.}^2}} = 1.01";$$

$$r_y = \sqrt{\frac{13.8 \times 10^6 \text{ mm}^4}{20.9 \times 10^3 \text{ mm}^2}} = 25.6 \text{ mm}$$

The 4×10 S4S section has a larger total cross-sectional area but a smaller radius of gyration (r_y); therefore, it will have a smaller axial compressive capacity than the 6×6 S4S column. A more detailed discussion about the effect of the radius of gyration on column design can be found in Chapter 10.

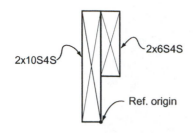

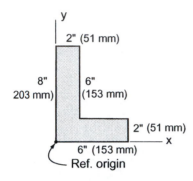

Problem 7.1.1

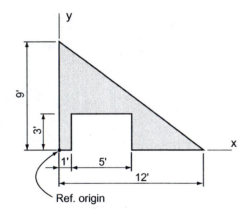

Problem 7.1.2

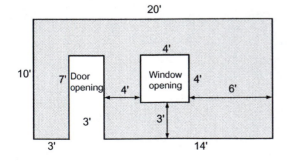

Problem 7.1.3

Supplementary Problems

Section 7.1—Center of Gravity and Centroids

7.1.1   A wood floor, supported by 2×10 S4S joists, are in need of bolstering to accommodate a larger load condition. One solution is to glue and nail a 2×6 S4S member to the existing 2×10s. Determine the new centroidal axis for this section.

7.1.2   Find the centroid of the L-shaped cross-sectional area shown. Use the reference at the lower left corner.

7.1.3   Determine the center of gravity (centroid) of the triangular wall panel shown. Use the negative area method for this problem.

7.1.4   A precast concrete wall panel with dimensions shown is to be hoisted into position at a building site. In hoisting the wall panel, it might be useful to know the location of its centroid. Determine the centroidal x and y axes referenced from the lower left corner.

Problem 7.1.4

7.1.5 A precast concrete T-section has the cross-sectional dimensions as shown. Using the reference origin as shown, determine the location of the centroidal x-axis for this composite section.

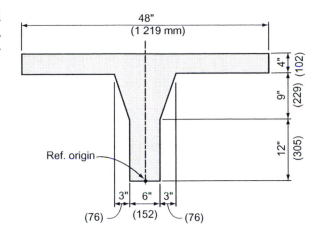

Problem 7.1.5

7.1.6 Find the centroid of the built-up steel section composed of a W12 × 87 (wide flange) with a $\frac{1}{2}$" × 14" cover plate welded to the top flange. See the steel table in the Appendix for information about the wide-flange section.

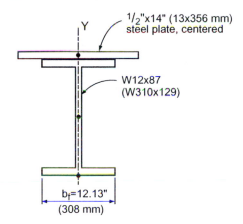

Problem 7.1.6

7.1.7 Determine the centroid of the built-up cross-section shown, assuming the reference origin is at the bottom of the channel section.

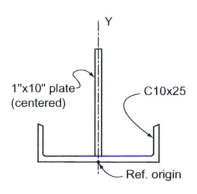

Problem 7.1.7

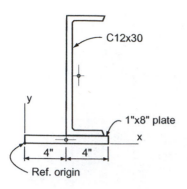

Problem 7.1.8

7.1.8 The bending capacity of a channel C12 × 30 steel beam is increased by welding on a 1" × 8" plate to the bottom of the channel. Using the reference origin shown, determine the location of the centroidal x and y axes for this composite section.

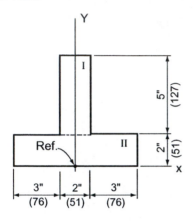

Problem 7.3.1

Section 7.3—Moment of Inertia

7.3.1 Determine the moment of inertia about the x and y axes for the composite cross-section shown.

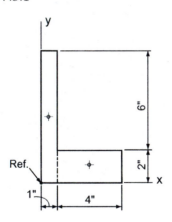

Problem 7.3.2

7.3.2 Find the I_x and I_y for the L-shaped cross-section shown.

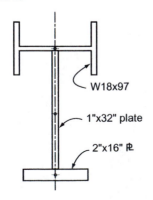

Problem 7.3.3

7.3.3 A built-up steel beam uses a W18 × 97 on its side with a 1" × 32" vertical plate and a 2" × 16" horizontal plate. Assuming the vertical plate is centered on the wide flange's web, calculate the I_x and I_y about the major centroidal axes of the cross-section.

7.3.4 A heavily loaded floor system uses a composite steel section as shown. A C15 × 40 channel section is attached to the top flange of the W18×50. Determine the I_x and I_y about the major centroidal axes using the cross-sectional properties given in the steel tables for standard rolled shapes (see Appendix).

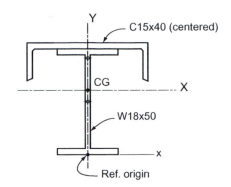

Problem 7.3.4

7.3.5 A built-up beam section (symmetrical about the y axis) is used to support large floor loads. Determine the moment of inertia about the major x axis.

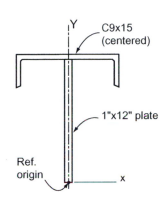

Problem 7.3.5

7.3.6 A built-up steel beam used to support a bridge road deck consists of a top cover plate welded to two C15 × 40 channel sections. Locate the major centroidal x axis and calculate the moment of inertia I_x and I_y about the major x and y axes.

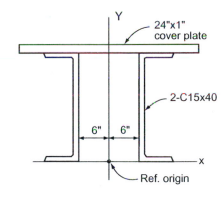

Problem 7.3.6

7.3.7 The bending capacity of a channel C10 × 15.3 steel beam is increased by welding on a C8 × 11.5 to the top flange as shown. Using the reference origin indicated, determine the location of the centroid of the composite section and solve for the I_x and I_y.

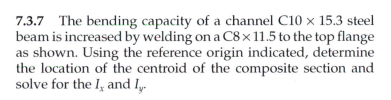
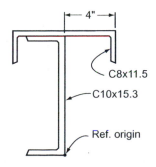

Problem 7.3.7

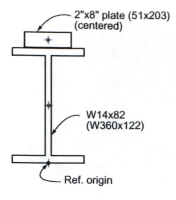

2"x8" plate (51x203)
(centered)

W14x82
(W360x122)

Ref. origin

Problem 7.3.8

7.3.8 Determine the I_x and I_y of the composite section shown.

8 Shear Forces and Bending Moments in Beams

Introduction

A beam is a long, slender structural member that resists loads usually applied transverse (perpendicular) to its longitudinal axis. These transverse forces cause the beam to bend in the plane of the applied loads, and internal stresses are developed in the material as it resists these loads. Beams are probably the most common type of structural member used in the roof and floors of a building of any size, as well as for bridges and other structural applications. Not all beams need to be horizontal; they may be vertical or inclined. They may have either one, two, or multiple reactions.

8.1 SHEAR AND BENDING MOMENT

The design of a beam entails the determination of size, shape, and material based on the bending stress, shear stress, and deflection due to the applied loads (see Figure 8.1). Six major

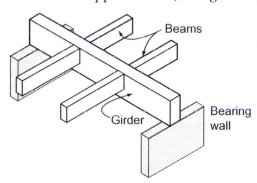

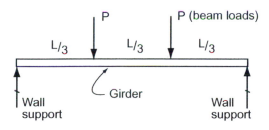

Figure 8.1 (a) Pictorial diagram of a beam/girder/wall arrangement. (b) FBD of the girder with beam loads.

Photo 8.1.1 Framing of a steel building under construction. Note the simple span and overhang beams.

Photo 8.1.2 Wood framing using glued-laminated beams and rafters.

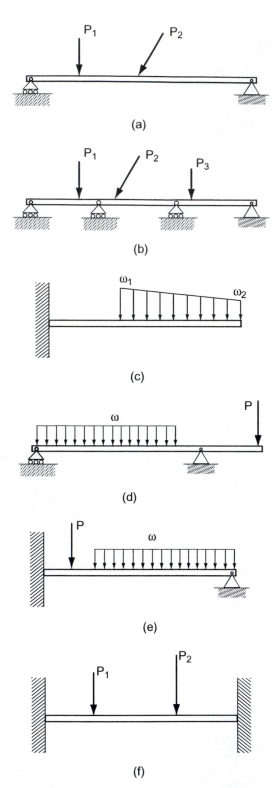

Figure 8.2   (a) Simply supported: two
supports. (b) Continuously supported: three or
more supports. (c) Cantilever: one end rigidly
supported. (d) Overhang: two supports—one
or both not located at the end. (e) Propped: two
supports—one end is fixed. (f) Restrained or
fixed: both supports are fixed, allowing no
rotation at the restrained ends.

beam classifications, based on their support conditions, are
illustrated in Figure 8.2.

Beam classification based on support conditions:

When a beam is subjected to any of these loading conditions, either singly or in any combination, the beam must
resist these loads and remain in equilibrium. In order for
the beam to remain in equilibrium, an internal force system must exist within the beam to resist the applied forces
and moments. Stresses and deflections in beams are functions of the internal reactions, forces, and moments. For
this reason, it is convenient to "map" these internal forces
and to construct diagrams that give a complete picture of
the magnitudes and directions of the forces and moments
that act throughout the beam length. These diagrams are
referred to as *load, shear (V),* and *moment (M)* diagrams
(see Figures 8.3 and 8.4).

■ The *load diagram* shown in Figure 8.3 is for a point
load at the free end of a cantilever beam. (The load
diagram is essentially the free-body diagram of the
beam.) In Figure 8.4, the load diagram is the FBD
of a double overhang beam with concentrated
loads at each end, plus a uniform load between
supports.

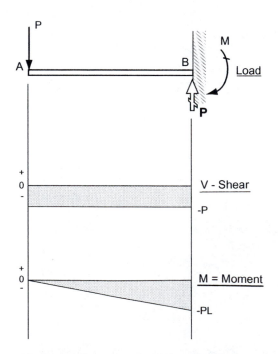

Figure 8.3   Load, shear (V), and moment (M)
diagrams.

- The *shear diagrams* shown in Figures 8.3 and 8.4 are graphs of the transverse shears along each respective beam's length.
- The *moment diagrams* shown in Figures 8.3 and 8.4 are graphs of the bending moment along the beam's length.

A shear diagram is a graph in which the abscissa (horizontal reference axis) represents distances along the beam length, and the ordinates (vertical measurements from the abscissa) represent the transverse shear at the corresponding beam sections. A moment diagram is a graph in which the abscissa represents distances along the beam, and ordinates represent the bending moment at the corresponding sections.

Shear (V) and moment (M) diagrams can be drawn by calculating values of shear and moment at various sections along the beam and plotting enough points to obtain a smooth curve. Such a procedure is rather time-consuming, and although it may be desirable for graphical solutions of certain structural problems, more rapid methods will be developed in Sections 8.3 and 8.4.

A *sign convention* is necessary for shear and moment diagrams if the results obtained from their use are to be interpreted conveniently and reliably.

By definition, the shear at a section is considered positive when the portion of the beam to the left of the section cut (for a horizontal beam) tends to be in the up position with respect to the portion to the right of the section cut, as shown in Figure 8.5.

Also by definition, the bending moment in a horizontal beam is positive at sections for which the top fibers of the

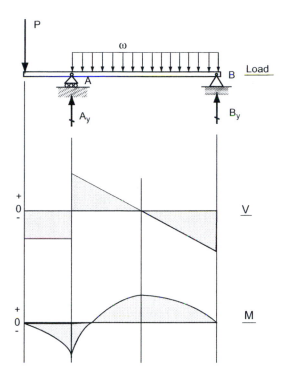

Figure 8.4   Load, shear, and moment diagrams for a single overhang beam.

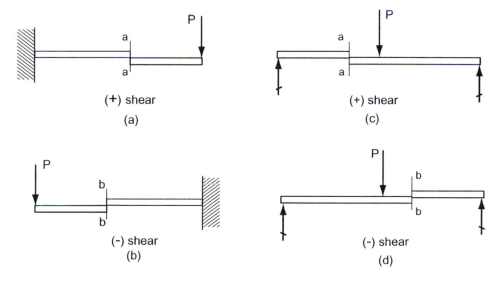

Figure 8.5   Sign convention for shear.

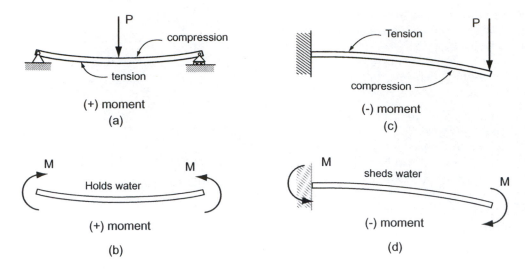

Figure 8.6   Sign convention for moment.

beam are in compression and the bottom fibers are in tension, as shown in Figure 8.6.

A positive moment generates a curvature that tends to hold water (concave-upward curvature), whereas a negative moment causes curvature that sheds water (concave-downward curvature).

This convention is a standard one for mathematics and is universally accepted. Since the convention is related to the probable deflected shape of the beam for a prescribed loading condition, it may be helpful to intuitively sketch the beam's deflected shape to assist in determining the appropriate moment signs (Figure 8.7).

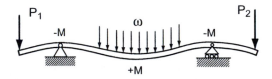

Figure 8.7   Exaggerated deflected shape due to loads on an overhang beam.

The overhang beam shown in Figure 8.7 exhibits a changing curvature that results in negative to positive to negative moments. The implication here is that there is a transverse section(s) in the beam span where the bending moment is zero to accommodate the required sign change. Such a section, termed the *inflection point(s)* or *point of inflection*, is almost always present in overhang and multiple-span beams.

An important feature of the sign convention used for shear and moment diagrams is that they differ from the conventions used in statics. When using the equations of equilibrium, forces directed up and to the right are positive, and counterclockwise moment tendencies are positive. The new sign conventions are used *only* for plotting the shear and moment diagrams. Make sure you do not confuse the two conventions.

8.2 EQUILIBRIUM METHOD FOR CONSTRUCTING SHEAR AND MOMENT DIAGRAMS

One basic method used in obtaining *shear* (*V*) and *moment* (*M*) diagrams is referred to as the *equilibrium method*. Specific values of *V* and *M* are determined from statics equations that are valid for appropriate sections of the member. (In these explanations, we shall assume that the member is a beam acted upon by downward loads, but actually the member could be turned at any angle.)

A convenient arrangement for constructing shear and moment diagrams is to draw a free-body diagram (FBD) of the entire beam and construct shear (*V*) and moment (*M*) diagrams directly below.

Unless the load is uniformly distributed or varies according to a known equation along the entire beam, no single elementary expression can be written for the shear (*V*) or moment (*M*) that applies to the entire length of the beam. Instead, it is necessary to divide the beam into intervals bounded by abrupt changes in the loading.

An origin should be selected (different origins may be used for different intervals), and positive directions should be indicated for the coordinate axes. Since *V* and *M* vary as a function of *x* along the beam length, equations for *V* and *M* can be obtained from free-body diagrams of portions of the beam (see Example Problem 8.1). Complete shear and moment diagrams should indicate values of shear and moment at each section where they are maximum positive and maximum negative. Sections where the shear and/or moment are zero should also be located.

Example Problem 8.1 (Equilibrium Method)

Draw the shear and moment diagram for a simply supported beam with a single concentrated load (Figure 8.8), using the equilibrium method.

Solution:

Solve for external reactions at *A* and *B*. Cut the beam through section *D-D*. Draw a FBD of each half of the beam.

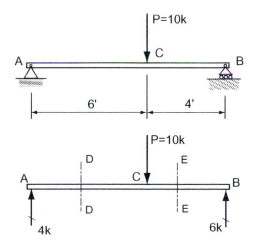

Figure 8.8   FBD of the entire beam (often referred to as the "Load" diagram).

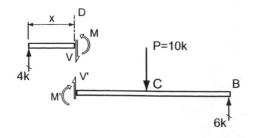

Figure 8.9   FBD of beam sections cut through D.

Examine segment *AD* from the FBD cut through *D* (Figure 8.9).

$$\sum F_y = +4k - V = 0; \quad V = 4\,k$$

$$\sum M_D = -(4k)(x) + M = 0; \quad M = 4\,k(x)$$

Note: *Shear* V *is a constant between* A *and* C. *The moment varies as a function of x (linearly) between* A *and* C.

$$@x = 0, \quad M = 0$$

$$@x = 6', \quad M = 24\text{ k-ft.}$$

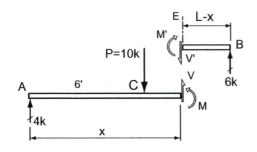

Figure 8.10   FBD of beam sections cut through E.

Examine segment *AE* from the FBD cut through *E* (Figure 8.10).

$$\sum F_y = +4k - 10\,k + V = 0; \quad V = 6\,k$$

$$\sum M_E = -4k(x) + (10\,k)(x - 6') + M = 0:$$

$$M = (60\text{ k-ft.}) - 6x$$

$$@x = 6', \quad M = 24\text{k-ft.}$$

$$@x = 10', \quad M = 0$$

Note: *Shear* V = 6 k *remains constant between* C *and* B. *The moment varies linearly, decreasing as x increases from* C *to* B.

Plotting the load, shear, and moment diagrams in order as shown in Figure 8.11, the following observations are made.

Shear (*V*) constant *A* to *C* (positive).
Shear constant *C* to *B* (negative).

$$V_{\max} = 6\,k \quad (-)\text{shear}$$

Moments are all positive.

Moments increase linearly from *A* to *C* ($x = 0$ to $x = 6'$).
Moment decreases linearly from *C* to *B* ($x = 6'$ to $x = 10'$).

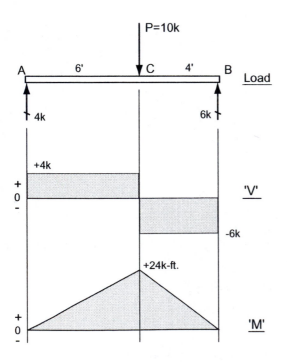

Figure 8.11   Load, shear, and moment diagrams.

Example Problem 8.2(Equilibrium Method)

Draw V and M diagrams for an overhang beam (Figure 8.12) loaded as shown. Determine the critical V_{max} and M_{max} locations and magnitudes.

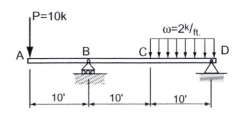

Figure 8.12   Loaded beam.

Draw a FBD of the beam and solve for the external support reactions. Based on your intuition, sketch the deflected shape of the beam (Figure 8.13) to assist in determining the signs for moment.

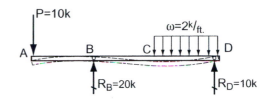

Figure 8.13   Load diagram.

Solution:

To find $V_{critical}$, examine sections (a) left and right of the concentrated loads and (b) at the beginning and end of the distributed loads (see Figures 8.14 to 8.19).

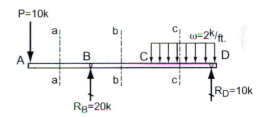

Figure 8.14   Cut sections a, b, and c between loads and reactions.

FBD at section cut *a-a:* (Figure 8.15)

Section *a-a*, $x = 0$ to $x = 10'$

$$\sum F_y = -10\,k + V = 0; \quad V = 10\,k \quad (-)\,shear$$

Just to the right of A, $V = 10\,k$

$$\sum M_{a-a} = +10k(x) - M = 0; \ M = 10x$$

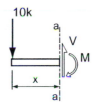

Figure 8.15   Section cut a-a.

FBD at section cut b-b: (Figure 8.16)

Section b-b, @ $x = 10'$ to $x = 20'$

Just right of B:

$$\sum F_y = -10\,k + 20\,k = 0; \ V = 10k \ (+)shear - constant$$

$$\sum M_B = +10k(x) - 20\,k(x-10') - M = 0;$$

$$M = -10\,k(x) + 200\ k\text{-}ft.$$

Just left of C; @ $x = 20'$

$$\sum F_y = -10\,k + 20\,k = 0; \quad V = 10\,k \quad (+)\,shear$$

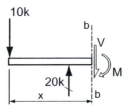

Figure 8.16   Section cut b-b.

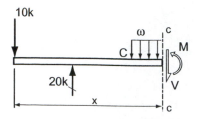

Figure 8.17   Section cut c-c.

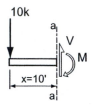

Figure 8.18

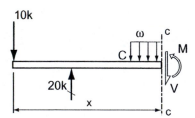

Figure 8.19

FBD at section cut c-c: (Figure 8.17)

Section c-c, @ $x = 20'$ to $x = 30'$

Just right of C:

$$\sum F_y = -10\,\text{k} + 20\,\text{k} = 0; \quad V = 10\,\text{k} \quad (+) \text{ shear}$$

Just left of D; @ $x = 30'$

$$\sum F_y = -10\,\text{k} + 20\,\text{k} - \left(2^{\,k}/_{\text{ft.}}\right)(x - 20') = 0$$

$$\therefore V = (50\,\text{k}) - 2x$$

Between C and D,

$$\sum M_{c-c} = +10\,\text{k}(x) - 20\,\text{k}(x-10')$$

$$+ \omega(x - 20')\left(\frac{x - 20'}{2}\right) + M = 0$$

$$M = -x^2 + 50x - 600\text{k-ft.}$$

M_{max} occurs at places where $V = 0$ or V changes sign. This occurs twice, at B and between C and D.

For M_{max} at B:

Examine a section cut just to the left or right of the concentrated load (Figure 8.18).

$$\sum M_a = +(10\,\text{k})(10') - M = 0;$$

$$M = -100\,\text{k-ft.}\ (-\text{moment})$$

For M_{max} between C and D:

Examine the shear equation at section cut c-c (Figure 8.19).

$$\sum F_y = -10\,\text{k} + 20\,\text{k} - \left(2^{\,k}/_{\text{ft.}}\right)(x - 20') - V = 0$$

$$\therefore V = 50 - 2x$$

But, M_{max} occurs at $V = 0$.

$$\therefore 0 = 50 - 2x; \quad x = 25'$$

$$\sum M_c = +(10\,\text{k})(25') - (20\,\text{k})(15')$$

$$+ \left(2^{\,k}/_{\text{ft.}}\right)(5')(25') + M = 0$$

$$M_{\text{max}} = 25\,\text{k-ft.}\ (+\text{moment})$$

Note: *Beams with one overhang end develop two possible* M_{max} *values.*

$$\therefore M_{\text{critical}} = 100\,\text{k-ft.}\ @\ B\ (-\text{moment})$$

Construct the resulting shear and moment diagrams (Figure 8.20).

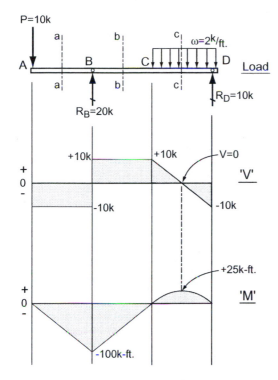

Figure 8.20   Load, shear, and moment diagrams.

8.3   RELATIONSHIP BETWEEN LOAD, TRANSVERSE SHEAR, AND BENDING MOMENT

The construction of shear and moment diagrams by the *equilibrium method* is quite time-consuming, particularly when a large number of section cuts must be considered. The mathematical relationships between loads, shears, and moments can be used to simplify the construction of such diagrams. These relationships can be obtained by examining a free-body diagram of an elemental length of a beam, as shown in Figures 8.21 and 8.22.

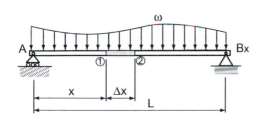

Figure 8.21   Beam with a generalized load.

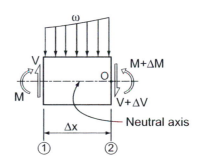

Figure 8.22   An elemental section of the beam.

In this example, we will assume a simply supported beam loaded with a varying distributed load (Figure 8.21). Detach a small length (Δx) of the beam between sections (1) and (2). Draw a FBD of the beam segment with an elemental length of the beam segment Δx.

V = shear at the left; (1)

$V + \Delta V$ = shear at the right; (2)

ΔV = change in shear between sections (1) and (2)

The beam element must be in equilibrium, and the equation $\left[\Sigma F_y = 0 \right]$ gives:

$$\Sigma F_y = +V - \omega(\Delta x) - (V + \Delta V) = 0$$

$$+V - \omega \Delta x - V - \Delta V = 0$$

$$\Delta V = -\omega \Delta x$$

$$\frac{\Delta V}{\Delta x} = -\omega$$

Note: *The negative sign represents a negative slope for this particular load condition.*

The preceding equation indicates that, at any section in the beam, the slope of the shear diagram is equal to the intensity of the loading.

If we examine the shear on the beam between points x_1 and x_2 (Figure 8.23), we obtain:

$$V = \omega \Delta x$$

But:

$$\Delta V = V_2 - V_1$$

and $\Delta x = x_2 - x_1$

$$\therefore (V_2 - V_1) = \omega(x_2 - x_1)$$

The interpretation of this equation says that:

The change in shear between sections at x_1 and x_2 is equal to the area under the load diagram between the sections (1) and (2).

Another equation of equilibrium about point 0 for Figure 8.22 can be written as:

$$\Sigma M_o = -V\Delta x - M + (M + \Delta M) + \omega(\Delta x)\frac{\Delta x}{2} = 0$$

$$-V\Delta x - M + M + \Delta M + \frac{\omega \Delta x^2}{2} = 0$$

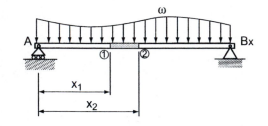

Figure 8.23   Section of beam between points ① and ②.

If Δx is a small value, the square of Δx becomes negligible.

$$\therefore \Delta M = V \Delta x$$

The preceding equation indicates that at any section in the beam the slope of the moment diagram is equal to the shear. Again, examining the beam between points (1) and (2) of Figure 8.22:

$$\Delta M = M_2 - M_1; \quad \text{where:} \quad \Delta x = x_2 - x_1$$

$$\therefore M_2 - M_1 = V(x_2 - x_1)$$

This equation states that:

The change in moment between any two sections, (1) and (2), is given by the area under the shear diagram between corresponding sections (1) and (2).

In Example Problems 8.1 and 8.2, the load, shear, and moment diagrams were plotted in that specific hierarchical order: Load $\rightarrow V \rightarrow M$. The V and M diagrams have a relationship derived from the previous diagram as shown above. Two additional diagrams, the slope and deflection diagrams, can be generated in a related fashion.

The relationship that exists between all five diagrams is shown in Figure 8.24.

Load: $\qquad \omega = \dfrac{\Delta V}{\Delta x}$

Shear V: $\qquad V = \dfrac{\Delta M}{\Delta x}$

Moment M: $\qquad M = \dfrac{\Delta \theta}{\Delta x} EI;$

where: E = modulus of elasticity; I = moment of inertia

Slope θ: $\qquad \theta = \dfrac{\Delta y}{\Delta x}$

Deflection y: $\qquad y$

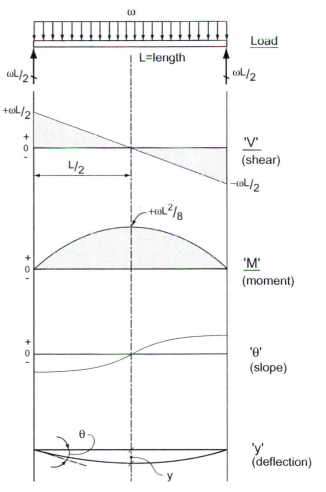

Figure 8.24 Relationship of load, shear, moment slope, and deflection diagrams.

8.4 SEMI-GRAPHICAL METHOD FOR CONSTRUCTING SHEAR AND MOMENT DIAGRAMS

The two expressions developed in the previous section, $(V_2 - V_1) = \omega(x_2 - x_1)$ and $M_2 - M_1 = V(x_2 - x_1)$, can be used to draw shear and moment diagrams and compute values of shear and moment at various sections along the beam as needed. This method is often referred to as the *semi-graphical method*.

Also necessary before attempting the semi-graphical method is an understanding of basic curves and curve relationships as shown in Figures 8.25 and 8.26.

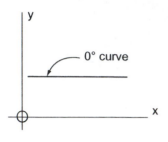

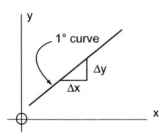

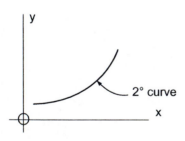

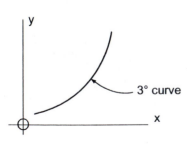

Figure 8.25   Basic family of curves.

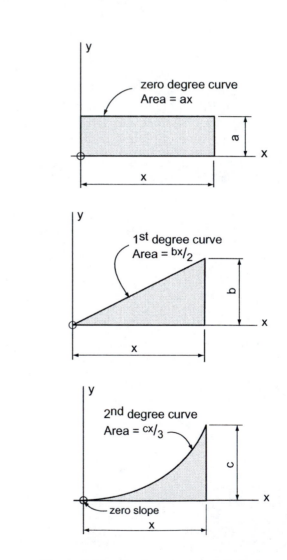

Figure 8.26   Basic curves and their properties.

Zero-degree curve

$y = c$

$c = $ constant

1st-degree curve

Straight line—may be uniformly increasing or decreasing.

$$\text{Slope} = \frac{\Delta y}{\Delta x}$$

$$y = cx$$

$$c = \text{constant}$$

2nd-degree curve

Parabolic—increasing (approaching verticality) or decreasing (approaching horizontality).

$y = kx^2 + c$

3rd-degree curve

Steeper than 2nd-degree curve.

$y = kx^3 + k'x^2 + \ldots$

Zero-degree curve

A zero-degree curve may represent a uniformly distributed load or the area of a shear diagram. $x = $ any point x along the beam.

1st-degree curve

A 1st-degree curve may represent triangular loading, the area under the shear diagram for a uniform load, or the area under the moment diagram for a concentrated load.

2nd-degree curve

A 2nd-degree curve usually represents the area of a shear diagram due to a triangular load distribution, or it could represent the moment diagram for a uniform load distribution.

General Considerations for Drawing V and M Diagrams:

1. When all loads and reactions are known, the shear and moment at the ends of the beam can be determined by inspection.
2. At a simply supported or pinned end, the shear must equal the end reaction, and the moment must be zero.
3. Both shear and moment are zero at a free end of a beam (cantilever beam or overhang beam).
4. At a built-in or fixed-end beam, the reactions are equal to the shear and moment values.
5. Load, shear, and moment diagrams are usually drawn in a definite sequence with the load diagram on top, followed by the shear diagram directly beneath it, and the moment diagram below the shear diagram.

6. When positive directions are chosen as upward and to the right, a uniformly distributed load acting down will give a negative slope in the shear diagram, and a positive distributed load (one acting upward) will result in a positive slope.

7. A concentrated force produces an abrupt change in shear.

8. The change in shear between any two sections is given by the area under the load diagram between the same two sections: $(V_2 - V_1) = \omega(x_2 - x_1)$

9. The change of shear at a concentrated force is equal to the concentrated force.

10. The slope at any point on the moment diagram is given by the shear at the corresponding point on the shear diagram; a positive shear represents a positive slope and a negative shear represents a negative slope.

11. The rate of increase or decrease in the moment diagram slope is determined by the increasing or decreasing areas in the shear diagram.

12. The change in moment between any two sections is given by the area under the shear diagram between corresponding sections:

$$M_2 - M_1 = V(x_2 - x_1)$$

13. A moment couple applied to a beam will cause the moment to change abruptly by an amount equal to the moment of the couple.

Example Problem 8.3 (Shear and Moment Diagrams)

Beam *ABC* is loaded with a single concentrated load as shown. This is the identical problem to Example Problem 8.1. Construct the shear (*V*) and moment (*M*) diagrams using the semi-graphical method (Figure 8.27).

Solution:

Load, shear, and moment diagrams come in a definite order because of their mathematical relationships (see Figure 8.25).

Draw a FBD of the beam and solve for the external support reactions. This FBD is the load diagram. By inspection, the shear at end *A* is +4 k.

Between *A* and *C*, there is no load shown on the load diagram. Therefore:

$$\omega = 0$$
$$V_2 - V_1 = \omega(x_2 - x_1)$$
$$\therefore V_2 - V_1 = 0$$

There is no change in shear between *A* and *C* (the shear is constant). At *C*, the 10 k concentrated load causes an abrupt change in shear, from +4 k to −6 k. The total shear change equals the magnitude of the concentrated load.

Between *C* and *B*, no load exists; therefore, there is no change in shear. The shear remains a constant −6 k.

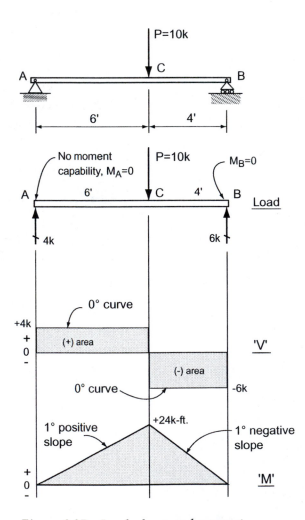

Figure 8.27   Load, shear, and moment diagrams.

At support B, an upward 6 k force returns the shear V to zero. There is no resultant shear at the very end of the beam.

The moment at pin and roller supports is zero; pins and rollers have no capacity to resist moment. The change in moment between any two points on a beam equals the area under the shear curve between the same two points:

$$M_2 - M_1 = V(x_2 - x_1)$$

Between A and C, the area under the shear curve is the area of a rectangle:

Area = 6' × 4 k = 24 k-ft.

Since the shear area is positive, the change in moment will occur along a positive, increasing curve. The change in moment is uniform (linearly increasing).

Shear diagram $\longrightarrow$	**Moment diagram**
(0° curve)	(1° curve)
(+ area)	(+ slope)

From C to B, the area of the shear diagram is:

Area = 4' × 6 k = 24 k-ft.

The change in moment from C to B is 24 k-ft.

Shear area $\longrightarrow$	**Moment diagram**
(0° curve)	(1° curve)
(− area)	(− slope)

Since the shear area is negative, the slope of the moment curve is negative.

The moment at B should go back to zero since no moment capability exists at the roller support.

Example Problem 8.4

Construct the V and M diagrams for the girder that supports three concentrated loads as shown in Figure 8.28.

Solution:

Draw the FBD of the girder and solve for the support reaction at each end. Construction lines should be drawn beneath the FBD at locations where loads occur.

In plotting the shear diagram, follow the direction for each force (including the support reactions) and maintain a constant shear until another concentrated load is encountered. The constant shear is a result of having no load areas that can cause a change in shear.

Beginning at the left support, the reaction pushes the shear to a magnitude of +11 k. No loads occur between A and B so the shear remains constant until the concentrated load at B pushes the shear down by 5 k to −6 k. Between B and C, the shear is constant until changed by the concentrated load of

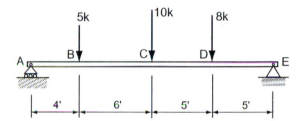

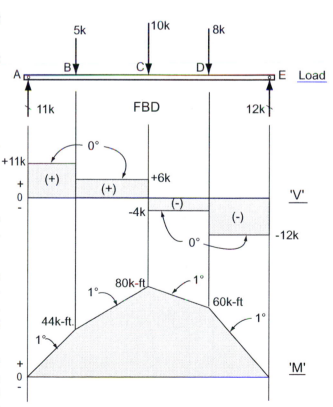

Figure 8.28   Load, V, and M diagrams.

-10 k. The resulting shear goes from $+6$ k to -4 k, reflecting a change in shear equal to 10 k (the concentrated load).

The shear remains constant at -4 k between C and D until the -8 k concentrated load pushes the shear down to -12 k. No load area exists between D and E, so the shear remains constant until the support reaction at E brings the shear diagram back to zero (a check on the condition of equilibrium).

Since the supports for the girder are a hinge and roller, the moment at the ends will be zero. Again, beginning from the left support, the moment changes from zero at A to the area contained under the shear diagram between A and B. The area is equal to: $A = (11\ \text{k}) \times (4') = 44$ k-ft. The shear area is a zero-degree curve, with a positive area. This results in a 1st-degree curve, with a positive slope in the moment diagram below.

Between B and C, the moment changes by the area $A = (6\ \text{k}) \times (6') = 36$ k-ft., with a 1st-degree curve, positive slope. The area under the shear diagram between C and D is equal to 20 k-ft., generating a 1st-degree curve with a negative slope (the shear area is negative). The last segment of shear area is 60 k-ft., producing a negative slope (1°) which brings the moment diagram back to zero (which it needs to be for a roller support).

Example Problem 8.5 (Semi-Graphical Method)

A cantilever beam supports a uniform load of $\omega = 2\ \text{kN}/\text{m}$ over its entire span, plus a concentrated load of 10 kN at the free end. Construct the V and M diagrams (Figure 8.29).

Solution:

Solve for the support reactions for the cantilever beam. Using the FBD of the beam, project construction lines from the free and supported ends.

Begin construction of the shear diagram by drawing the 10 kN shear force at the left (free) end. Between the left end at A and the support at B, the shear will be changing at a rate of $2\ \text{kN}/\text{m}$. The uniform load represents a negative area (producing a negative slope) with a zero-degree curve. In the shear diagram below, the curve will be a 1st-degree curve with a negative slope. The area under the load diagram between A and B is equal to:

$$A = \left(2\ ^{\text{kN}}/_{\text{m}}\right) \times (3\ \text{m}) = 6\ \text{kN}$$

A change of 6 kN, from the initial 10 kN shear at end A, results in a shear of -16 kN at support B. This shear value is then brought back to zero by the support reaction

$$B_y = +16\ \text{kN}$$

The free end of a cantilever has no capacity to develop a moment so the moment curve begins with $M_A = 0$. Examination of the shear diagram above reveals a 1st-degree curve with an increasing negative area. This increasing negative area

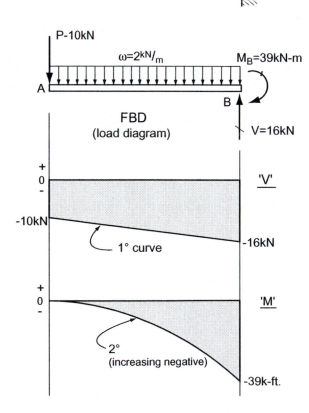

Figure 8.29   Load, V, and M diagrams.

(from left to right) will generate a 2nd-degree curve with an increasing negative slope in the moment diagram below. The area of the trapezoid in the shear diagram is:

$$A = \frac{(10 \text{ kN} + 16 \text{ kN})}{2} \times (3 \text{ m}) = 39 \text{ kN-m}$$

In plotting the moment curve, begin with $M_A = 0$ and draw a 2nd-degree curve that gets steeper (a tangent line to the curve becomes more vertical) as you go from left to right. The end point should result in a moment equal to $M_B = -39$ kN-m. This moment value corresponds to the moment computed earlier using the equation of equilibrium.

Example Problem 8.6 (Semi-Graphical Method)

Construct V and M diagrams for the simply supported beam *ABC*, which is subjected to a partial uniform load (Figure 8.30).

Solution:

Draw a FBD of the beam and solve for the external reactions. This is the *load diagram*.

By inspection, we see that at A, the reaction of 15 k is the shear. The shear at the end reaction point is equal to the reaction itself. Between A and B, there is a downward $(-)$ uniform load of 2 k/ft. The change in shear between A and B equals the area under the load diagram between A and B.

$$\text{Area} = 2 \text{ k/ft. } (10 \text{ ft.}) = 20 \text{ k}$$

Therefore, shear changes from $+15$ k to -5 k. V goes to zero at some distance x from A.

$$\underbrace{V_2 - V_1}_{15 \text{ k} \rightarrow 0} = \underbrace{\omega}_{2 \text{ k/ft.}} \underbrace{(x_2 - x_1)}_{x}; \qquad \therefore \; 15 \text{ k} = 2^{\text{k}}\!/_{\text{ft.}} (x);$$

$$x = 7.5'$$

Between B and C, no load exists on the beam, so no change in shear occurs. Shear is constant between B and C.

The moment at the pin and roller is 0.

Compute the area under the shear diagram between A and x.

$$\text{Area} = (^1\!/_2)(7.5 \text{ ft.})(15 \text{ k}) = 56.25 \text{ k-ft.}$$

The change in moment between A and x equals 56.25 k-ft., and since the shear area is $(+)$, the slope of the moment curve is $(+)$. A 1st-degree shear curve results in a 2nd-degree moment curve. The shear curve is positive, but since the area is decreasing, the corresponding moment slope is positive but decreasing.

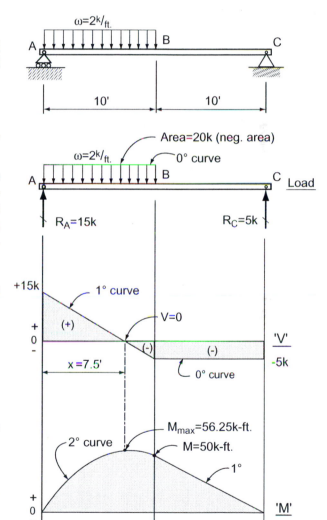

Figure 8.30   Load, V, and M diagrams.

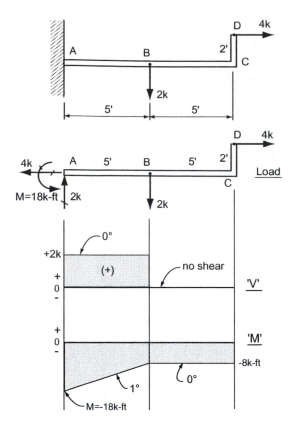

Figure 8.31   Load, V, and M diagrams.

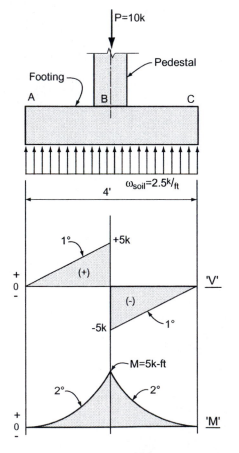

Figure 8.32   Load, V, and M diagrams.

Example Problem 8.7 (Figure 8.31)

For a cantilever beam with an upturned end, draw the load, shear, and moment diagrams.

Solution:

Determine the support reactions. Then move the horizontal 4-k force at C to align with the beam axis $A\,B\,C$.

Since the 4-k force is moved to a new line of action, a moment $M = 8$ k-ft. must be added to point C.

The V diagram is very simple in this example. The left support pushes up with a force of 2 k and remains constant until B, since no other loads are present between A and B. At B a 2-k downward-acting force brings the V back to 0, and it remains 0 all the way to C (no vertical loads occur between B and C).

The moment diagram starts with a moment at the left end because of the presence of the support moment $M = 18$ k-ft. Imagine the beam curvature in determining whether the $M = 18$ k-ft. is plotted in the positive or negative direction. Since the curvature due to bending results in tension on the top surface of the beam, the sign convention says this is a negative moment condition. Between A and B, the moment remains negative but with a positive slope of the 1st degree. There is no change in moment between B and C; therefore, the magnitude remains -8 k-ft., which corresponds to the applied moment at C.

Example Problem 8.8 (Figure 8.32)

The diagram shows a bearing load on a spread footing. Draw the load, shear, and moment diagrams of the figure shown.

Solution:

A typical spread footing supporting a column load develops unique V and M diagrams.

Between A and B, the shear changes from 0 to 5 k (which is the area under the upward-acting load of the soil bearing from A to B). Since the load envelope is positive, the slope of the shear diagram is positive. The column load at B causes an abrupt change to occur in the shear diagram. A positive load envelope between B and C again generates a 1st-degree positive slope to 0.

The moment at the left end of the footing is 0 and increases positively to a magnitude of 5 k-ft. at B. A decreasing negative slope is generated between B and C as the curve diminishes to 0 at C.

$$M_2 - M_1 = V(x_2 - x_1)$$

Area of triangle:

$$A = \tfrac{1}{2}(2')(5\text{ k}) = 5\text{ k-ft.}$$

Example Problem 8.9 (Figure 8.33)

A header beam spanning a large opening in an industrial building supports a triangular load as shown. Construct the V and M diagrams and label the peak values.

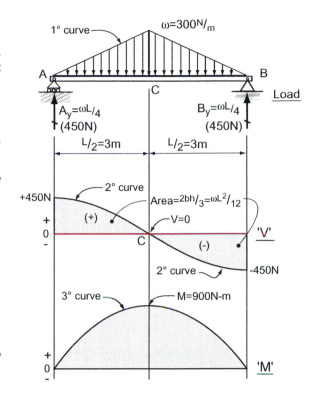

Figure 8.33   Load, V, and M diagrams.

Solution:

The triangular load is symmetrical and has a maximum load intensity at midspan equal to $\omega = 300\,^{N}/_{m}$.

Due to the symmetrical loading, the support reactions are equal.

$$A_y = B_y = \tfrac{1}{2} \times \left(\frac{L}{2}\right) \times (\omega) = \frac{\omega L}{4}$$

$$= \frac{\left(300\,^{N}/_{m}\right)(6\text{ m})}{4} = 450\text{ N}$$

Construct the V diagram:

- At the left support at A, the shear is equal to $\dfrac{\omega L}{4} = 450$ N

- Between A and the mid span at C, the load is changing at a linear rate (1st-degree curve). This triangular area (negative) produces a 2nd-degree curve in the shear diagram below with an increasing negative slope. The change in shear is equal to $\Delta V = \tfrac{1}{2}(3\text{ m})\left(300\,^{N}/_{m}\right) = 450$ N.

- Since the shear at A is 450 N and the change in shear between A and C is 450 N, the resulting shear at C is zero (0).

- From C to the right support at B, the shear changes by $\Delta V = \tfrac{1}{2}(3\text{ m})\left(300\,^{N}/_{m}\right) = 450$ N. Therefore, the shear at B is equal to -450 N. The curve generated between C and B is 2nd degree, with a negative-decreasing slope. Since the load area approaches zero at B, the slope of a line tangent to the shear curve at B should be horizontal.

In constructing the M diagram:

- Between A and C, the moment diagram changes by the area under the 2nd-degree curve in the shear diagram.

$$\Delta M = A = \tfrac{2}{3}\left(\frac{L}{2}\right)\left(\frac{\omega L}{4}\right) = \frac{\omega L^2}{12}$$

$$= \frac{\left(300\,^{N}/_{m}\right)(6\text{ m})^2}{12} = 900\text{ N-m}$$

- The moment curve between A and C is 3rd degree, with a positive, decreasing slope.

■ From C to the right support at B, the moment changes by

$$\Delta M = \frac{\omega L^2}{12} = \frac{\left(300\,^{N}\!/_{m}\right)(6\text{ m})^2}{12} = 900 \text{ N-m}$$

■ The moment curve between C and support B is 3rd degree, with a negative, increasing slope (slope is getting steeper).

Example Problem 8.10 (Figures 8.34 and 8.35)

A major compound roof beam over a classroom is composed of two beams joined by a mechanical hinge at D. Solve for the support reactions at A, C, and E, then construct the V and M diagrams.

Solution:

Support reactions at A, C, and E can be solved by separating the compound beam into two simpler beams: AD and DE.

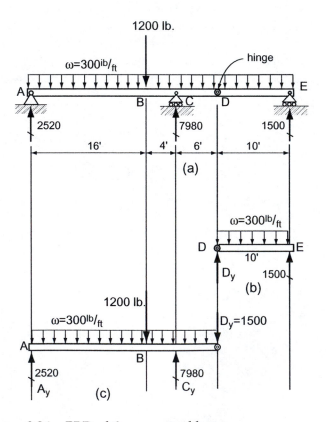

Figure 8.34   FBD of the compound beam.

Using the FBD of DE (Figure 8.34b), solve for the vertical support reactions D_y and E_y.

$$\sum M_D = -\left(300^{\text{lb.}}\!/_{\text{ft.}}\right)(10')(5') + E_y(10') = 0$$

$$\therefore E_y = 1500 \text{ lb.}$$

$$\sum F_y = +D_y - \left(300^{\text{lb.}}\!/_{\text{ft.}}\right)(10') + 1500 \text{ lb.} = 0$$

$$\therefore D_y = 1500 \text{ lb.}$$

Hinge reaction $D_y = 1500$ lb. is now shown as an applied force at D for beam ACD in Figure 8.34c. Summing moments about the support at A, the reaction at C_y can be solved.

$$\sum M_A = -\left(300^{\text{lb.}}\!/_{\text{ft.}}\right)(26')(13') - (1200 \text{ lb.})(16')$$

$$- (1500 \text{ lb.})(26') + C_y(20') = 0$$

$$\therefore C_y = 7980 \text{ lb.}$$

$$\sum F_y = +A_y - \left(300^{\text{lb.}}\!/_{\text{ft.}}\right)(26') - (1200 \text{ lb.})$$

$$- (1500 \text{ lb.}) + \left(\underset{C_y}{7980 \text{ lb.}}\right) = 0$$

$$\therefore A_y = +2520 \text{ lb.}$$

Reassemble the two beams into the original compound beam, including the solved support reactions. Since all external forces are now known, shear (V) and moment (M) diagrams can be constructed.

Draw construction lines from each support position, the concentrated load, and the hinge at D. These locations are key positions where changes in the V and M diagrams are likely to occur.

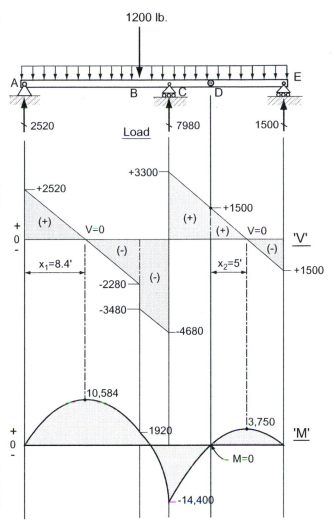

Figure 8.35   Load, V, and M diagrams.

For constructing the V diagram:

- Record a $+2520$ lb. shear at support A.
- Between A and the concentrated load at B, the shear diagram will change by
  $\Delta V = A = \left(300^{\text{lb.}}\!/_{\text{ft.}}\right)(16') = 4800 \text{ lb.}$
- From the $+2520$ lb. at A, the shear changes by 4800 lb. to position at -2280 lb. at B.
- At point B, the concentrated load causes an abrupt change of -1200 lb. with $V = -3480$ lb.
- Between B and C, the shear changes an additional $\Delta V = A = \left(300^{\text{lb.}}\!/_{\text{ft.}}\right)(4') = 1200 \text{ lb.}$ The shear is now equal to $V = -4680$ lb.
- At the support C, the reaction of $C_y = 7980$ lb. results in a shear of $V = +3300$ lb.
- The change in shear between C and D (the hinge) is equal to: $\Delta V = A = \left(300^{\text{lb.}}\!/_{\text{ft.}}\right)(6') = 1800 \text{ lb.}$ The shear changes from $V_C = +3300$ lb. to $V_D = +1500$ lb.

- Continuing from D to E, the change in shear is:
$\Delta V = A = \left(300\,{}^{lb.}\!/_{ft.}\right)(10') = 3000$ lb.
- The shear at E is $V_E = -1500$ lb., which is equal to the support reaction at E.

Constructing the M diagram:

- Begin with a zero moment at the left roller support at A.
- Between A and B, the shear diagram crosses the zero axis, indicating that a peak point in the moment diagram will occur.
- Determine the distance x where the shear crosses the zero axis by using the general equation: $\Delta V = \omega x$ (the area under the distributed load).

$$\Delta V = 2520 \text{ lb.} = \left(300\,{}^{lb.}\!/_{ft.}\right)(x)$$

$$\therefore x = \frac{2520 \text{ lb.}}{300\,{}^{lb.}\!/_{ft.}} = 8.4 \text{ ft.}$$

- The change in moment between A and distance $x = 8.4'$ is $\Delta M = \left({}^{1}\!/_{2}\right)(2520 \text{ lb.})(8.4') = 10{,}584$ lb.-ft. The moment curve in this section is 2nd degree with a decreasing positive slope.
- From $V = 0$ to point B, the moment changes by:

$$\Delta M = \left({}^{1}\!/_{2}\right)(2280 \text{ lb.})\left(16' - \underset{x}{8.4'}\right) = 8664 \text{ lb.-ft.}$$

The moment at B is equal to $M_B = 1920$ lb.-ft., with a 2nd degree, increasing negative slope.
- Between B and the support at C, the change in moment is equal to the area of the trapezoid:

$$\Delta M = \left(\frac{3480 + 4680}{2} \text{ lb.}\right)(4') = 16{,}320 \text{ lb.-ft.}$$

A 2nd-degree, increasing negative slope develops with the moment at C equal to:

$$M_C = (16{,}320 - 1920) \text{ lb.-ft.} = 14{,}400 \text{ lb.-ft.}$$

- The change in moment from C to D is:

$$\Delta M = \left(\frac{3300 \text{ lb.} + 1500 \text{ lb.}}{2}\right)(6')$$

$$= 14{,}400 \text{ lb.-ft.}$$

The moment at the hinge D is zero, as it should be.
- The change in shear from V_D to where it crosses the zero axis is 1500 lb. Distance x_2 is,

$$\Delta V = 1500 \text{ lb.} = \omega x_2 = \left(300\,{}^{lb.}\!/_{ft.}\right)x_2$$

$$\therefore x_2 = 5 \text{ ft.}$$

- The moment changes from $M_D = 0$ to

$$M = \left(\tfrac{1}{2}\right)(1500\ \text{lb.})(5') = 3750\ \text{lb.-ft.}$$

This is another peak in the moment diagram since V crossed the zero axis.

- For the last 5 ft. the moment changes by 3750 lb.-ft., which brings the moment back to zero at E. The curve is a 2nd-degree curve, with an increasing negative slope.

Example Problem 8.11 (Figures 8.36 and 8.37)

A compound beam with internal hinges is loaded as shown. Solve for the support reactions and draw the V and M diagrams.

Solution:

Separate the compound beam into the three simpler beams and draw an individual FBD for each. Begin the support re-action analysis of the middle section carrying the single concentrated load.

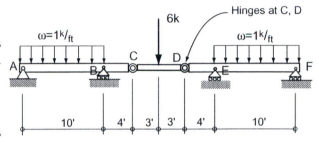

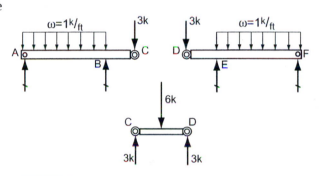

Figure 8.36 FBD of the compound beam.

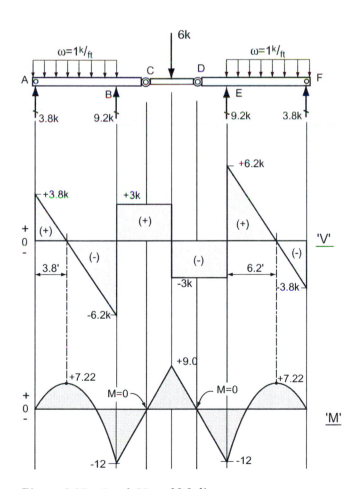

Figure 8.37 Load, V, and M diagrams.

Beam CD represents a simple beam with two pin supports, and beams ABC and DEF are single overhang beams. Beam section CD develops pin reactions of 3 k at each end. Each 3-k reaction force is applied equal and opposite to the end of the overhang beam as a concentrated load. The overhang beam's support reactions are then obtained by using the equations of equilibrium.

Once support reactions have been determined, reassemble the beam into its original condition with the solved reaction forces and begin construction of the V and M diagrams.

Remember, the hinges at C and D have no moment capability ($M = 0$).

The shear diagram crosses the 0 axis in five places; therefore, five peak points develop in the moment diagram. Moment is most critical at the support points B and E.

Supplementary Problems

Section 8.2—Equilibrium Method for V and M Diagrams

Construct the load, shear, and moment diagrams for the following beam conditions using the equilibrium method.

8.2.1

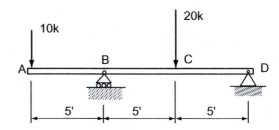

Problem 8.2.1

8.2.2

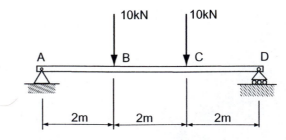

Problem 8.2.2

8.2.3

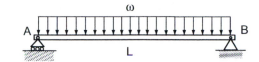

Problem 8.2.3

8.2.4

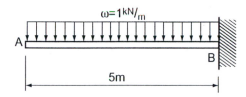

Problem 8.2.4

Section 8.4—Semi-Graphical Method for Constructing V
and M *Diagrams*

Construct the load, shear, and moment diagrams for the following beam conditions using the semi-graphical method.

8.4.1

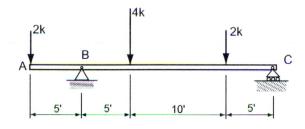

Problem 8.4.1

8.4.2

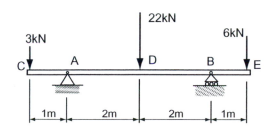

Problem 8.4.2

8.4.3

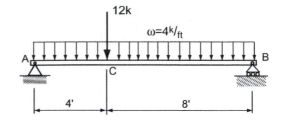

Problem 8.4.3

8.4.4

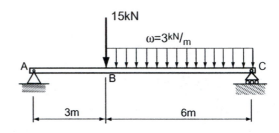

Problem 8.4.4

8.4.5

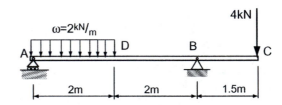

Problem 8.4.5

8.4.6

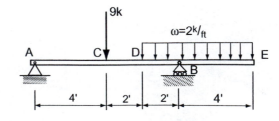

Problem 8.4.6

8.4.7

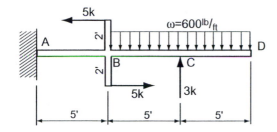

Problem 8.4.7

8.4.8

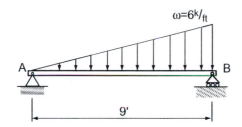

Problem 8.4.8

8.4.9

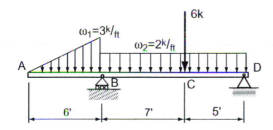

Problem 8.4.9

8.4.10

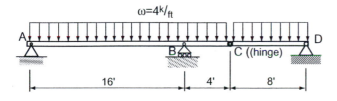

Problem 8.4.10

8.4.11

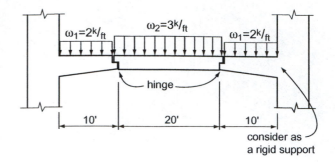

Problem 8.4.11

8.4.12

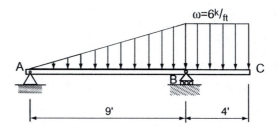

Problem 8.4.12

9

Bending and Shear Stresses in Beams

Introduction

One of the earliest studies concerned with the strength and deflection of beams was carried out by *Galileo Galilei*. Galileo was the first to discuss the bending strength of a beam (Figure 9.1). He thus became the founder of an entirely new branch of science: the theory of the strength of materials, which has played a vital part in modern engineering science.

Galileo started with the observation of a cantilever beam subjected to a load at the free end. He equated the statical moments of the external load with that of the resultant of the tensile stresses in the beam (which he assumed to be uniformly distributed over the entire cross-section of the beam—as in Figure 9.2) in relation to the axis of rotation (assumed to be located at the lower edge of the embedded cross-section). Galileo concluded that the bending strength of a beam was directly proportional to its width but proportional to the square of its height. However, as Galileo based his proposition merely on considerations of statics and did not yet introduce the notion of elasticity, propounded by Robert Hooke half a century later, he erred in the evaluation of the magnitude of the bending strength in relation to the tensile strength of the beam.

Two centuries after Galileo Galilei's initial beam theory, Charles-Augustin de Coulomb (1736–1806) and Louis-Marie-Henri Navier (1785–1836) finally succeeded in finding the correct answer. In 1773, Coulomb published a paper that discarded the fulcrum concept and proposed the triangular distribution shown in Figure 9.3, in which both the tensile and compressive stresses have the same linear distribution.

Figure 9.1   Cantilever loaded at the free end.
From Galileo Galilei Discorsi e Demo strazioni Matematiche, *Leyden, 1638. Drawing based on illustration in Schweizerische Bauzertung, Vol. 119.*

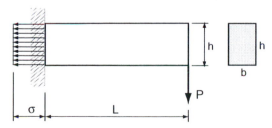

Figure 9.2   Flexure according to Galileo.
Redrawn from an illustration in Schweizerische Bauzertung, *Vol. 116.*

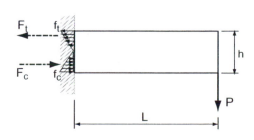

Figure 9.3   Flexure according to Coulomb.

297

9.1 FLEXURAL STRAIN

The accuracy of *Coulomb's theory* can be demonstrated by examining a simply supported beam subjected to bending. The beam is assumed to be (a) initially straight and of constant cross-section, (b) elastic and have equal moduli of elasticity in tension and compression, and (c) homogeneous—of the same material throughout. It is further assumed that a plane section before bending remains a plane after bending (see Figures 9.4 and 9.5).

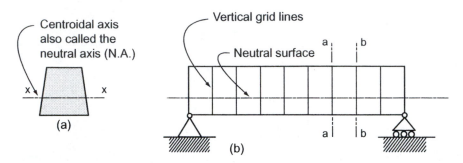

Figure 9.4   (a) Beam cross-section. (b) Beam elevation before loading.

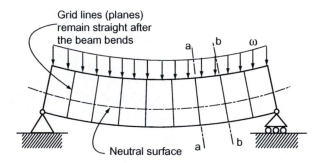

Figure 9.5   Beam bending under load.

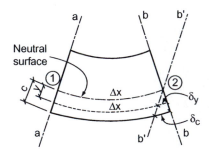

Figure 9.6   Beam section after loading.

For this to be strictly true, it is necessary that the beam be bent only with couples (no shear on transverse planes). The beam must be proportioned so that it will not buckle, and the loads must be applied so that no twisting (torsion) occurs.

Examining a portion of the bent beam between sections a-a and b-b (Figure 9.6), one observes that at some distance *c* above the bottom of the beam, the longitudinal elements (sometimes called *fibers*) undergo no change in length.

The curved surface (*1 - 2*) formed by these elements is referred to as the *neutral surface*, and the intersection of this surface with any cross-section is called the neutral axis of the cross-section. The neutral axis corresponds to the centroidal axis of a cross-section. All elements (fibers) on one side of the neutral surface are compressed, and those on the opposite side are in tension. For the simple beam shown in Figure 9.4, the portion of the beam above the neutral surface experiences compression, while the lower portion is undergoing tensile stressing.

The assumption is made that all longitudinal elements (fibers) have the same length initially before loading.

Referring again to Figure 9.6,

$$\frac{\delta_y}{y} = \frac{\delta_c}{c} \qquad \text{or} \qquad \delta_y = \frac{y}{c}\delta_c$$

where:

δ_y = deformation developed along fibers located a distance below the neutral surface

δ_c = deformation at the bottom surface of the beam—a distance c below the neutral surface

Since all elements had the same initial length, Δx, the strain of any element can be determined by dividing the deformation by the length of the element; the strain becomes

$$\varepsilon_y = \frac{y\varepsilon_c}{c}; \qquad \text{and} \qquad \varepsilon = \frac{f}{E}$$

which indicates that the strain of any fiber is directly proportional to the distance of the fiber from the neutral surface.

With the premise that the longitudinal strains are proportional to the distance from the neutral surface accepted, the assumption is now made that Hooke's law applies (which restricts stresses to magnitudes within the proportional limit of the material). Then the equation becomes

$$\frac{\varepsilon_y}{y} = \frac{f_y}{E_y y} = \frac{f_c}{E_c c}$$

The final result, if $E_c = E_y$ (constant), is

$$\frac{f_y}{y} = \frac{f_c}{c}$$

which verifies Coulomb's conclusion.

Redrawing the diagram shown in Figure 9.6 to include the compressive as well as the tensile deformations due to bending stress, we can use Hooke's law to explain the stress variations occurring on the cross-section (see Figure 9.7).

The deformation at the neutral axis is zero after bending; therefore, the stress at the neutral axis (N.A.) is zero. At the top fiber, the maximum shortening (compressive deformation) occurs from the development of maximum compressive stresses. Conversely, the maximum tensile stress occurs at the bottom fibers, resulting in a maximum elongation deformation (see Figure 9.8).

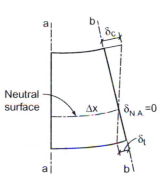

Figure 9.7   Deformed section on beam due to bending.

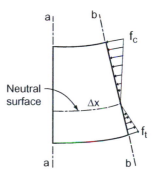

Figure 9.8   Bending stresses on section b-b.

9.2 FLEXURAL (BENDING) STRESS EQUATION

Consider a portion of a beam that is subjected to pure bending only by couples (designated by **M**) at each end, as shown in Figure 9.9. Since the beam is in equilibrium, the moments at each end will be numerically equal, but of opposite sense. Due to the moment couples, the beam is bent from its original straight position to the curved (deformed) shape indicated in Figure 9.9.

Due to this bending action, we find that the lengths of the upper parts of the beam decrease, while the bottom parts of the beam undergo lengthening. This action has the effect of placing the upper portion of the beam in compression and the lower portion of the beam in tension. An equation must be obtained that will relate bending stress to the external moment and the geometric properties of the beam. This can be done by examining a segment of the beam whose internal force system at any given transverse section is a moment **M**, as shown in Figure 9.10. The development of the flexure formula can be found in Appendix C.

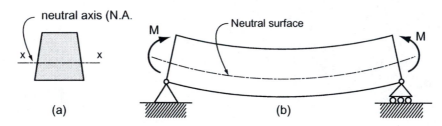

Figure 9.9   (a) Elevation of beam in bending. (b) Beam cross section.

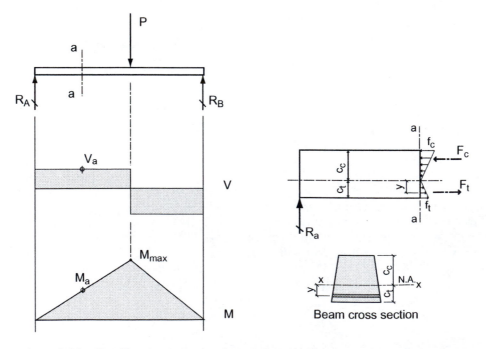

Figure 9.10   Bending stresses on a beam cross-section.

c_c = distance from neutral axis (N.A.) to the extreme compressive fiber

c_t = distance from N.A. to the extreme tensile fiber

y = distance from N.A. to some area ΔA

ΔA = small strip of area on the beam cross-section

Flexural stresses are directly proportional to the beam's bending moment (load, span, and location along the beam's length) and inversely proportional to the beam's moment of inertia relative to the neutral axis. Bending stress will vary from zero at the neutral axis and reaches a maximum at the top and/or bottom fibers.

The general flexure formula is expressed as:

$$f_y = \frac{My}{I_x}$$

where:

f_y = bending stress at level y on the beam cross-section. (psi, ksi) or (kPa, MPa)

M = bending moment at a location along the length of the beam. Generally obtained from the M diagram. (lb.-in., lb.-ft., k-in., k-ft.) or (kN-m)

y = vertical distance measured from the neutral axis to point y on the cross-section. (in.) or (m)

I_x = moment of inertia of the beam cross-section referenced to the neutral axis (N.A.). (in.^{4}) or (m^4)

Generally, in beam design, the maximum bending stress is desired. Since the flexural stresses are highest at the extreme fibers of the cross-section (very top and/or bottom), the value of c is substituted for y and the equation becomes:

$$f_b = \frac{Mc}{I}$$

where:

f_b = bending stress at the extreme fiber, top or bottom

c = distance from the N.A. to the extreme fiber

I = moment of inertia of the cross-section about its centroidal (or N.A.) axis

M = moment at some point along the beam length

Note: *Bending stress f is directly proportional to the value c; therefore, the largest bending stress on a cross-section is obtained by selecting the largest c value for unsymmetrical cross-sections (see Figure 9.11).*

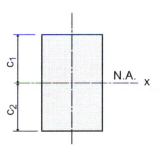

(a) Rectangular cross section

$c_1 = c_2$; $f_{top} = f_{bottom}$

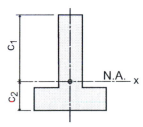

(b) Unsymmetrical cross section

$c_1 > c_2$; $f_{top} > f_{bottom}$

Figure 9.11   *Distances to the extreme fiber for beam cross-sections.*

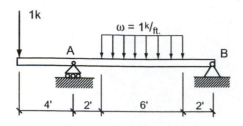

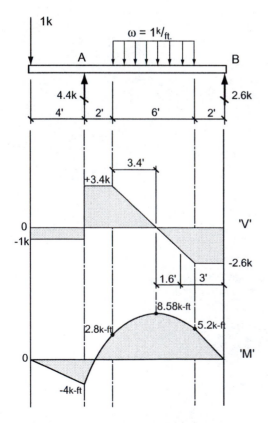

Figure 9.12

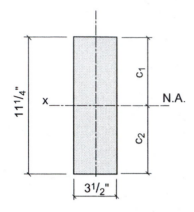

Figure 9.13   Beam cross-section.

Figure 9.14   Load, V, and M diagrams.

Example Problem 9.1: Bending Stress (Figures 9.12 to 9.14)

A 4×12 S4S Douglas fir beam is loaded and supported as shown.

    a. Calculate the maximum bending stress developed in the beam.
    b. What is the magnitude of the bending stress developed 3' to the left of support B?

Solution:

The maximum bending stress developed in the beam occurs where the bending moment is largest. To determine the maximum moment, plot the V and M diagrams.

V at 3' to the left of B:

$$V = \omega(1.6')$$

$$V = 1.6\ k$$

4×12 S4S

$$A = 39.4\ in.^2$$

$$I_x = \frac{bh^3}{12} = \frac{(3.5\ in.)(11.25\ in.)^3}{12} = 415.3\ in.^4$$

$$c_1 = c_2 = \frac{11.25\ in.}{2} = 5.63\ in.$$

a. $M_{max} = 8.58\ k\text{-ft.}$

$$f_b = \frac{Mc}{I_x} = \frac{(8.58\ k\text{-ft.})(12\ ^{in.}/_{ft.})(5.63\ in.)}{415.3\ in.^4}$$

$$= 1.4\ k/in.^2$$

b. Moment at 3' left of B:

$$M = 8.58\ k\text{-ft.} - {}^1/_2(1.6')(1.6\ k)$$

$$M = 8.58\ k\text{-ft.} - 1.28\ k\text{-ft.}$$

$$M = 7.3\ k\text{-ft.}$$

$$f_b = \frac{Mc}{I_x} = \frac{(7.7\ k\text{-ft.})(12\ ^{in.}/_{ft.})(5.63\ in.)}{415.3\ in.^4}$$

$$= 1.19\ k/in.^2$$

Example Problem 9.2 (Figures 9.15 to 9.18)

A beam must span a distance of 12' and carry a uniformly distributed load of 120 lb./ft. Determine which cross-section would be the least stressed: a, b, or c.

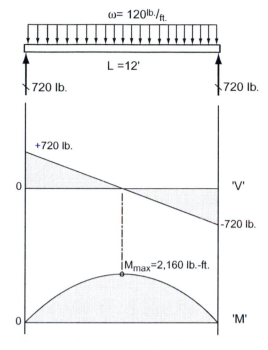

Figure 9.15   Load, V, and M diagrams.

Solution:

   a. Plank

$$A = 20 \text{ in.}^2; I_x = 6.7 \text{ in.}^4; c = 1''$$

$$f_{max} = \frac{Mc}{I} = \frac{(2160 \text{ lb.-ft.})(12 \,^{in.}/_{ft.})(1 \text{ in.})}{6.7 \text{ in.}^4}$$

$$= 3870 \,^{lb.}/_{in.^2}$$

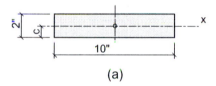

Figure 9.16

   b. Rectangular beam

$$A = 20 \text{ in.}^2; I_x = 41.7 \text{ in.}^4; c = 2.5''$$

$$f_{max} = \frac{Mc}{I} = \frac{(2160 \text{ lb.-ft.})(12 \,^{in.}/_{ft.})(2.5 \text{ in.})}{41.7 \text{ in.}^4}$$

$$= 1550 \,^{lb.}/_{in.^2}$$

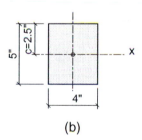

Figure 9.17

   c. I-beam

$$A = 20 \text{ in.}^2; I_x = 154.7 \text{ in.}^4; c = 4''$$

$$f_{max} = \frac{Mc}{I} = \frac{(2160 \text{ lb.-ft.})(12 \,^{in.}/_{ft.})(4 \text{ in.})}{154.7 \text{ in.}^4}$$

$$= 670 \,^{lb.}/_{in.^2}$$

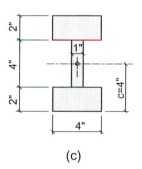

Figure 9.18

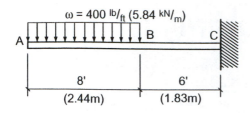

Figure 9.19

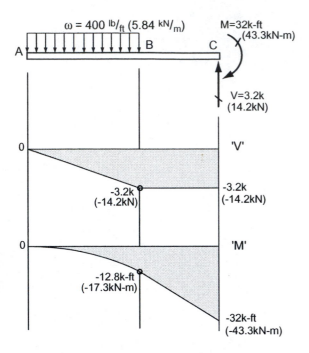

Figure 9.20   Load, V, and M diagrams.

Example Problem 9.3 (Figures 9.19 and 9.20)

A W8×28 (W200×42) steel beam is loaded and supported as shown. Determine the maximum bending stress. A36 steel.

Solution:

$$V = \omega \times 8' = (400 \text{ lb./ft.}) \times (8') = 3200 \text{ lb.}$$

$$V_{SI} = (5.84 \text{ kN/m}) \times (2.44 \text{ m}) = 14.2 \text{ kN}$$

$$M = 3200 \text{ lb.} \times (10') = 32{,}000 \text{ lb.-ft.} = 32 \text{ k-ft.}$$

$$M_{SI} = 14.2 \text{ kN} \times (3.05 \text{ m}) = 43.3 \text{ kN-m}$$

$$f_{max} = \frac{Mc}{I} \text{ (Using } M_{max} \text{ from the moment diagram)}$$

$$I = 98.0 \text{ in.}^4 \quad (40.8 \times 10^6 \text{ mm}^4)$$

$$c = \frac{d}{2} = \frac{8.06''}{2} = 4.03'' \quad (c_{SI} = \frac{d}{2} = \frac{205 \text{ mm}}{2} = 103 \text{ mm})$$

$$f_{max} = \frac{(32 \text{ k-ft.})(12 \text{ in.}/\text{ft.})(4.03'')}{98.0 \text{ in.}^4} = 15.8 \text{ k/in.}^2$$

$$f_{SI} = \frac{(4.33 \times 10^4 \text{ kN-mm})(103 \text{ mm})}{(40.8 \times 10^6 \text{ mm}^4)} = 0.109 \text{ kN/mm}^2$$

$$= 109 \text{ MN/m}^2$$

$$F_b = 22 \text{ k/in.}^2 \quad (150 \text{ MPa})$$

What is the bending stress that would result if the steel beam were replaced by a 6×16 S4S Southern pine No. 1 beam? (Obtain section properties from timber tables in Appendix A.)

Solution:

$$f = \frac{Mc}{I}$$

$$I = 1707 \text{ in.}^4$$

$$c = \frac{d}{2} = \frac{15.5''}{2} = 7.75''$$

$$f = \frac{(384{,}000 \text{ lb.-in.})(7.75 \text{ in.})}{1{,}707 \text{ in.}^4} = 1740 \text{ lb./in.}^2$$

$$F_{allow} = 1550 \text{ psi} < 1740 \text{ psi}$$

$\therefore$ Not good (NG), overstressed

Example Problem 9.4 (Figures 9.21 to 9.23)

Determine the maximum tensile and compressive bending stresses in the beam shown.

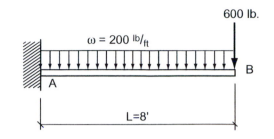

Figure 9.21

Solution:

$$\Sigma A = 10 \text{ in.}^2; \quad \Sigma Ay = 26 \text{ in.}^3$$

$$\bar{y} = \frac{\Sigma Ay}{\Sigma A} = \frac{26 \text{ in.}^3}{10 \text{ in.}^2} = 2.6''$$

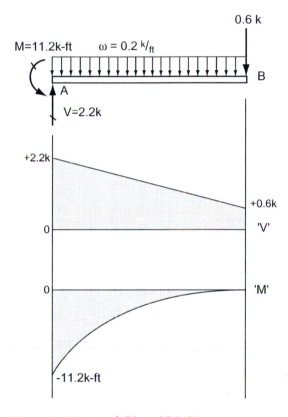

Figure 9.22   Load, V, and M diagrams.

Component	A (in.²)	y (in.)	$y\Delta A$ (in.³)
	6	4	24
	4	0.5	2

Component	I_{xc} (in.⁴)	d_y (in.)	Ad_y^2 (in.⁴)
	18	1.4	11.76
	0.33	2.1	17.64

$$\Sigma I_{xc} = 18.33 \text{ in.}^4; \quad \Sigma Ad_y^2 = 29.4 \text{ in.}^4$$

$$I_x = 18.33 \text{ in.}^4 + 29.4 \text{ in.}^4 = 47.73 \text{ in.}^4$$

$$\underset{\text{(top)}}{f_b} = \frac{Mc_t}{I} = \frac{(11{,}200 \text{ lb.-ft.})(12 \text{ in.}/\text{ft.})(4.4'')}{47.73 \text{ in.}^4}$$

$$= 12{,}390 \text{ lb.}/\text{in.}^2$$

$$\underset{\text{(bottom)}}{f_b} = \frac{Mc_t}{I} = \frac{(11{,}200 \text{ lb.-ft.})(12 \text{ in.}/\text{ft.})(2.6'')}{47.73 \text{ in.}^4}$$

$$= 7320 \text{ lb.}/\text{in.}^2$$

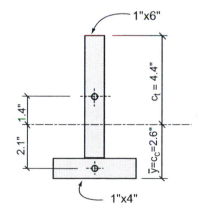

Figure 9.23   Beam cross-section.

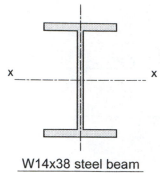

W14x38 steel beam

$A = 11.20 \text{ in.}^2$
$I_x = 385 \text{ in.}^4$
$S_x = 54.6 \text{ in.}^3$

Figure 9.24

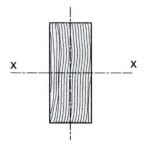

4x10 S4S timber beam

$A = 32.38 \text{ in.}^2$
$I_x = 230.8 \text{ in.}^4$
$S_x = 49.9 \text{ in.}^3$

Figure 9.25

Section Modulus

The majority of the structural shapes used in practice (structural steel, timber, aluminum, etc.) are standard shapes (shown in Figures 9.24 and 9.25) that are normally available in industry. Cross-sectional properties such as area *(A)*, moment of inertia *(I)*, and dimensional size (depth and width) for standard shapes are usually listed in handbooks and tables.

The properties of nonstandard sections and built-up sections may be calculated by the methods outlined in Chapter 7.

As a means of expanding the basic flexure equation into a design form, the two section properties *I* and *c* are combined as *I/c*, which is called the *section modulus*.

$$f_b = \frac{Mc}{I} = \frac{M}{I/c}$$

Section modulus: $S = I/c$;

therefore

$$f_b = \frac{Mc}{I} = \frac{M}{S}$$

where:

> S = section modulus (usually about the *x* axis), (in.^{3}) or (mm^3)
> M = bending moment in the beam (usually M_{max})

Since *I* and *c* of standard sections are known, their section moduli *(S)* are also listed in handbooks. For nonstandard sections and for regular geometric shapes, the section modulus may be obtained by calculating the moment of inertia *I* of the area and then dividing *I* by *c*, the distance from the neutral axis to the extreme fiber. In symmetrical sections, *c* has only one value, but in unsymmetrical sections *c* will have two values, as shown in Figure 9.11. In the analysis and design of beams, however, we are usually interested only in the maximum stress that occurs in the extreme fiber. In all such problems, the greatest value of *c* must be used.

If we rewrite the basic flexure equation into a design form:

$$S_{required} = \frac{M_{max}}{F_b}$$

where:

> F_b = allowable bending stress (ksi or psi); (MPa or kPa)
> M = maximum bending moment in the beam (k-in. or lb.-in.); (kN-mm)

the usefulness of the section modulus becomes quite apparent, since only one unknown exists rather than two (I and c).

Example Problem 9.5: Section Modulus (Figures 9.26 to 9.28)

Two C10×15.3 steel channels are placed back to back to form a 10"-deep beam. Determine the permissible P if $F_b = 30$ ksi. Assume A572 grade 50 steel.

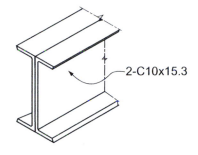

Solution:

$$I_x = 67.4 \text{ in.}^4 \times 2 = 134.8 \text{ in.}^4$$

$$M_{max} = \tfrac{1}{2}(5)(5) + (P/2)(5)$$

$$M_{max} = 12.5 + 2.5P$$

$$= (12.5 \text{ k-ft.} + 2.5P) \times (12 \text{ }^{in.}/_{ft.})$$

$$f = \frac{Mc}{I} = \frac{M}{S}; \quad \therefore M = F_b \times S_x$$

$$S_x = 2 \times 13.5 \text{ in.}^3 = 27 \text{ in.}^3$$

Figure 9.26   Two steel channels.

Equating both M_{max} equations:

$$M = (30 \text{ k/in.}^2) \times (27 \text{ in.}^3) = 810 \text{ k-in.}$$

$$(12.5 \text{ k-ft.} + 2.5P)(12 \text{ }^{in.}/_{ft.}) = 810 \text{ k-in.}$$

Dividing both sides of the equation by 12 in./ft.:

$$(12.5 \text{ k-ft.}) + (2.5 \text{ ft.})P + 67.5 \text{ k-ft.}$$

$$2.5P = 55 \text{ k}$$

$$\therefore P = 22 \text{ k}$$

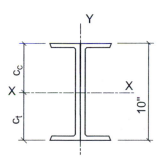

Figure 9.27   Beam cross-section.

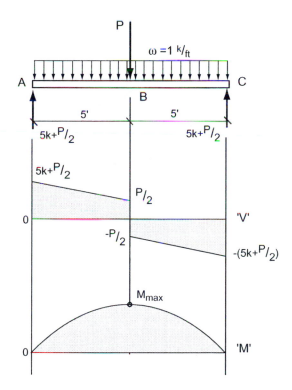

Figure 9.28   Load, V, and M diagrams.

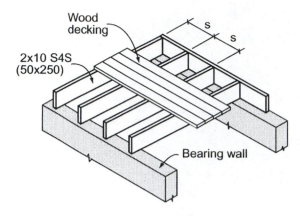

Figure 9.29   Timber floor framing.

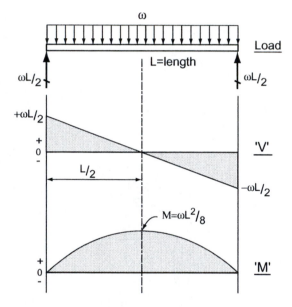

Figure 9.30   Load, V, and M diagrams.

Example Problem 9.6 (Figures 9.29 and 9.30)

A timber floor system utilizing 2×10 S4S (50×250 S4S) joists spans a length of 14' (4.27 m), simply supported. The floor carries a DL + LL load of 50 psf (2.39 kPa). At what spacing should the joists be placed? Assume Douglas fir–larch No. 2 (F_b = 1450 psi) or (10 MPa).

Solution:

2×10 S4S (50×250 S4S); S_x = 21.39 in.^{3} (350 × 10^3 mm^3)
Based on the allowable stress criteria:

$$f = \frac{Mc}{I} = \frac{M}{S}$$

$$M_{max} = S \times F_b = (21.4 \text{ in.}^3)(1.45 \text{ k/in.}^3) = 31 \text{ k-in.}$$

$$M = \frac{(31 \text{ k-in.})}{12 \text{ in./ft.}} = 2.58 \text{ k-ft.}$$

$$M_{SI} = (350 \times 10^{-6} \text{ m}^3)(10 \text{ MN/m}^2)$$

$$= 3.5 \times 10^{-3} \text{ MN-m} = 3.5 \text{ kN-m}$$

Based on the bending moment diagram:

$$M_{max} = \frac{\omega L^2}{8}; \quad \therefore \ \omega = \frac{8M}{L^2}$$

Substituting for M obtained previously,

$$\omega = \frac{8(2.58 \text{ k-ft.})}{(14')^2} = 0.105 \text{ k/ft.} = 105 \text{ lb./ft.}$$

$$\omega_{SI} = \frac{8(3.5 \text{ kN-m})}{(4.27 \text{ m})^2} = 1.54 \text{ kN/m}$$

But,

$$\omega = \text{lb./ft.}^2 \times \text{tributary width (joist spacing } s)$$

$$\omega = \text{N/m}^2 \times \text{tributary width (joist spacing } s)$$

$$s = \frac{\omega}{50 \text{ lb./ft.}^2} = \frac{105 \text{ lb./ft.}}{50 \text{ lb./ft.}^2} = 2.1'$$

$$s_{SI} = \frac{1.54 \text{ kN/m}}{2.39 \text{ kN/m}^2} = 0.64 \text{ m}$$

$$s = 25'' \text{ spacing}$$

Use 24" o.c. spacing.

Note: Spacing is more practical for plywood subflooring, based on a 4 ft. module of the sheet.

Example Problem 9.7 (Figures 9.31 to 9.33)

Design the roof and second-floor beams if $F_b = 1550$ psi (Southern pine No. 1).

Solution:

Load conditions:

Roof: Snow + DL (roof) = 200 lb./ft.

Walls: 400 lb. concentrated load on beams at second floor

Railing: 100 lb. concentrated load on beam overhang

Second Floor: DL + LL = 300 lb. (also on the deck)

Roof beam design:

$$M_{max} = 3600 \text{ lb.-ft.}$$

$$S_{required} = \frac{(3.6 \text{ k-ft.})(12 \text{ in./ft.})}{1.55 \text{ k/in.}^2} = 27.9 \text{ in.}^3$$

From the timber table in Appendix A:

Use: 4×8 S4S ($S_x = 30.7$ in.^{3})

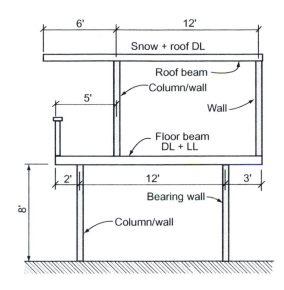

Figure 9.31   Building section.

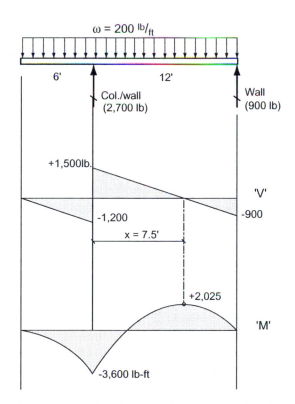

Figure 9.32   Load, V, and M diagrams for the roof beam.

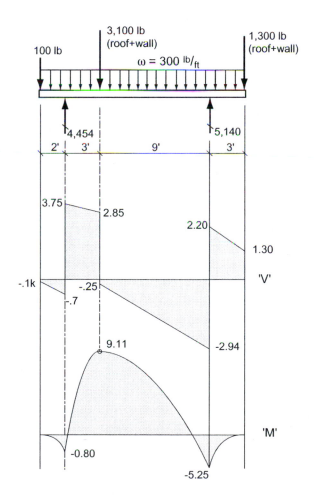

Figure 9.33   Load, V, and M diagrams for the second-floor beam.

second-floor beam design:

$$M_{max} = 9110 \text{ lb.-ft.}$$

$$S_{required} = \frac{M_{max}}{F_b} = \frac{(9.110 \text{ k-ft.})(12 \text{ in./ft.})}{1.55 \text{ k/in.}^2} = 70.5 \text{ in.}^3$$

From the timber tables in Appendix A:

Use: 4×12 S4S ($S_x = 73.8 \text{ in.}^3$)

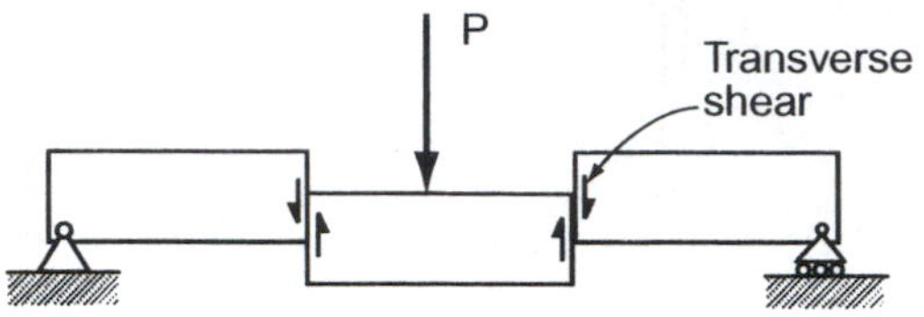

9.3 SHEARING STRESS—LONGITUDINAL AND TRANSVERSE

In addition to the internal bending moment present in beams, a second important factor to be considered in the determination of the strength of beams is shear. There is generally present an internal shear force, *V*, which may in some cases govern the design of beams. Many materials (wood, for example) are primarily weak in shear; for this reason, the load that can be supported may depend on the ability of the material (beam) to resist shearing forces.

Since beams are normally horizontal and the cross-sections upon which bending stresses are investigated are vertical, these shearing stresses in beams are generally referred to as vertical (transverse) and horizontal (longitudinal).

Transverse shear action (Figure 9.34) is a pure shearing condition and occurs even where there is no bending of the beam. However, beams do bend and, when they bend, fibers on one side of the neutral axis are placed in compression and those on the other side are placed in tension. In effect, the fibers on either side of the neutral surface tend to slip in directions opposite to one another.

The existence of horizontal (longitudinal) shearing stresses in a bent beam can readily be visualized by bending a deck of cards. The sliding of one surface over another, which is plainly visible, is a shearing action, which, if prevented, will set up horizontal shearing stresses on those surfaces (see Figure 9.35b).

If one constructs a beam by stacking one 4" × 4" member on top of another without fastening them together, and then loads this beam in a direction normal to the beam length, the resulting deformation will appear somewhat like that shown in Figure 9.36. The fact that a solid beam does not exhibit this relative movement of longitudinal elements, as shown in Figure 9.37, indicates the presence of shearing stresses on longitudinal planes. The evaluation of these shearing stresses will now be studied by means of free-body diagrams and the equilibrium approach.

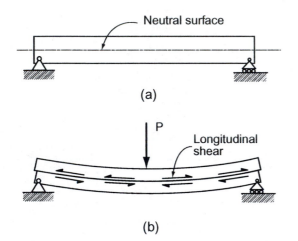

Figure 9.34   Transverse shear of a beam.

*Figure 9.35   (a) Beam with no load.
(b) Loaded beam with longitudinal shear.*

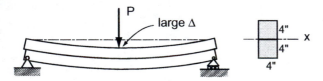

Figure 9.36   Two 4" × 4" members (unfastened)—large deflection.

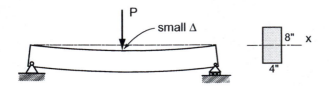

Figure 9.37   4"×8" solid section (smaller deflection under load).

Relationship Between Transverse and Longitudinal Shearing Stress

In Chapter 8 we developed a method of plotting shear (V) diagrams based on beams experiencing transverse shearing action. This section will now show that at any point in a deflected beam, the vertical and horizontal shearing stresses are equal. Therefore, the V diagram is a representation of both transverse and longitudinal shear along the beam.

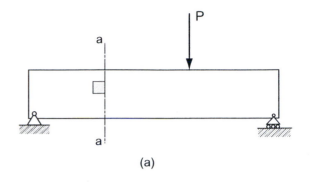

Consider a simply supported beam as shown in Figure 9.38a. When a section a-a is passed through the beam, a shear force V, representing the sum total of all unit transverse shearing stresses on the cut section, develops as shown in Figure 9.38b. If we now isolate a small, square element of this beam, the following relationship develops:

$$V = \Sigma f_v A$$

where:

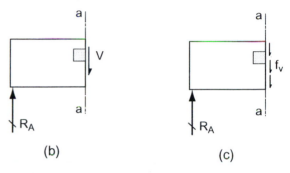

f_v = unit shearing stress
A = cross-sectional area of beam

Figure 9.38 *(a) Simply supported beam. (b) Transverse shear force (V). (c) Transverse shearing stress.*

Removing the small elemental square from the beam, we draw a free-body diagram showing the forces acting on it (Figures 9.39 to 9.41).

Examine the shear stress along section cut a-a.

Assume: $\Delta y = \Delta x$ and that the elemental square is very small.

f_v = (transverse shear stress)

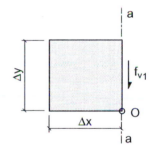

Figure 9.39 *Elemental square.*

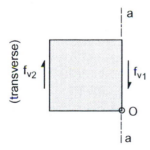

Figure 9.40 *Transverse shear stresses.*

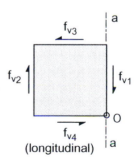

Figure 9.41 *Longitudinal shear stresses.*

For equilibrium vertically,

$$\sum F_y = 0; \quad f_{v_1} = f_{v_2} \quad \text{(forms a moment couple)}$$

To place the elemental square in rotational equilibrium, sum moments about point O.

$$\sum M_O = 0; \quad (f_{v_1})(\Delta x) = (f_{v_3})(\Delta y)$$

But:

$$\Delta y = \Delta x$$

Therefore; $f_{v_1} = f_{v_2} = f_{v_3} = f_{v_4}$

Shears f_{v_3} and f_{v_4} form a counterclockwise couple.

From the preceding example, we can conclude that:

$f_{\text{transverse}} = f_{\text{longitudinal}}$ at a given point along the beam length.

9.4   GENERAL SHEAR STRESS EQUATION

Horizontal (longitudinal) shearing stresses are not evenly distributed over the beam's cross-section (see Figure 9.42). In the development of the general shear stress equation in Appendix C, it was observed that the shear stresses are largest at the neutral axis and are zero at the top and bottom fibers. This appears to be the reverse of the bending stresses, which were maximum at the extreme top and/or bottom fibers and zero at the neutral axis.

The general formula for determining the horizontal (longitudinal) shearing stress in beams is:

$$f_v = \frac{VQ}{Ib}$$

where:

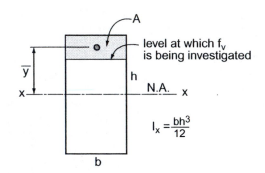

Figure 9.42   Rectangular beam cross-section.

f_v = unit shearing stress; transverse or longitudinal (psi or ksi); (kPa or MPa)

V = shear in the beam at a given point along the beam length, usually obtained from the shear (V) diagram (lb. or k); (kN)

$Q = A\bar{y}$ = first moment (in.³); (mm³ or m³)

A = area above or below the level at which the shear stress is desired (in.²); (mm² or m²)

$\bar{y}$ = distance from the beam cross-section's neutral axis (N.A.) to the centroid of the area above or below the desired plane where the shear stress is being examined (in.); (mm or m)

I = moment of inertia of the entire beam cross-section (in.⁴); (mm⁴ or m⁴)

b = width of the beam at the plane where the shear stress is being examined (in.); (mm or m)

Example Problem 9.8: Shear Stress
(Figures 9.43 to 9.47)

Calculate the maximum bending and shear stress for the beam shown.

Solution:

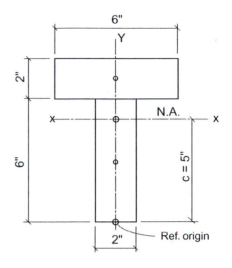

Figure 9.43   Beam cross-section.

Component	I_{xc} (in.⁴)	A (in.²)	d_y (in.)	Ad_y^2 (in.⁴)
[horizontal rectangle]	4	12	2	48
[vertical rectangle]	36	12	2	48

$$I_x = \sum I_{xc} + \sum Ad_y^2 = 40 \text{ in.}^4 + 96 \text{ in.}^4$$

$$= 136 \text{ in.}^4$$

$$f_{b_{max}} = \frac{M_{max}c}{I_x} = \frac{(5000 \text{ lb.-ft})(12 \text{ in./ft.})(5'')}{136 \text{ in.}^4}$$

$$= 2200 \text{ }^{lb.}/_{in.^2}$$

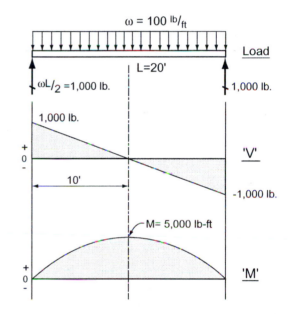

Figure 9.44   Load, V, and M diagrams.

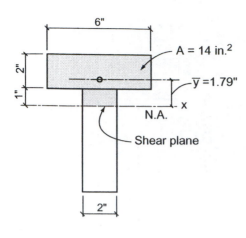

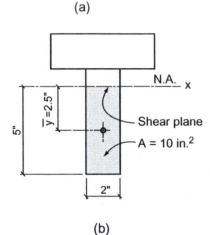

(b)

Figure 9.45   (a) Section above the N.A. (b) Section below the N.A.

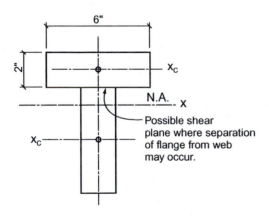

Figure 9.46   Shear plane between the flange and web.

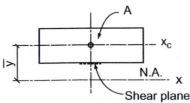

Figure 9.47   Flange section above the shear plane.

Component	A (in.^{2})	y (in.)	$y\Delta A$ (in.^{3})
	12	2	24
	2	0.5	1

From Figure 9.45a

$$V_{max} = 1000 \text{ lb. (from } V \text{ diagram)}$$

$$\bar{y} = \frac{25 \text{ in.}^3}{14 \text{ in.}^2} = 1.79"$$

$$Q = \sum Ay = A\bar{y} = (14 \text{ in.}^2)(1.79 \text{ in.}) = 25 \text{ in.}^3$$

$$I_x = 136 \text{ in.}^4 \text{(for the entire cross-section)}$$

$$b = 2"$$

$$\therefore f_v = \frac{VQ}{Ib} = \frac{(1000 \text{ lb.})(25 \text{ in.}^3)}{(139 \text{ in.}^4)(2 \text{ in.})} = 92 \text{ }^{lb.}/_{in.^2}$$

From Figure 9.45b

$$V = 1000 \text{ lb.}$$

$$Q = A\bar{y} = 25 \text{ in.}^3$$

$$I_x = 136 \text{ in.}^4$$

$$b = 2"$$

$$\therefore f_v = \frac{(1000 \text{ lb.})(25 \text{ in.}^2)}{(136 \text{ in.}^4)(2 \text{ in.})} = 92 \text{ }^{lb.}/_{in.^2}$$

Note: Pick the easier half of the cross-section in calculating $Q = A\bar{y}$.

What shear stress develops at the base of the flange? (This calculation would give an indication as to what kind of shear stress must be resisted if glue, nails, or any other fastening device is used to join the flange to the stem.) (Figures 9.46 and 9.47.)

$$V = 1000 \text{ lb.}$$

$$I = 136 \text{ in.}^4$$

$b = 2"$ or 6" but a smaller b gives a larger f_v.

$$\therefore \text{Use } b = 2"$$

$$Q = A\bar{y} = (12 \text{ in.}^2)(2 \text{ in.}) = 24 \text{ in.}^3$$

$$\therefore f_v = \frac{(1000 \text{ lb.})(24 \text{ in.}^2)}{(136 \text{ in.}^4)(2 \text{ in.})} = 88.3 \text{ }^{lb.}/_{in.^2}$$

Example Problem 9.9 (Figures 9.48 to 9.51)

Determine the maximum shear stress developed on the beam cross-section shown below.

Solution:

Component	I_{xc} (in.^{4})
	$\dfrac{2(6)(4)^3}{12} = 64$
	$\dfrac{2(12)^3}{12} = 288$

$$I_x = \sum I_{xc} = 352 \text{ in.}^4$$

Two locations will be examined to determine the maximum shear stress. One shear plane is through the neutral axis (normally the critical location), and the other will be where the section necks down to 2".

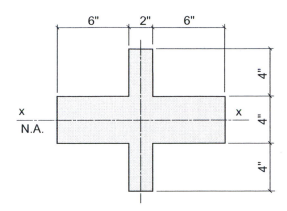

Figure 9.48   Beam cross-section.

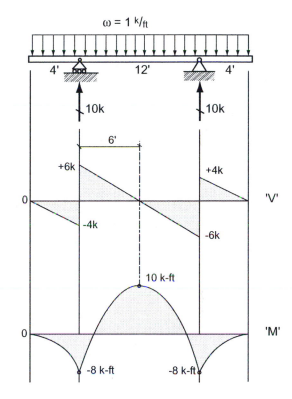

Figure 9.49   Load, V, and M diagrams.

Component	A (in.^{2})	y (in.)	$y\Delta A$ (in.^{3})
	8	4	32
	28	1	28

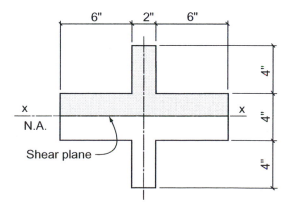

Figure 9.50   Beam cross-section.

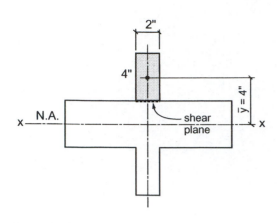

Figure 9.51   Beam cross-section.

$$\sum Ay = 60 \text{ in.}^3$$

$$Q = A\bar{y} = \sum Ay = 60 \text{ in.}^3$$

$$f_v = \frac{VQ}{Ib} = \frac{(6000 \text{ lb.})(60 \text{ in.}^3)}{(352 \text{ in.}^4)(14 \text{ in.})} = 73.1 \text{ }^{lb.}/_{in.^2}$$

(at the neutral axis)

Component	A (in.²)	y (in.)	yΔA (in.³)
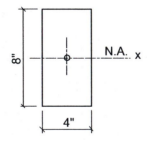	8	4	32

$$Q = A\bar{y} = \sum Ay = 32 \text{ in.}^3$$

$$f_v = \frac{VQ}{Ib} = \frac{(6000 \text{ lb.})(32 \text{ in.}^3)}{(352 \text{ in.}^4)(2 \text{ in.})} = 272.7 \text{ }^{lb.}/_{in.^2}$$

$$\therefore f_{v_{max}} = 272.7 \text{ psi}$$

This is where longitudinal shear failure would probably occur.

Example Problem 9.10 (Figures 9.52 to 9.58)

For the beam cross-section shown, determine the longitudinal shear stress that develops at the N.A. and at 1" increments above the N.A. Use V_{max} for your calculations.

Figure 9.52   Beam cross-section.

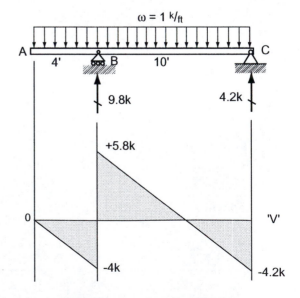

Figure 9.53   Load and V diagrams.

Solution:

For the general shear stress equation, a determination must be made for the V_{max} and the cross-sectional properties I, b, and Q. The V_{max} value is most conveniently obtained directly from the shear diagram.

Since the moment of inertia I is constant for a given cross-section, it may be calculated as:

$$I_x = \frac{bh^3}{12} = \frac{(4")(8")^3}{12} = 171 \text{ in.}^4$$

The width of the cross-section (shear plane) is also constant; therefore, $b = 4"$.

The values of $Q = A\bar{y}$ for each of the four shear planes, including the neutral axis at which the horizontal (longitudinal) shear is desired, are shown in Figures 9.54 to 9.57 and are tabulated as follows:

$$I_x = 171 \text{ in.}^4; \quad b = 4"$$

$$Q = A\bar{y} = (16 \text{ in.}^2)(2 \text{ in.}) = 32 \text{ in.}^3$$

$$\underset{\text{(N.A.)}}{f_v} = \frac{VQ}{Ib} = \frac{(5800 \text{ lb.})(32 \text{ in.}^3)}{(171 \text{ in.}^4)(4 \text{ in.})} = 271 \text{ }^{\text{lb.}}/_{\text{in.}^2}$$

$$I_x = 171 \text{ in.}^4; \quad b = 4"$$

$$Q = A\bar{y} = (4 \text{ in.})(3 \text{ in.})(2.5 \text{ in.}) = 30 \text{ in.}^3$$

$$\underset{\text{(1" above N.A.)}}{f_v} = \frac{VQ}{Ib} = \frac{(5800 \text{ lb.})(30 \text{ in.}^3)}{(171 \text{ in.}^4)(4 \text{ in.})} = 254 \text{ }^{\text{lb.}}/_{\text{in.}^2}$$

$$I_x = 171 \text{ in.}^4; \quad b = 4"$$

$$Q = A\bar{y} = (4 \text{ in.})(2 \text{ in.})(3 \text{ in.}) = 24 \text{ in.}^3$$

$$\underset{\text{(2" above N.A.)}}{f_v} = \frac{VQ}{Ib} = \frac{(5800 \text{ lb.})(24 \text{ in.}^3)}{(171 \text{ in.}^4)(4 \text{ in.})} = 204 \text{ }^{\text{lb.}}/_{\text{in.}^2}$$

Plotting the shear stress values on a graph, adjacent to the beam cross-section, we obtain a parabolic curve as shown in Figure 9.58. Had the values of shearing stress been obtained for the corresponding points below the neutral axis, we should have found corresponding magnitudes. By completing the curve, it will be noted that the maximum value of horizontal shearing stress occurs at the neutral plane (surface), where $A\bar{y}$ is a maximum and the bending stresses are equal to zero.

Because of its frequent use in design, an expression for the maximum horizontal shearing stress occurring in solid rectangular beams (primarily timber beams) may be derived.

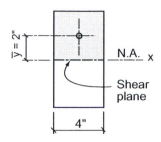

Figure 9.54   Shear at the N.A.

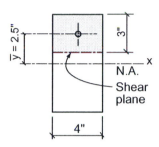

Figure 9.55   Shear 1" above the N.A.

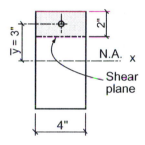

Figure 9.56   Shear 2" above the N.A.

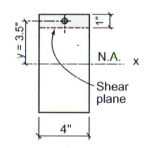

Figure 9.57   Shear 3" above the N.A.

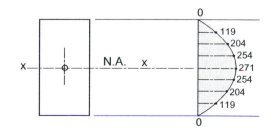

Figure 9.58   Graph of shear stress intensities at various locations on the beam cross-section.

The shear plane is maximum at the neutral axis, as found in Figure 9.58.

$$f_v = \frac{VQ}{Ib}; \quad I_x = \frac{bh^3}{12}; \quad A = b \times \frac{h}{2} = \frac{bh}{2}$$

$$b = b; \quad \bar{y} = \frac{h}{4}; \quad Q = A\bar{y}$$

Therefore,

$$f_v = \frac{V\left(\dfrac{bh}{2} \times \dfrac{h}{4}\right)}{\left(\dfrac{bh^3}{12}\right)(b)} = \frac{12Vbh^2}{8b^2h^3} = \frac{3V}{2bh}$$

However, *(bh)* = area of the *entire* beam cross-section.

Simplifying:

$$f_{v_{max} \atop (N.A.)} = \frac{3V}{2A} = \frac{1.5V}{A} \quad \text{(For solid rectangular cross-sections)}$$

where:

$$f_{v_{max} \atop (N.A.)} = \text{maximum shearing stress at the N.A.}$$

$$V = \text{maximum shear on the loaded beam}$$

$$A = \text{cross-sectional area of the beam}$$

From the equation just developed, we find that the maximum (design) shear stress for a rectangular beam is 50% larger than the average shear value (see Figure 9.59).

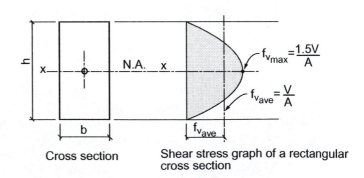

Cross section   Shear stress graph of a rectangular cross section

Figure 9.59   Shear stress distribution—key points.

Example Problem 9.11 (Figure 9.60)

Determine the adequacy of a wood floor joist (2×10's @ 16" o.c.) supporting a $DL = 10$ psf and $LL = 40$ psf. Assume that the allowable stresses for the Hem-fir No. grade material are as follows: $F_b = 1150$ psi, $F_v = 75$ psi.

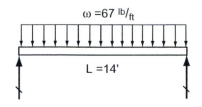

Figure 9.60   Floor joist.

Solution:

2×10 S4S joist ($A = 13.9$ in.²; $S_x = 21.4$ in.³)

$$\omega = (50 \text{ lb./ft.}^2) \times \left(\frac{16 \text{ in.}}{12 \text{ in.}/\text{ft.}}\right) = 67 \text{ lb./ft.}$$

$$V_{\max} = \frac{\omega L}{2} = \frac{(67 \text{ lb./ft.})(14 \text{ ft.})}{2} = 469 \text{ lb.}$$

$$M_{\max} = \frac{\omega L^2}{8} = \frac{(67 \text{ lb./ft.})(14 \text{ ft.})^2}{8} = 1642 \text{ lb.-ft.}$$

$$f_b = \frac{M_{\max}}{S_x} = \frac{(1642 \text{ lb./ft.})(12 \text{ in./ft.})}{21.4 \text{ in.}^3}$$

$$= 921 \text{ lb.}/\text{in.}^2 < F_b = 1150 \text{ psi}$$

Therefore, OK in bending

$$f_v = \frac{VQ}{Ib} = \frac{3V}{2A} = \frac{1.5(469 \text{ lb.})}{(13.9 \text{ in.}^2)}$$

$$= 51 \text{ lb./in.}^2 < F_v = 75 \text{ psi}$$

The section is also adequate in shear.

Shearing Stress Variations in Beams

Beams must be designed to safely withstand the maximum stresses due to bending and shear. The variation of tensile and compressive bending stresses over a cross-sectional area was discussed in Section 9.2. As in bending stress, shear stress also varies on a cross-section, as illustrated for a rectangular cross-section in Figure 9.60. Except for a few exceptions, the maximum shearing stress generally occurs at the neutral axis.

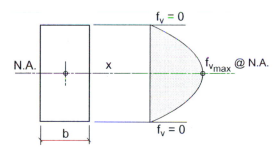

Figure 9.61   Rectangular beam.

Shearing stress variation over the cross-section of a T-beam, C-beam, and wide-flange section is illustrated in Figures 9.61 to 9.64. The shaded curve in Figure 9.62 indicates what the stress variation would be if the beam area had remained rectangular with a constant width **b.** This

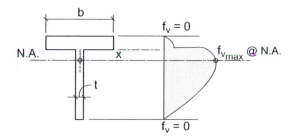

Figure 9.62   T-beam.

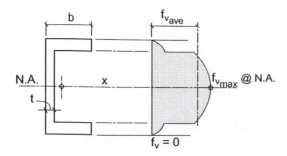

Figure 9.63   Channel section.

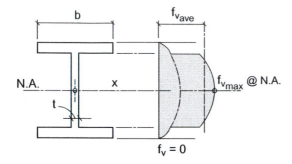

Figure 9.64   Wide flange.

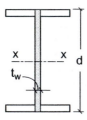

Figure 9.65   Wide-flange—W.

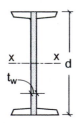

Figure 9.66   Standard—S.

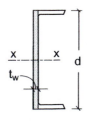

Figure 9.67   Channel—C.

variation would be similar to that shown in Figures 9.63 and 9.64. The sudden increase in shear stress at the under-side of the top flange comes from the change of the width from **b** to **t** in

$$f_v = \frac{VQ}{Ib}$$

A similar change occurs at the flange-to-web transition of a T-beam in Figure 9.62, but here the curve below the neutral axis follows the usual pattern for a rectangular beam.

Upon examination of the shear stress distribution for a wide-flange section, we find that most of the shear is resisted by the web, and very little resistance is offered by the flanges. The opposite is true in the case for flexural stresses—the flanges resist most of the bending stress, and the web offers little resistance to bending (see Figure 9.64).

The calculation of the exact maximum stress magnitude us-ing **VQ/Ib** can become difficult because of the presence of fil-lets (rounding) where the flange joins the web. A high level of accuracy is even harder to achieve in channels or standard I-shapes that have sloping flange surfaces. Accordingly, the American Institute of Steel Construction (AISC) recommends the use of a much simpler approximate formula for the com-mon steel shapes:

$$f_{v_{average}} = \frac{V_{max}}{t_w d}$$

where:

V = shear force

d = beam depth

t_w = web thickness

This formula gives the average unit shearing stress for the web over the full depth of the beam, ignoring the contri-bution of the flange (see Figures 9.65 to 9.67).

Webs resist approximately 90% of the total shear for struc-tural shapes. In contrast, flanges resist 90% of the bending stresses.

Depending on the particular steel shape, the average shear stress formula:

$$f_{v_{\text{average}}} = \frac{V_{\text{max}}}{t_w d}$$

Figure 9.68   Web buckling in steel beams.

can be as much as 20% in error in the nonconservative direction. This means that when a shearing stress computed from this equation gets within 20% of the maximum allowable shear stress, the actual maximum stress (VQ/Ib) might be exceeding the allowable stress by a small amount.

Fortunately, this low level of accuracy is seldom a problem for two reasons:

1. Structural steels are very strong in shear.
2. Most beams and girders in buildings, unlike those in some machines, have low shearing stresses.

High shearing stress may be present in short-span, heavily loaded beams, or if large concentrated loads are applied adjacent to a support. In determining the size of a steel beam, flexural stresses or deflection will usually govern. When shearing stresses do become excessive, steel beams do not fail by ripping along the neutral axis, as might occur in timber beams. Rather, it is the compression buckling of the relatively thin web that constitutes a shear failure (see Figure 9.68). The AISC has provided several design formulas for determining when extra bearing area must be provided at concentrated loads or when web stiffeners are needed to prevent such failures (see Figure 9.69).

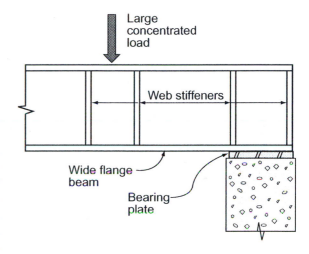

Figure 9.69   Web stiffeners.

Example Problem 9.12: Shearing Stress (Figure 9.70)

An American Standard S12×31.8 beam resists a shear $V = 12$ k at the supports. Determine the average web shear stress. $F_v = 14.5$ ksi (A36 steel).

Solution:

$$f_{v_{\text{ave}}} = \frac{V}{t_w d}$$

$$V = 12 \text{ k}$$

$$t_w = 0.35"$$

$$d = 12"$$

$$f_{v_{\text{ave}}} = \frac{V}{t_w d} = \frac{(12 \text{ k})}{(0.35 \text{ in.})(12 \text{ in.})}$$

$$= 2.86 \text{ k/in.}^2 < F_v = 14.5 \text{ ksi}$$

$\therefore$ OK

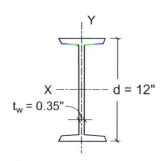

Figure 9.70   S12×31.8 steel beam.

Example Problem 9.13 (Figure 9.71)

A W12×45 (310×67) beam is loaded as shown. Calculate the critical $f_{v_{\text{average}}}$.

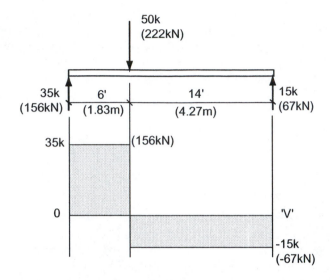

Figure 9.71   Load and V diagrams.

Solution:

$$F_v = 14.5 \text{ ksi (A36 steel)}$$

$$V = 35 \text{ k (156 kN)}$$

$$t_w = 0.335'' \text{ (9 mm)}$$

$$d = 12.06'' \text{ (306 mm)}$$

$$f_{v_{\text{ave}}} = \frac{V_{\text{max}}}{t_w d} = \frac{(35 \text{ k})}{(0.335'')(12.06'')}$$

$$= 8.66 \text{ k/in.}^2 < F_v = 14.5 \text{ ksi}$$

$$f_{v_{\text{ave}}}_{\text{(si)}} = \frac{(156 \text{ kN})}{(9 \times 10^{-3} \text{ m})(306 \times 10^{-3} \text{ m})}$$

$$= 56.6 \times 10^3 \text{ kN/m}^2$$

$$= 57 \text{ MPa} < F_v = 100 \text{ MPa}$$

$$\therefore \text{ OK in shear}$$

9.5 DEFLECTION IN BEAMS

As discussed in earlier sections of Chapter 9, the design of beams for a particular load and support condition requires the strength investigation of bending stress and shear stress. Quite frequently, however, the design of a beam is governed by its permissible *deflection*. In design, deformation (called deflection in beams) often shares an equivalent importance with strength considerations, especially in long-span structures.

Deflection, a *stiffness* requirement, represents a change in the vertical position of a beam due to the applied loads. Load magnitude, beam span length, the moment of inertia of the beam cross-section, and the beam's modulus of elasticity are all factors in the amount of deflection that results. Generally, the amount of *allowable* or *permissible deflection* is limited by building codes or by practical considerations such as minimizing plaster cracking in ceiling surfaces or reducing the springiness of a floor.

Wood as a structural material is less stiff (lower E-value) than steel or concrete; hence, deflection is always a concern. Detrimental effects from large deflections can include nail popping in gypsum ceilings, cracking of horizontal plaster surfaces, and visible sagging of ceilings and floors. In some design situations (primarily longer spans), a wood member satisfying the strength requirements will not necessarily satisfy deflection criteria.

Steel beams, although stronger relative to wood, still need to be checked for deflection. Particular care must be given in long-span situations because of the likelihood of objectionable sag or *ponding* of water. Ponding is potentially one of the most dangerous conditions for flat roofs. It occurs when a flat roof deflects enough to prevent normal water runoff. Instead, some water collects in the midspan and, with the added weight of accumulated water, the roof deflects a little more, allowing even more water to collect, which in turn causes the roof to deflect more. This progressive cycle continues until structural damage or collapse occurs. Building codes require that all roofs be designed with sufficient slope to ensure drainage after long-term deflection, or that roofs be designed to support maximum roof loads, including the possible effects of ponding.

The *allowable deflection limits* for beams are given in Table 9.1. These limits are based on the American Institute of Timber Construction (AITC), American Institute of Steel Construction (AISC), and Uniform Building Code (UBC) standards.

The calculation of *actual beam deflections* is often approached from a mathematical viewpoint that requires the solution of a second-order differential equation subject to the loading and the type of end supports of the beam. While this method is mathematically straightforward, it presents formidable problems associated with the evaluation of the proper boundary

Table 9.1   Recommended Allowable Deflection Limits.

Use Classification	LL only	DL+LL
Roof beams:		
Industrial	1/180	1/120
Commercial and institutional		
• without plaster ceiling	1/240	1/180
• with plaster ceiling	1/360	1/240
Floor beams:		
Ordinary usage*	1/360	1/240

*Ordinary usage is for floors intended for construction in which walking comfort and the minimizing of plaster cracking are primary considerations.

conditions, as well as in the mathematics required to obtain the solution.

There are many ways to approach the problem of beam deflections: the moment-area method, conjugate beam, double integration, and formulas. This section will deal exclusively with the use of established deflection formulas found in standard handbooks such as the AISC manual, timber design manuals, and so on. Also, deflections will automatically be calculated for most beam designs done on a computer. The intent of this section is to present a few fundamental concepts dealing with deflection and its role in beam design rather than to explore the many sophisticated mathematical techniques that may be employed in obtaining deflection values. An understanding of the basics of deflection will enable the user of computer software to better understand the results obtained.

Deflection Formulas

Many loading patterns and support conditions occur so frequently in construction that reference manuals (AISC, AITC, etc.) and engineering handbooks tabulate the appropriate formulas for their deflections. A few of the more common cases are shown in Table 9.2. More often than not, the required deflection values in a beam design situation can be determined via these formulas, and one does not need to resort to deflection theory. Even when the actual loading situation does not match one of the tabulated cases, it is sufficiently accurate for most design situations to approximate the maximum deflection by using one or more of the formulas.

Computed actual deflections must be compared against the allowable deflections permitted by the building codes.

$$\Delta_{\text{actual}} \leq \Delta_{\text{allowable}}$$

Table 9.2   Common Cases of Beam Loading and Deflection.

Beam Load and Support	Actual Deflection
(a) Uniform load, simple span	$\Delta_{max} = \dfrac{5\omega L^4}{384EI}$ (at the centerline)
(b) Concentrated load at midspan	$\Delta_{max} = \dfrac{PL^3}{48EI}$ (at the centerline)
(c) Two concentrated loads at third points	$\Delta_{max} = \dfrac{23PL^3}{648EI} = \dfrac{PL^3}{28.2EI}$ (at the centerline)
(d) Three concentrated loads at quarter points	$\Delta_{max} = \dfrac{PL^3}{20.1EI}$ (at the centerline)
(e) Uniform load, both ends fixed	$\Delta_{max} = \dfrac{\omega L^4}{384EI}$ (at the centerline)
(f) Uniform load, cantilever beam	$\Delta_{max} = \dfrac{\omega L^4}{8EI}$ (at the centerline)
(g) Cantilever with concentrated load at the end	$\Delta_{max} = \dfrac{PL^3}{3EI}$ (at the free end)

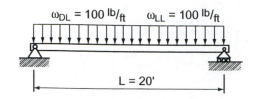

$\omega_{DL} = 100\ ^{lb}/_{ft}$ $\omega_{LL} = 100\ ^{lb}/_{ft}$

L = 20'

Figure 9.72

Example Problem 9.14: Deflection in Beams (Figure 9.72)

Using DF-L No. 1, design the simply supported floor beam shown to meet bending, shear, and deflection criteria.

$$\tfrac{1}{4} < b/h < \tfrac{1}{2}$$

$$\Delta_{allow.(DL+LL)} = L/240; \quad \Delta_{allow(LL)} = L/360$$

$$F_b = 1300\ psi;\ F_v = 85\ psi;\ E = 1.6 \times 10^6\ psi$$

Solution:

Bending:

$$M_{max} = \frac{\omega L^2}{8} = \frac{(200\ lb./ft.)(20')^2}{8} = 10,000\ lb.\text{-}ft.$$

$$S_{required} = \frac{M_{max}}{F_b} = \frac{(10\ k\text{-}ft)(12\ in./ft.)}{(1.3 \times 10^3\ k/in.^2)} = 92.3\ in.^3$$

Shear:

$$V_{max} = \frac{\omega L}{2} = \frac{(200\ lb./ft.)(20')}{2} = 2000\ lb.$$

$$A_{required} = \frac{1.5 V_{max}}{F_v} = \frac{1.5(2000\ lb.)}{(85\ lb./in.^2)} = 35.3\ in.^2$$

Deflection (Allowable):

$$\Delta_{allow(DL+LL)} = \frac{L}{240} = \frac{(20')(12\ in./ft.)}{240} = 1"\ or$$

$$\Delta_{allow(LL)} = \frac{L}{360} = \frac{(20')(12\ in./ft.)}{360} = 0.67"$$

Note that the $S_{required}$ and $A_{required}$ values calculated do not account for the beam's own weight.

Try: 6 × 12 S4S.

($A = 63.25\ in.^2$; $S_x = 121.23\ in.^3$; $I_x = 697.07\ in.^4$)

Economical (efficient) beams usually have width-to-depth $\tfrac{1}{4} < b/h < \tfrac{1}{2}$

Check the effect of the beam's weight as it affects the bending and shear stress condition.

Bending:

$$S_{additional} = S_{add} = \frac{M_{add}}{F_b}$$

Where $M_{add.}$ = additional bending moment due to the beam's weight

$$M_{add.} = \frac{(\omega_{\text{beam wt.}})(L)^2}{8}$$

Conversion for wood density of 35 pcf (Douglas fir and Southern pine) to pounds per lineal foot of beam is:

$$\omega_{beam} = 0.252 \times \text{cross-sectional area of the beam}$$

$$\omega_{beam} = 0.252 \times (63.25) = 16 \text{ lb./ft.}$$

$$\therefore M_{add.} = \frac{(16 \text{ lb./ft.})(20')^2}{8} = 800 \text{ lb.-ft.}$$

$$S_{add.} = \frac{M_{add.}}{F_b} = \frac{(800 \text{ lb./ft.})(12 \text{ in./ft.})}{1300 \text{ lb./in.}^2} = 7.4 \text{ in.}^3$$

$$\therefore S_{total} = 92.3 \text{ in.}^3 + 7.4 \text{ in.}^3$$

$$= 99.7 \text{ in.}^3 < 121.2 \text{ in}^3 \quad \therefore \text{ OK}$$

Shear:

$$V_{add.} = \text{Additional shear developed due to the beam's own weight}$$

$$\therefore V_{add.} = \frac{\omega_{beam}L}{2} = \frac{(16 \text{ lb./ft.})(20')}{2} = 160 \text{ lb.}$$

$$A_{add.} = \frac{1.5V_{add.}}{F_v} = \frac{1.5(160 \text{ lb.})}{85 \text{ lb./in.}^2} = 2.8 \text{ in.}^2$$

$$\therefore A_{total} = 35.3 \text{ in.}^2 + 2.8 \text{ in.}$$

$$= 38.1 \text{ in.}^2 < 63.25 \text{ in.}^2 \quad \therefore \text{ OK}$$

Deflection (Actual):

$$\Delta_{actual} = \frac{5\omega_{LL}L^4}{384EI}$$

$$= \frac{5(100 \text{ lb./ft.})(20')^4(1728 \text{ in.}^3/\text{ft.}^3)}{384(1.6 \times 10^6 \text{ lb./in.}^2)(697.1 \text{ in.}^4)} = 0.32"$$

$$\Delta_{actual(LL)} = 0.32" < \Delta_{allow.(LL)} = 0.67"$$

$$\Delta_{actual} = \frac{5\omega_{total}L^4}{384EI}$$

$$= \frac{5(216 \text{ lb./ft.})(20')^4(1728 \text{ in.}^3/\text{ft.}^3)}{384(1.6 \times 10^6 \text{ lb./in.}^2)(697.1 \text{ in.}^4)} = 0.70"$$

Note: $\omega_{total} = 216$ lb./ft. *includes the beam weight.*

$$\Delta_{actual(DL+LL)} = 0.70'' < \Delta_{allow(DL+LL)} = 1.0''$$
$$\therefore OK$$

Use: 6×12 S4S.

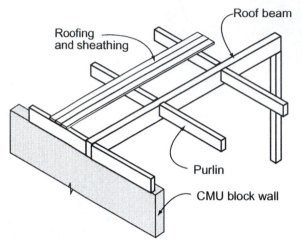

Figure 9.73   Roof framing.

Example Problem 9.15 (Figures 9.73 to 9.75)

Design a Southern pine No. 1 beam to carry the loads shown (roof beam, no plaster). Assume the beam is supported at each end by an 8" block wall. $F_b = 1550$ psi; $F_v = 110$ psi; $E = 1.6 \times 10^6$ psi.

Solution:

Bending:

$$S_{required} = \frac{M_{max}}{F_b} = \frac{(12.8 \text{ k-ft.})(12 \text{ in./ft.})}{1.55 \text{ k/in.}^2} = 99.1 \text{ in.}^3$$

Shear:

$$A_{required} = \frac{1.5V_{max}}{F_v} = \frac{1.5(2750 \text{ lb.})}{110 \text{ lb./in.}^2} = 37.5 \text{ in.}^2$$

Deflection (Allowable):

$$\Delta_{allow.} = \frac{L}{240} = \frac{(15')(12 \text{ in./ft.})}{240} = 0.75''$$

Try 6×12 S4S.

$$(A = 63.3 \text{ in.}^2; S_x = 121 \text{ in.}^3; I_x = 697 \text{ in.}^4)$$
$$\omega_{beam} \approx 0.252 \times 63.3 = 16 \text{ lb./ft.}$$

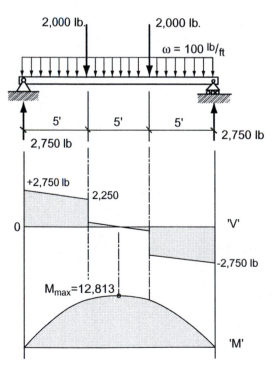

Figure 9.74   Load, V, and M diagrams.

Bending:

$$M_{add.} = \frac{\omega_{beam}L^2}{8} = \frac{(16 \text{ lb./ft.})(15')^2}{8} = 450 \text{ lb.-ft.}$$

$$S_{add.} = \frac{M_{add.}}{F_b} = \frac{(450 \text{ lb.-ft.})(12 \text{ in./ft.})}{1550 \text{ lb./in.}^2} = 3.5 \text{ in.}^3$$

$$S_{total} = 99.1 \text{ in.}^3 + 3.5 \text{ in.}^3$$

$$= 102.5 \text{ in.}^3 < 121 \text{ in.}^3 \quad \therefore \text{ OK}$$

Note: $S_{add.}$ *is usually 3–5% of* $S_{required}$.

Shear:

$$V_{add.} = \frac{\omega_{beam}L}{2} = \frac{(16 \text{ lb./ft.})(15')}{2} = 120 \text{ lb.}$$

$$A_{add.} = \frac{1.5V_{add.}}{F_v} = \frac{1.5(120 \text{ lb.})}{110 \text{ lb./in.}^2} = 1.6 \text{ in.}^2$$

$$A_{total} = 37.5 \text{ in.}^2 + 1.6 \text{ in.}^2$$

$$= 39.1 \text{ in.}^2 < 63.3 \text{ in.}^2 \quad \therefore \text{ OK}$$

Actual Deflection:

Using *superposition* (the combination, or *superimposing*, of one load condition onto another):

$$\Delta_{actual} = \frac{5\omega_{LL}L^4}{384EI} + \frac{23PL^3}{648EI} \quad \text{(at the center span)}$$

$$\Delta_{actual} = \frac{5(100 + 16)(15')^4(1728)}{384(1.6 \times 10^6)(697)}$$

$$+ \frac{23(2000)(15')^3(1728)}{684(1.6 \times 10^6)(697)}$$

$$\Delta_{actual} = 0.12" + 0.35" = 0.47" < 0.75" \quad \therefore \text{ OK}$$

Check the bearing stress between the beam and the block wall support.

$$f_{bearing} = f_p = \frac{P}{A_{brg}} = \frac{2870 \text{ lb.}}{(8" \times 5.5")} = 65.2 \text{ }^{lb.}/_{in.^2}$$

The allowable bearing stress perpendicular to the grain for Southern pine No. 1 is:

$$F_{c\perp} = 440 \text{ psi} \quad \therefore \text{ OK}$$

Use 6×12 S4S.

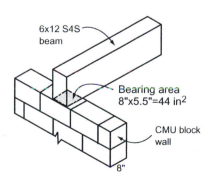

Figure 9.75   Beam bearing on CMU wall.

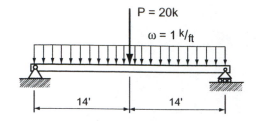

Figure 9.76

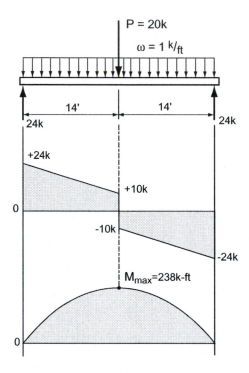

Figure 9.77   Load, V, and M diagrams.

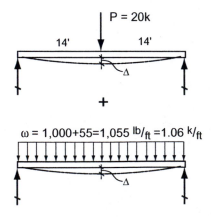

Figure 9.78   Load/deflection conditions.

Example Problem 9.16 (Figures 9.76 to 9.78)

A steel beam (A572/50) is loaded as shown. Assuming a deflection requirement of $\Delta_{\text{total}} = L/240$ and a depth restriction of 18" nominal, select the most economical section.

$$F_b = 30 \text{ ksi}; \ F_v = 20 \text{ ksi}; \ E = 30 \times 10^3 \text{ ksi}$$

Solution:

$$V_{\text{max}} = 24 \text{ k}; \ M_{max} = 238 \text{ k-ft.}$$

Steel beams are usually designed for bending. Once a trial section has been selected, shear and deflection are checked.

Bending:

$$S_{\text{req'd}} = \frac{M}{F_b} = \frac{(238 \text{ k-ft})(12 \text{ in./ft.})}{30 \text{ k/in.}^2} = 95.2 \text{ in.}^3$$

Try W18×55

$(S_x = 98.3 \text{ in.}^3, I_x = 890 \text{ in.}^4, t_w = 0.390", d = 18.11")$

$$M_{\text{add.}} = \frac{\omega_{\text{beam}} L^2}{8} = \frac{(55 \text{ lb./ft.})(28')^2}{8} = 5390 \text{ lb.-ft.}$$

$$S_{\text{add.}} = \frac{M_{\text{add.}}}{F_b} = \frac{(5.39 \text{ k-ft.})(12 \text{ in./ft.})}{30 \text{ k/in.}^2} = 2.2 \text{ in.}^3$$

$$S_{\text{total}} = 95.2 \text{ in.}^3 + 2.2 \text{ in.}^3$$
$$= 97.4 \text{ in.}^3 < 98.3 \text{ in.}^3 \quad \therefore \text{ OK}$$

Shear Check:

$$V_{\text{add.}} = \frac{\omega_{\text{beam}} L}{2} = \frac{(55 \text{ lb./ft.})(28')}{2} = 770 \text{ lb.}$$

$$f_{v_{\text{average}}} = \frac{V_{\text{max}}}{t_w d} = \frac{(24 \text{ k} + 0.77 \text{ k})}{(0.39")(18.11")}$$

$$= 3.5 \text{ ksi} < 20 \text{ ksi} \quad \therefore \text{ OK}$$

Deflection Check:

$$\Delta_{\text{allow.}} = \frac{L}{240} = \frac{(28')(12 \text{ in./ft.})}{240} = 1.4"$$

$$\Delta_{\text{actual}} = \frac{PL^3}{48EI} = \frac{5\omega L^4}{384EI}$$

$$\Delta_{\text{actual}} = \frac{(20 \text{ k})(28')^3(1728)}{48(30 \times 10^3)(890)}$$

$$+ \frac{5(1.06 \text{ k/ft.})(28')^4(1728)}{384(30 \times 10^3)(890)}$$

$$\Delta_{\text{actual}} = 0.59" + 0.55" = 1.14" < 1.4" \quad \therefore \text{ OK}$$

Use W18×55.

9.6 LATERAL BUCKLING IN BEAMS

In the previous discussion on beams, it was implied that making a beam as deep (large I_x) as possible was generally advantageous since the I_x and S_x values are maximized. There are, however, limits on how deep a beam should be when used in the context of the building. When a simply supported beam is subjected to a load, the top flange or surface is in compression while the bottom flange or surface is in tension. At the compression side of the beam, there is a tendency for it to buckle (deflect sideways), just as a column can buckle under axial loading. In a cantilever or overhang beam, the buckling or *sidesway* will develop due to the compression on the bottom surface of the beam (Figures 9.79 and 9.80). Very narrow, deep beams are particularly susceptible to lateral buckling, even at relatively low stress levels.

To resist the tendency of a beam to displace laterally, either the compression surface needs to be braced by other framing members or the beam needs to be reproportioned to provide a larger I_y. The vast majority of beams, such as floor and roof beams in buildings, are laterally supported by the floor or roof structures attached to and supported by them.

Steel decking welded to the beams, beams with the top flange embedded in the concrete slab, or composite construction (steel beams mechanically locked to the steel decking and concrete slab) are examples of lateral support for steel beams.

Wood framing typically employs continuous support along the top compression surface through sheathing nailed at a relatively close spacing and solid blocking to provide restraint against rotation at the ends. Depending on the span of the wood beam, bridging or solid blocking is provided at intervals to resist lateral buckling.

Some roof beams that support relatively lightweight roof sheathing are not considered to be laterally supported.

Certain beams are inherently stable against any lateral buckling tendency by virtue of their cross-sectional shapes. For example, a rectangular beam with a large width-to-depth ratio (I_y and I_x are relatively close) and loaded in the vertical plane should have no lateral stability problem (Figures 9.81 to 9.84). A wide-flange beam having a

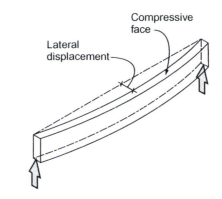

Figure 9.79   *Simply supported beam.*

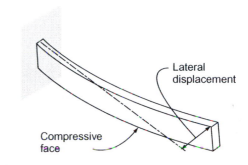

Figure 9.80   *Cantilever beam.*

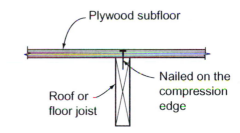

Figure 9.81   *Typical wood floor or roof joist with continuous nailing on the compression edge.*

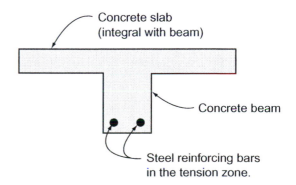

Figure 9.82   *Concrete slab/beam cast monolithically.*

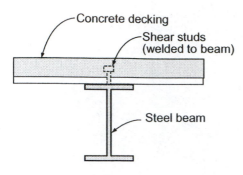

Figure 9.83   Composite concrete slab with steel beam.

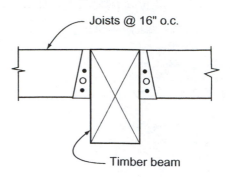

Figure 9.84   Timber beam with joist framing.

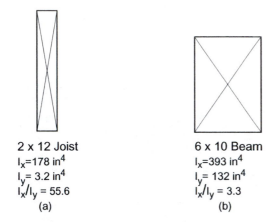

Figure 9.85   (a) Poor lateral resistance. (b) Good lateral resistance.

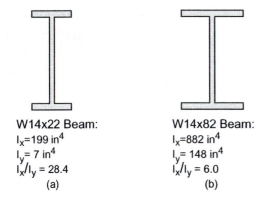

Figure 9.86   (a) Poor lateral resistance. (b) Good lateral resistance.

compression flange that is both wide and thick so as to provide a resistance to bending in a horizontal plane (relatively large I_y) will also have considerable resistance to buckling (Figures 9.85 and 9.86).

The problem of lateral instability in unbraced steel beams (W shapes) is amplified because the cross-sectional dimensions are such that relatively slender elements are stressed in compression. Slender elements have large width-to-thickness ratios, and these elements are particularly susceptible to buckling.

A beam that is not laterally stiff in cross-section must be braced every so often along its compressive side in order to develop its full moment capacity. Sections not adequately braced or laterally supported by secondary members could fail prematurely.

In Section 9.2, the design of steel beams assumed an allowable bending stress of $F_b = 0.6F_y$ (where $F_b = 22$ ksi for A36 steel). Steel beams laterally supported along their compression flanges, meeting the specific requirements of the AISC, are allowed to use $F_b = 0.66F_y$ (where $F_b = 24$ ksi for A36 steel). When the unsupported lengths of the compression flanges become large, allowable bending stresses may be reduced below the $F_b = 0.6F_y$ level.

For the purposes of preliminary sizing of steel beams in architectural practice, and in particular for this text, the allowable bending stress will be taken as:

$$F_b = 0.6F_y$$

In the case of timber beams, the dimensions of the cross-sections are such that the depth-to-width ratios are relatively small. A common method of dealing with the lateral stability issue is to follow rules of thumb that have developed over time. These rules apply to sawed lumber beams and joists/rafters (see Table 9.3). The beam depth-to-width ratios are based on nominal dimensions.

Table 9.3   Lateral bracing requirements for timber beams.

Beam Depth/ Width Ratio	Type of Lateral Bracing Required	Example
2 to 1	None	
3 to 1	The ends of the beam should be held in position	End blocking / Joist or beam
5 to 1	Hold the compression edge in line (continuously)	Sheathing or decking / Nailing / Joist or rafter
6 to 1	Diagonal bracing should be used	Nailed sheathing/decking / Bridging / Joist
7 to 1	Both edges of the beam should be held in line	Nailed sheathing/decking, top and bottom / Bridging / Joist

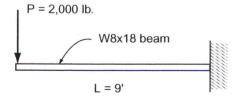

P = 2,000 lb.

W8x18 beam

L = 9'

Problem 9.1.1

Supplementary Problems

9.1.1   A cantilever beam has a span of 9 feet with a concentrated load of 2000 lb. at its unsupported end. If a W8×18 is used (F_b = 22 ksi), is it safe?

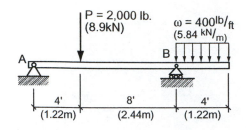

Problem 9.1.2

9.1.2 The single overhang beam uses a 4×12 S4S (100 × 300 mm) Douglas fir–larch No. 1 member. Determine the maximum bending stress developed. Is it safely designed? (F_b = 1300 psi or 8.97 MPa)

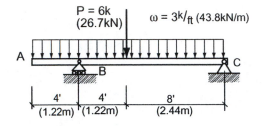

Problem 9.1.3

9.1.3 A 16-foot-long (4.88-m) single overhang beam is loaded as shown. Assuming a W8×35 (W200×52), determine the maximum bending stress developed. (F_b = 22 ksi or 150 MPa).

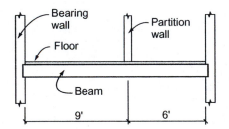

Problem 9.1.4

9.1.4 A beam as shown supports a floor and partition where the floor load is assumed to be uniformly distributed (500 lb/ft.) and the partition contributes a 1000-lb. concentrated load. Select the lightest W8 steel section if F_b = 22 ksi.

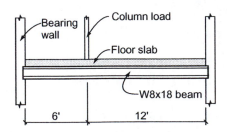

Problem 9.1.5

9.1.5 A W8×18 floor beam supports a concrete slab and a column load of 2400 lb. from above. The slab load on the beam is equal to 400 lb./ft. Draw V and M diagrams and determine the adequacy of the beam based on bending stress. (F_b for A36 steel is 22 ksi.)

9.1.6   A lintel beam over a doorway opening 10' wide supports a triangular load as shown. Assuming the lintel beam
to be a W8×15 (W200×22), A36 steel, determine the bending
stress developed. What size timber beam, 8" (200 mm)
nominal width, could be used if F_b = 1600 psi or 11 MPa?

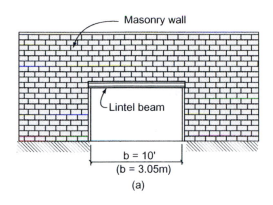

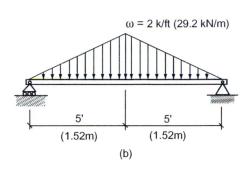

(a)   (b)

Problem 9.1.6

9.1.7   Glu-laminated beams are used to support the roof
and pulley load at a warehouse. Beams span 24' (7.32 m)
plus an 8' (2.44 m) overhang over the loading area. Determine the bending stress adequacy of the beam.

Properties of the glu-lam:

b = 6.75" (171 mm)

h = 12" (305 mm)

S = 162 in.^{3} (2.66 × 10^6 mm^3)

I = 974 in.^{4} (404 × 10^6 mm^4)

F_b = 2400 psi (16.5 MPa)

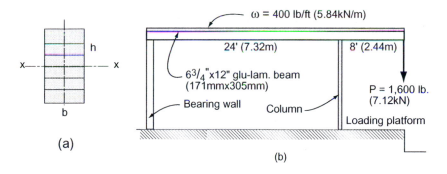

(a)

(b)

Problem 9.1.7

9.1.8   A W8×28 (W200×42) steel beam is loaded and supported as shown. Determine the maximum bending stress
developed at the wall. What is the bending stress at a point
4 feet to the right of the free end of the beam? (Construct
the *V* and *M* diagrams.)

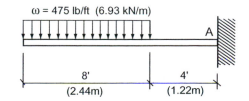

Problem 9.1.8

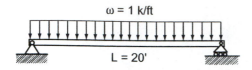

ω = 1 k/ft

L = 20'

Problem 9.1.9

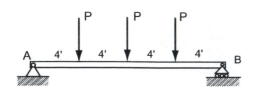

P   P   P

A   4'   4'   4'   4'   B

Problem 9.1.10

9.1.9 Select the lightest 14" nominal depth W beam to carry the load shown. Assume A36 steel (F_b = 22 ksi).

9.1.10 A W18×40 (A36) beam is used to support three concentrated loads of magnitude P. Determine the maximum permissible P. Draw V and M diagrams as an aid.

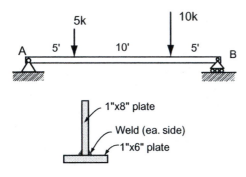

5k   10k

A   5'   10'   5'   B

1"x8" plate

Weld (ea. side)

1"x6" plate

Problem 9.1.11

9.1.11 Two steel plates (A572, F_y = 50 ksi) are welded together to form an inverted T-beam. Determine the maximum bending stress developed. Also determine the maximum shear stress at the neutral axis (N.A.) of the cross-section and at the intersection where the stem joins the flange.

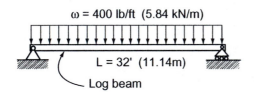

ω = 400 lb/ft (5.84 kN/m)

L = 32' (11.14m)

Log beam

Problem 9.1.12

9.1.12 A log of diameter D is available to be used as a beam carrying a uniformly distributed load of 400 lb./ft. (5.84 kN/m) over a length of 32 feet (11.14 m). Determine the required diameter D necessary if F_b = 1200 psi (8.27 MPa) and F_v = 100 psi (690 kPa).

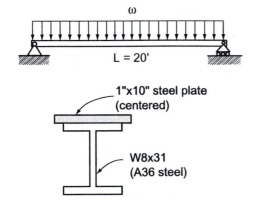

ω

L = 20'

1"x10" steel plate
(centered)

W8x31
(A36 steel)

Problem 9.1.13

9.1.13 The 20-foot beam shown in the figure has a cross-section built up from a 1" × 10" steel plate welded onto the top of a W8×31 section. Determine the maximum load w the beam can sustain when the steel section reaches a maximum allowable bending stress of F_b = 22 ksi. For the w calculated, determine the shear stress f_v developed between the plate and the top flange surface (use the flange width for b).

9.1.14 A lintel beam 12' long is used in carrying the imposed loads over a doorway opening. Assuming that a built-up box beam is used with a 12" overall depth as shown, determine the maximum bending stress and shear stress developed.

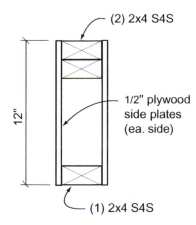

(2) 2x4 S4S

1/2" plywood side plates (ea. side)

12"

(1) 2x4 S4S

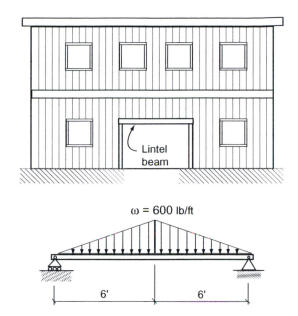

Lintel beam

ω = 600 lb/ft

6' 6'

Problem 9.1.14

9.1.15 The cross-section of the rough-cut timber beam shown is loaded with ω over 6 ft. (1.83 m) of the span. Determine the maximum value of ω if the allowable bending stress is F_b = 1600 psi (11 MPa) and the allowable shear stress is F_v = 85 psi (586 kPa).

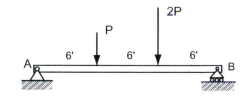

ω = (lb/ft or kN/m)

2' (.61m) 2' (.61m)

L = 10' (3.05m)

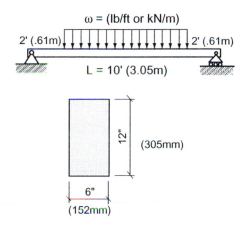

12" (305mm)

6" (152mm)

Problem 9.1.15

9.1.16 A 4×12 S4S beam carries two concentrated loads as shown. Assuming F_b = 1600 psi and F_v = 85 psi, determine:

    a. The maximum permissible load P.
    b. The bending and shear stress 4 ft. to the right of support A.

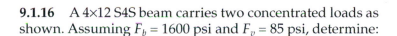

2P

P

6' 6' 6'

A B

Problem 9.1.16

9.1.17 The beam shown is constructed by welding cover plates to two channel sections. What maximum uniformly distributed load can this beam support on a 20-ft. span if F_b = 22 ksi.

Check the shear stress where the plate attaches to the channel flange.

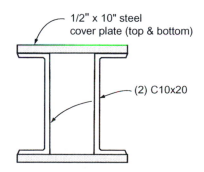

1/2" x 10" steel cover plate (top & bottom)

(2) C10x20

Problem 9.1.17

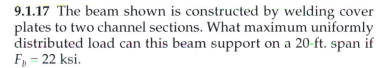

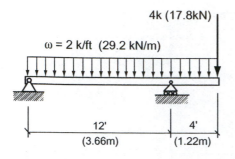

Problem 9.1.18

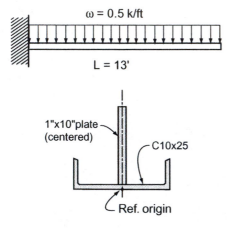

Problem 9.1.19

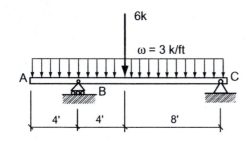

Problem 9.1.20

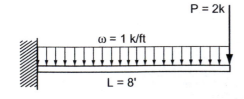

Problem 9.1.21

9.1.18 Select the lightest W section steel beam based on the bending condition. Check $fv_{average}$ for the beam selected. $F_b = 22$ ksi (152 MPa); $F_v = 14.5$ ksi (100 MPa).

9.1.19 A built-up beam section is used to support a uniformly distributed load as shown. Determine the maximum bending stress developed in the beam. Check the shear stress (based on V_{max}) at the intersection of the channel and plate.

9.1.20 A 16' span beam is loaded as shown. Determine the following:

    a. Support reactions at B and C.
    b. V diagram—label the peak points.
    c. M diagram—label the peak points.
    d. Maximum bending stress developed assuming a W8×35 beam. (A36 steel)
    e. Is the beam overstressed?
    f. What would you select as the most "economical" (lightest) section for the condition shown?

9.1.21 Assuming A36 steel, select the most economical W8 section. Check the shear stress and determine the deflection at the free end.

$$F_b = 22 \text{ ksi}$$

$$F_v = 14.5 \text{ ksi}$$

$$E = 29 \times 10^3 \text{ ksi}$$

9.1.22 Design a Douglas fir–larch No. 1 beam to support the load shown.

$$F_b = 1300 \text{ psi}$$

$$F_v = 85 \text{ psi}$$

$$E = 1.6 \times 10^6 \text{ psi}$$

$$\Delta_{\text{allow}(LL)} = L/360$$

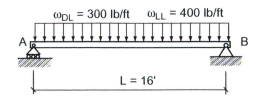

Problem 9.1.22

9.1.23 Design the beam shown assuming the loads are due to dead and live loads.

$$\Delta_{\text{allow (DL+LL)}} = L/240$$

Assuming the beam is Southern pine No. 1 supported at both ends by girders as shown, calculate and check the bearing stress developed between the 6 ×___ beam and a

6×12 S4S girder.

$F_b = 1550$ psi; $F_v = 110$ psi;

$E = 1.6 \times 10^6$ psi; $F_{C \text{ (perpendicular)}} = 410$ psi

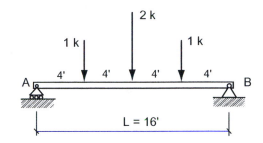

Problem 9.1.23

9.1.24 Design B-1 and SB-1 assuming A36 steel. Maximum depth for each is restricted to 16" nominal.

LL	40 psf
Concrete	150 pcf
Curtail wall on spandrel beam SB-1	300 plf
Suspended plaster ceiling	5 psf
Metal deck	4 psf

$\Delta_{\text{allow}(LL)} < L/360$ for B-1;

$\Delta_{\text{allow}(DL+LL)} < L/240$ for SB-1

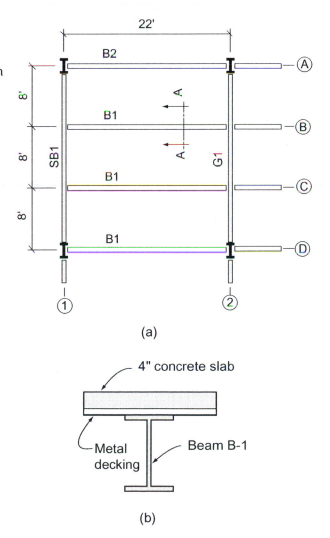

Problem 9.1.24

10 Column Analysis and Design

Introduction

Columns are essentially vertical members responsible for supporting compressive loads from roofs and floors, transmitting the vertical forces to the foundations and on to the subsoil. The structural work performed by the column is somewhat simpler than that of the beam because the applied loads are in the same vertical orientation. Although columns are normally considered to be vertical elements, they can actually be positioned in any orientation. Columns are defined by their length dimension between support ends and can be either very short (e.g., footing piers) or very long (e.g., bridge and freeway piers). They are used as major elements in trusses, building frames, and substructure supports for bridges (Figures 10.1 and 10.2). Loads are typically applied at member ends, producing axial compressive stresses.

Common terms used to identify column elements include *studs*, *struts*, *posts*, *piers*, *piles*, and *shafts*. Virtually every common construction material is used for column construction, including steel, timber, concrete (reinforced and prestressed), and masonry. Each material possesses characteristics (material and production) that present opportunities and limitations on the shapes of cross-sections and profiles chosen. Columns are major structural components that significantly affect the building's overall performance and stability and thus are designed with larger safety factors than other structural components. Failure of a joist or beam may be localized and may not severely affect the building's integrity; however, failure of a strategic column may be catastrophic for a large area of the structure. Safety factors for columns adjust for the uncertainties of material irregularities, support fixity at the column ends, and take into consideration construction inaccuracies, workmanship, and unavoidable eccentric (off-axis) loading.

Figure 10.1 Maritime Museum, Astoria, Oregon.

Figure 10.2 Fisheries Building, UW Campus.

10.1 SHORT AND LONG COLUMNS— MODES OF FAILURE

Large slabs of stone used at *Stonehenge* were extremely massive and tended to be stabilized by their own weight. Massive stone columns continued to be used in Greek and Roman structures, but with the development of wrought iron, cast iron, steel, and reinforced concrete, columns began to take on much more slender proportions. Column slenderness greatly influences a column's ability to carry load. Because a column is a compression member, it would be reasonable to assume that one would fail due to crushing or excessive shortening once the stress level exceeded the elastic (yield point) limit of the material. However, for most columns failure occurs at a lower level than the column's material strength because most are relatively slender (long in relation to their lateral dimension) and fail due to buckling (lateral instability). *Buckling* is the sudden uncontrolled lateral displacement of a column, at which point no additional load can be supported. The sideways deflection or buckle will eventually fail in bending if loads are increased. Very short, stout columns fail by crushing due to material failure (Figure 10.3); long, slender columns fail by buckling—a function of the column's dimensions and its modulus of elasticity (Figure 10.4).

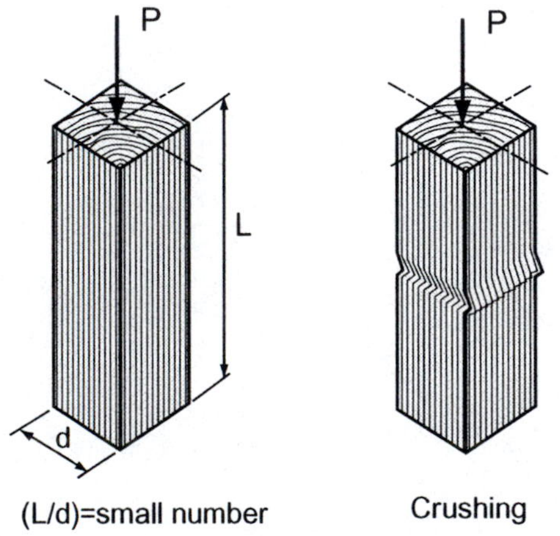

Short Columns

Stress computations for short columns are very simple and rely on the basic stress equation developed at the beginning of Chapter 6. If the load and column size are known, the actual compressive stress may be computed as:

$$f_a = \frac{P_{\text{actual}}}{A} \leq F_a$$

where:

f_a = actual compressive stress (psi, ksi, Pa, kPa)

A = cross-sectional area of column (in.^{2}, mm^2)

P_{actual} = actual load on the column (pounds, kips, N, kN)

F_a = allowable compressive stress per codes (psi, ksi, Pa, kPa)

This stress equation can be easily rewritten into a design form when determining the requisite short column size when the load and allowable material strength are known:

$$A_{\text{required}} = \frac{P_{\text{actual}}}{F_a}$$

where:

A_{required} = minimum cross-sectional area of the column

(L/d)=small number   Crushing

Figure 10.3   Short columns—High stress level.

Long Columns—Euler Buckling

The buckling phenomenon in slender columns is due to the inevitable eccentricities in loading and the likelihood of irregularities in a material's resistance to compression. Buckling could be avoided (theoretically) if the loads applied were absolutely axial and the column material was totally homogeneous with no imperfections. Obviously, this is not possible; hence, buckling is a fact of life for any slender column. The load capacity of a slender column is directly dependent on the dimension and shape of the column as well as the stiffness of the material (E), but is independent of the strength of the material (yield stress).

The buckling behavior of slender columns, within their elastic limit, was first investigated by a Swiss mathematician named Leonhard Euler (1707–1783). Euler's equation presents the relationship between the load that causes buckling of a pinned end column and the stiffness properties of the column. The critical buckling load can be determined by the equation:

$$P_{\text{critical}} = \frac{\pi^2 E I_{\text{min.}}}{L^2}$$

where:

P_{critical} = critical axial load that causes buckling in the column (pounds, kips, Newtons, kilonewtons)

E = modulus of elasticity of the column material (psi, ksi, Pascal, kPa)

$I_{\text{min.}}$ = smallest moment of inertia of the column cross-section (in.⁴, mm⁴)

L = column length between pinned ends (in., mm)

Note that as the column length becomes very long, the critical load becomes very small, approaching zero as a limit. Conversely, very short column lengths require extremely large loads to cause the member to buckle. High loads result in high stresses, which cause crushing rather than buckling.

The *Euler equation* demonstrates the susceptibility of the column to buckling as a function of the column length squared, the stiffness of the material used (E), and the cross-sectional stiffness as measured by the moment of inertia (I).

To understand the phenomenon of buckling further, let's examine a slender column with a slight initial bow to it before loading (Figure 10.5).

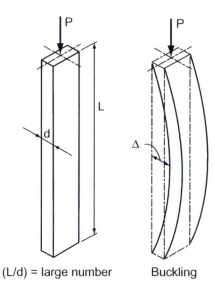

(L/d) = large number   Buckling

Figure 10.4   Long columns—Elastic instability.

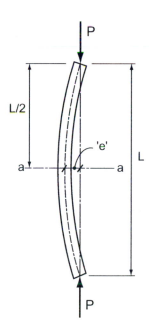

Figure 10.5   Column with an eccentricity e.

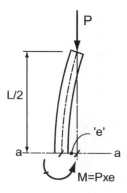

Figure 10.6   The offset load P produces a moment M = Pxe.

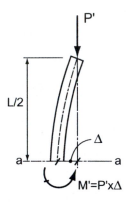

Figure 10.7   P'>P: Increased load with an increased displacement Δ>e.

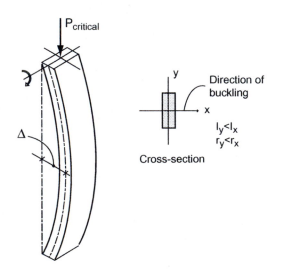

Figure 10.8   Column buckling about the weak axis.

Since the load **P** is offset (eccentric to the central axis of the column), a moment **M = Pxe**, shown in Figure 10.6, results in bending stresses being present in addition to the compressive stress $f = P/A$. If the load is increased, an additional moment results in bending the column further, and thus results in a larger eccentricity or displacement. This moment $M' = P'x\Delta$, shown in Figure 10.7, results in an increased bending that causes more displacement, thus creating an even larger moment (**P − Δ effect**). A progressive bending moment and displacement continues until the stability of the column is compromised. The critical load at which the column's ability to resist uncontrolled, progressive displacement has been reached is referred to as *Euler's critical buckling load*.

It is important to note again that the Euler equation (which contains no safety factors) is valid only for long, slender columns that fail due to buckling, and in which stresses are well within the elastic limit of the material. Short columns tend to fail by crushing at very high stress levels, well beyond the elastic range of the column material.

Slenderness Ratios

The geometric property of a cross-section called the *radius of gyration* was introduced briefly in Chapter 7. This dimensional property is being recalled in connection with the design of columns. Another useful form of the Euler equation can be developed by substituting the radius of gyration for the moment of inertia in which:

$$r = \sqrt{\frac{I}{A}} \quad \text{and} \quad I = Ar^2$$

where:

r = radius of gyration of the column cross-section (in.)
I = least (minimum) moment of inertia (in.⁴)
A = cross-sectional area of the column (in.²)

The critical stress developed in a long column at buckling can be expressed as:

$$f_{\text{critical}} = \frac{P_{\text{critical}}}{A} = \frac{\pi^2 E(Ar^2)}{AL^2} = \frac{\pi^2 E}{\left(L/r\right)^2}$$

The (L/r) is known as the *slenderness ratio*. The critical buckling stress of a column depends inversely on the square of the slenderness ratio. The slenderness ratio is a primary indicator of the mode of failure one might expect for a column under load. Column sections with high r values are more resistant to buckling (see Figure 10.8). Since the radius of gyration is derived from the moment of inertia, we can deduce that cross-sectional configuration is critical in generating higher r values.

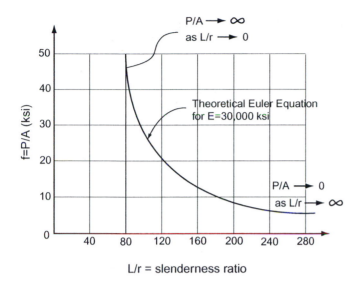

Figure 10.9 *Buckling stress versus slenderness ratio.*

High slenderness ratios mean lower critical stresses (Figure 10.9) that will cause buckling; conversely, lower slenderness ratios result in higher critical stress (but still within the elastic range of the material).

As a comparison of steel column sections often found in buildings, note the difference in $r_{min.}$ values for three sections shown in Figure 10.10. All three sections have relatively equal cross-sectional areas but very different radii of gyration about the critical buckling axis. If all three columns were assumed as 15 feet in length and pin connected at both ends, the corresponding slenderness ratios are quite different indeed.

In general, the most efficient column sections for axial loads are those with almost equal r_x and r_y values. Circular pipe sections and square tubes are the most effective shapes since the radii of gyration about both axes are the same ($r_x = r_y$). For this reason, these types of sections are often used as columns for light to moderate loads. However, they are not necessarily appropriate for heavy loads and where many beam connections must be made. The practical considerations and advantages of making structural connections to easily accessible wide-flange shapes often outweigh the pure structural advantages of closed cross-sectional shapes (like tubes and pipes).

Special wide-flange sections are specifically manufactured to provide relatively symmetrical columns (r_x/r_y ratios approaching 1.0) with large load-carrying capability. Most of these column sections have depth and flange widths approximately equal ("boxy" configuration) and are generally in the 10", 12", and 14" nominal depth category.

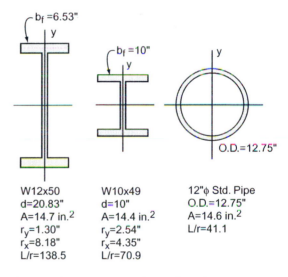

W12x50
d=20.83"
A=14.7 in.²
r_y=1.30"
r_x=8.18"
L/r=138.5

W10x49
d=10"
A=14.4 in.²
r_y=2.54"
r_x=4.35"
L/r=70.9

12"φ Std. Pipe
O.D.=12.75"
A=14.6 in.²
L/r=41.1

Figure 10.10 *Comparison of steel cross-sections with equivalent areas.*

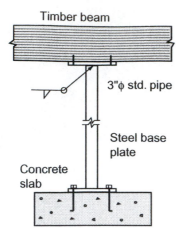

Figure 10.11   Steel pipe column.

Figure 10.12   FBD of column.

Example Problem 10.1: Short and Long Columns— Modes of Failure (Figures 10.11 and 10.12)

Determine the critical buckling load for a 3" ϕ standard weight steel pipe column that is 16 ft. tall and pin connected. Assume that $E = 29 \times 10^6$ psi

Solution:

From the Euler buckling equation:

$$P_{critical} = \frac{\pi^2 E I_{min.}}{L^2}$$

The least (smallest) moment of inertia is normally used in the Euler equation to produce the critical buckling load. In this example, however, $I_x = I_y$ for a circular pipe:

$$I = 3.02 \text{ in.}^4 \text{ (There is no weak axis for buckling.)}$$

$$P_{critical} = \frac{(3.14)^2 \left(29 \times 10^6 \, ^{lb.}\!/_{in.^2}\right)(3.02 \text{ in.}^4)}{\left(16 \text{ ft.} \times 12 \, ^{in.}\!/_{ft.}\right)^2}$$

$$= 23{,}420 \text{ lb.}$$

The accompanying critical stress can be evaluated as:

$$f_{critical} = \frac{P_{critical}}{A} = \frac{23{,}420 \text{ lb.}}{2.23 \text{ in.}^2} = 10{,}500 \text{ psi}$$

This column buckles at a relatively low stress level compared with the allowable compressive stress.

$$F_{compression} = 22 \text{ ksi}$$

Example Problem 10.2 (Figure 10.13)

Determine the critical buckling stress for a 30-foot-long, W12×65 steel column. Assume simple pin connections at the top and bottom.

$$F_y = 36 \text{ ksi (A36 steel)}; \quad E = 29 \times 10^3 \text{ ksi}$$

Solution:

For a W12×65, $r_x = 5.28"$, $r_y = 3.02"$, $A = 19.1 \text{ in.}^2$

Compute the slenderness ratio L/r for each of the two axes. Substitute the larger of the two values into the Euler equation because it will yield the more critical stress value.

$$\frac{L}{r_x} = \frac{30 \text{ ft.} \times 12 \, ^{in.}\!/_{ft.}}{5.28 \text{ in.}} = 68.2$$

$$\frac{L}{r_y} = \frac{30 \text{ ft.} \times 12 \, ^{in.}\!/_{ft.}}{3.02 \text{ in.}} = 119.2 \leftarrow \text{ governs}$$

(produces a smaller stress value at buckling)

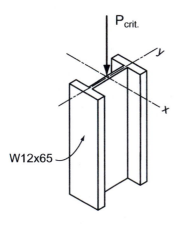

Figure 10.13   Critical load on a W column.

The use of L/r_x would clearly yield a much larger stress value. This indicates that the column would buckle about the y-axis under a much smaller load than would be required to make it buckle the other way. In practical terms, this means that, in case of overload, the column would not be able to reach the critical load necessary to make it buckle about its strong axis; it would have failed at a lower load value by buckling about its weak axis. Therefore, in computing critical load and stress values, always use the greater L/r value.

$$f_{cr.} = \frac{\pi^2 E}{(^L/_r)^2} = \frac{(3.14)^2(29 \times 10^3 \text{ ksi})}{(119.2)^2} = 20.1 \text{ ksi}$$

$$P_{cr.} = f_{cr} \times A = (20.1 \text{ }^k/_{in^2}) \times (19.1 \text{ in.}^2) = 384 \text{ k}$$

Example Problem 10.3 (Figure 10.14)

Evaluate the Euler buckling load of a long column, W250×149 (W10×100), pin connected at the top and bottom. $E = 207 \times 10^3$ MPa (30×10^3 ksi), and $F_y = 345$ MPa (50 ksi).

Solution:

The Euler buckling formula for long columns is:

$$P_{critical} = \frac{\pi^2 E I_{min.}}{L^2}$$

$$L = 6 \text{ m}; \quad L^2 = 36 \text{ m}^2$$

Looking up the moment of inertia in Appendix Table 10 (Table 6 for U.S. customary)

$$I_{min.} = I_y = 86.1 \times 10^6 \text{ mm}^4 \text{ (207 in.}^4)$$

For dimensional unit compatibility, the I value will be converted to meter units ($1 \text{ mm}^4 = 10^{-12} \text{ m}^4$) and

$$\text{MPa} = {}^{\text{MN}}/_{m^2}$$

Thus: $\quad I_{min.} = 86.1 \times 10^{-6} \text{ m}^4$

$$P_{critical} = \frac{\pi^2 \left(207 \times 10^3 \text{ }^{MN}/_{m^2}\right)(86.1 \times 10^{-6} \text{ m}^4)}{36 \text{ m}^2}$$

$$= 4.88 \text{ MN}$$

Solving this same problem using U.S. customary units;

$$P_{crit.} = \frac{\pi^2 \left(30 \times 10^3 \text{ }^k/_{in.^2}\right)(207 \text{ in.}^4)}{\left(19.7 \text{ ft.} \times 12 \text{ }^{in.}/_{ft.}\right)^2}$$

$$= 1096 \text{ k (4880 kN)}$$

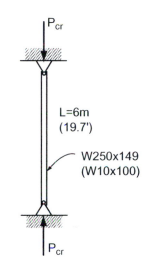

Figure 10.14   Buckling load on a long column.

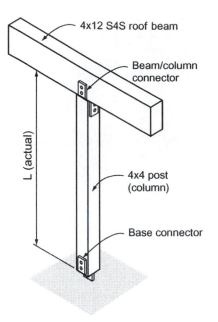

Figure 10.15   Wood beam/column framing.

10.2   END SUPPORT CONDITIONS AND LATERAL BRACING

In the previous analysis of Euler's equation, each column was assumed to have pinned ends in which the member ends were free to rotate (but not translate) in any direction at their ends. If, therefore, a load is applied vertically until the column buckles, it will do so in one smooth curve (see Figure 10.8). The length of this curve is referred to as the *effective* or *buckled length*. In practice, however, this is not always the case, and the length free to buckle is greatly influenced by its end support conditions.

The assumption of pinned ends is an important one since a change of end conditions imposed on such a column may have a marked effect upon its load-carrying capacity. If a column is solidly connected at the top and bottom, it is unlikely to buckle under the same load assumed for a pinned-end column. Restraining the ends of a column from translation and a free-rotation condition generally increases the load-carrying capacity of a column. Allowing translation as well as rotation at the ends of a column generally reduces its load-carrying capacity.

Column design formulas generally assume a condition in which both ends are fixed in translation but free to rotate (pin connected). When other conditions exist, the load-carrying capacity is increased or decreased, so the allowable compressive stress must be increased or decreased, or the slenderness ratio must be increased. For example, in steel columns, a factor, *K*, is used as a multiplier for converting the actual length to an effective buckling length based on end conditions (Figure 10.15). The *theoretical K* values listed in Figures 10.16 to 10.21 are *less conservative* than the actual values often used in structural design practice.

Case A: Both Ends Pinned—Structure adequately braced against lateral (wind and earthquake) forces.

$$L_e = L$$

$$K = 1.0$$

$$P_{\text{critical}} = \frac{\pi^2 EI}{L^2}$$

Examples:
Timber column nailed top and bottom.

Timber columns using standard, stamped, sheet metal beam/column and base connectors.

Steel column with simple clip angle connections top and bottom.

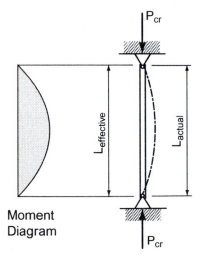

Figure 10.16   Case A—effective buckling length, both ends pinned.

Case B: Both Ends Fixed—Structure adequately braced against lateral forces.

$$L_e = 0.5L$$

$$K = 0.5$$

$$P_{critical} = \frac{\pi^2 EI}{(0.5L)^2} = \frac{4\pi^2 EI}{L^2}$$

Examples:
Concrete column rigidly (monolithically cast) connected to large beams top and bottom.

Steel column rigidly connected (welded) to large steel beams top and bottom.

Steel column rigidly connected to a beam above and substantial base plate on the bottom, generally welded.

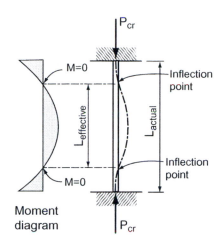

Figure 10.17   Case B—effective buckling length, both ends fixed.

Case C: One End Pinned and One End Fixed—Structure adequately braced against lateral forces.

$$L_e = 0.7L$$

$$K = 0.7$$

$$P_{critical} = \frac{\pi^2 EI}{(0.7L)^2} \approx \frac{2\pi^2 EI}{L^2}$$

Examples:
Concrete column rigidly connected to concrete slab at the base and attached to light-gauge roofing at the top.

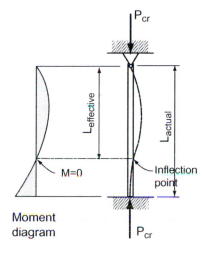

Figure 10.18   Case C—effective buckling length, one end pinned and the other fixed.

Case D: One End Free and One End Fixed—Lateral translation possible (develops eccentric column load).

$$L_e = 2.0L$$

$$K = 2.0$$

$$P_{critical} = \frac{\pi^2 EI}{(2L)^2} = \frac{\tfrac{1}{4}\pi^2 EI}{L^2}$$

Examples:
Water tank mounted on a simple pipe column.

Flagpole analogy.

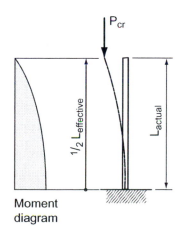

Figure 10.19   Case D—effective buckling length, one end free and one end fixed.

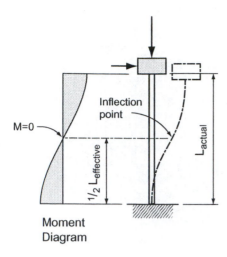

Figure 10.20   Case E—effective buckling length, both ends fixed with some lateral translation.

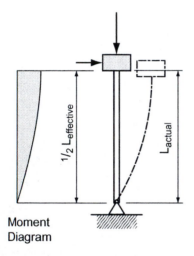

Figure 10.21   Case F—effective buckling length, base pinned, top fixed with some lateral translation.

Case E: Both Ends Fixed with Some Lateral Translation

$$L_e = 1.0L$$

$$K = 1.0$$

$$P_{critical} = \frac{\pi^2 EI}{L^2}$$

Examples:
Columns in a relatively flexible rigid frame structure (concrete or steel).

Case F: Base Pinned, Top Fixed with Some Lateral Translation

$$L_e = 2.0L$$

$$K = 2.0$$

$$P_{critical} = \frac{\pi^2 EI}{(2L)^2} = \frac{{}^1\!/_4\,\pi^2 EI}{L^2}$$

Examples:
Steel column with a rigid connection to a beam above and a simple pin connection at the base. There is some flexibility in the structure, allowing column loads to be positioned eccentrically.

Intermediate Lateral Bracing

In the previous section we found that the selection of the type of end connection used directly influenced the buckling capacity of a column. Fixed connections seem to be an obvious solution to minimizing column sizes; however, the cost associated with achieving rigid connections is high, and such connections are difficult to make. Also, timber columns are generally assumed to be pin connected since the material strength generally precludes the construction of true rigid joints. So what other methods are there to achieve an increase in column capacity without specifying larger column sizes?

A common strategy used to increase the effectiveness of a column is to introduce *lateral bracing* about the weak axis of buckling (Figure 10.22). Infill wall panels, window headers, girts for curtain walls, and other systems provide lateral bracing potentials that can be used to reduce the buckling length of the column. Bracing provided in one plane does not, however, provide resistance to buckling in the perpendicular plane (Figure 10.23). Columns must be checked in both directions to determine the critical slenderness ratio to be used in analysis or design. Lateral restraint provided by the bracing is for buckling resistance about the *y axis* only. The column is still susceptible to buckling in one smooth curve about the *x axis*.

Very slender sections can be used for columns if they are adequately braced or stiffened against buckling between floors (Figure 10.24). Slenderness ratios must be calculated for both axes to determine which direction governs.

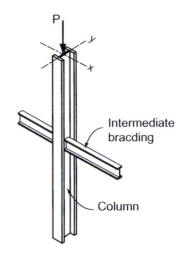

Figure 10.22   Wide-flange column section braced about the weak axis of buckling.

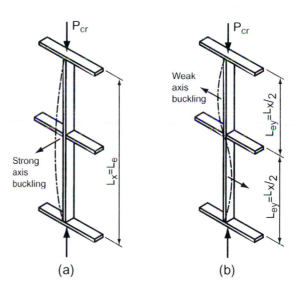

Figure 10.23   (a) Strong axis. (b) Weak axis.

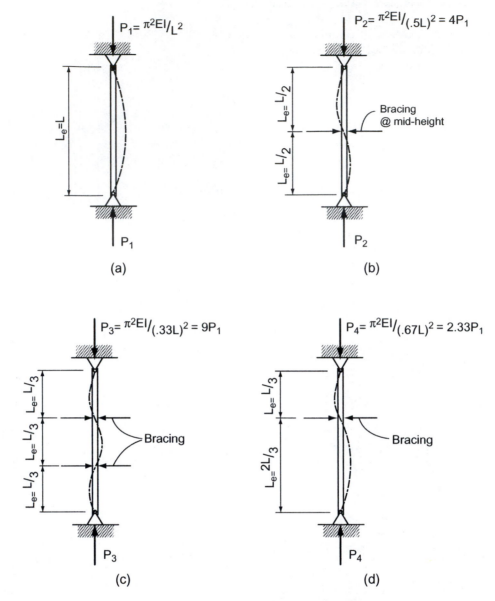

*Figure 10.24 Effective buckling lengths for various lateral support conditions.
(a) No bracing. (b) Braced at midpoint. (c) Third-point bracing. (d) Asymmetric
bracing.*

Example Problem 10.4 (Figure 10.25)

Determine the Euler buckling capacity of a timber pole, $L = 24$ ft. (7.32 m) used in carrying a major roof load. The pole has a 10" (254-mm) diameter and is considered pin connected at the top and bottom. Assume $E = 1.0 \times 10^6$ psi or $E = 6.895 \times 10^3$ MPa.

Solution:

$$P_{critical} = \frac{\pi^2 EI}{L^2_{effective}} = \frac{\pi^2 EI}{(KL)^2}$$

Since the top and bottom are assumed to be pin connected, $K = 1.0$.

For a circular cross-section, the area and moment of inertia are computed as:

$$A = \frac{\pi D^2}{4} = \frac{\pi (10 \text{ in.})^2}{4} = 78.5 \text{ in.}^2$$

$$I = \frac{\pi D^4}{64} = \frac{\pi (10 \text{ in.})^4}{64} = 491 \text{ in.}^4$$

In SI units,

$$A = \frac{\pi D^2}{4} = \frac{\pi (254 \text{ mm})^2}{4} = 50.6 \times 10^3 \text{ mm}^2$$

$$I = \frac{\pi D^4}{64} = \frac{\pi (254 \text{ mm})^4}{64} = 2.04 \times 10^8 \text{ mm}^4$$

$$= 2.04 \times 10^{-4} \text{ m}^4$$

$$KL = (1.0)(24 \text{ ft.}) = 24 \text{ ft.} \quad \text{or} \quad KL = (1.0)(7.32 \text{ m})$$

$$= 7.32 \text{ m}$$

$$P_{cr.} = \frac{\pi^2 \left((1.0 \times 10^3 \text{k}/_{\text{in.}^2})(491 \text{ in.}^4) \right)}{\left(24 \text{ ft.} \times 12^{\text{in.}}/_{\text{ft.}} \right)^2} = 58.4 \text{ k}$$

In SI units,

$$P_{cr.} = \frac{\pi^2 \left(6.895 \times 10^3 \text{MN}/_{\text{m}^2} \right)(2.04 \times 10^{-4} \text{ m}^4)}{(7.32 \text{ m})^2}$$

$$= 0.26 \text{ MN} = 260 \text{ kN}$$

The critical stress in the pole can be computed as:

$$f_{critical} = \frac{P_{cr.}}{A} = \frac{58.4 \text{ k}}{78.5 \text{ in.}^2} = 0.744^{\text{k}}/_{\text{in.}^2} = 744 \text{ psi}$$

(A relatively low stress value)

$$f_{cr.} = \frac{P_{cr.}}{A} = \frac{260 \text{ kN}}{50.6 \times 10^{-3} \text{ m}^2}$$

$$= 5138^{\text{kN}}/_{\text{m}^2} = 5.14 \text{ MPa}$$

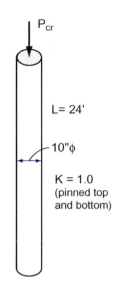

Figure 10.25   Circular cross-section pole.

P_{cr}

$L = 24'$

10" φ

K = 1.0
(pinned top
and bottom)

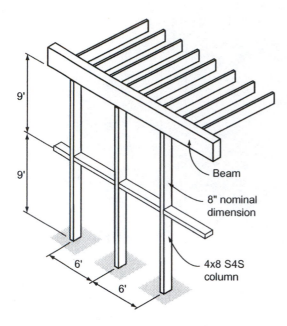

9'

9'

6'

6'

Beam

8" nominal
dimension

4x8 S4S
column

Figure 10.26   Timber roof/wall framing.

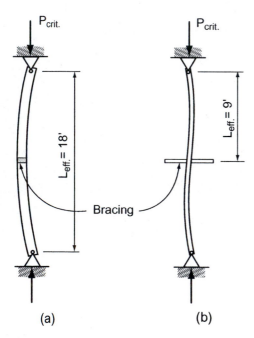

P_crit.

P_crit.

$L_{eff.} = 18'$

$L_{eff.} = 9'$

Bracing

(a)

(b)

*Figure 10.27   (a) Strong axis buckling.
(b) Weak axis buckling.*

Example Problem 10.5 (Figures 10.26 and 10.27)

Determine the critical buckling load for a 4×8 S4S Douglas fir column that is 18' long and braced at midheight against the weak direction of buckling. $E = 1.3 \times 10^6$ psi.

Solution:

$$4\times8 \text{ S4S } (I_x = 111.2 \text{ in.}^4, I_y = 25.9 \text{ in.}^4,$$
$$A = 25.38 \text{ in.}^2)$$

$$r_x = \sqrt{\frac{I_x}{A}} = \sqrt{\frac{111.2 \text{ in.}^4}{25.38 \text{ in.}^2}} = 2.1 \text{ in.}$$

$$r_y = \sqrt{\frac{I_y}{A}} = \sqrt{\frac{25.9 \text{ in.}^4}{25.38 \text{ in.}^2}} = 1.01 \text{ in.}$$

Load causing buckling about the *x* axis:

$$P_{cr.} = \frac{\pi^2 E I_x}{L_x^2} = \frac{(3.14)^2(1.3 \times 10^6 \text{ lb.}/\text{in.}^2)(111.2 \text{ in.}^4)}{(18' \times 12 \text{ in.}/\text{ft.})^2}$$

$$= 30,600 \text{ lb.}$$

Load causing buckling about the *y* axis:

$$P_{cr.} = \frac{\pi^2 E I_y}{L_y^2} = \frac{(3.14)^2(1.3 \times 10^6 \text{ lb.}/\text{in.}^2)(25.9 \text{ in.}^4)}{(9' \times 12 \text{ in.}/\text{ft.})}$$

$$= 28,500 \text{ lb.}$$

Since the load required to cause the member to buckle about the weaker *y* axis is less than the load that is associated with buckling about the stronger *x* axis, the critical buckling load for the entire column is 28.5 k. In this case, the member will buckle in the direction of the least dimension.

When columns are actually tested, there is usually a difference found between actual buckling loads and theoretical predictions. This is particularly true for columns near the transition between short- and long-column behavior. The result is that buckling loads are often slightly lower than predicted, particularly near the transition zone, where failure is often partly elastic and partly inelastic (crushing).

Example Problem 10.6 (Figures 10.28 to 10.30)

A W8×40 steel column supports trusses framed into its web, which serve to fix the weak axis and light beams that attach to the flange, simulating a pin connection about the strong axis. If the base connection is assumed as a pin, determine the critical buckling load the column is capable of supporting.

Solution:

$$W8\times40; \quad (A = 11.7 \text{ in.}^2, r_x = 3.53'', I_x = 146 \text{ in.}^4,$$
$$r_y = 2.04'', I_y = 49.1 \text{ in.}^4)$$

The first step is to determine the critical axis for buckling (i.e., which one has the larger KL/r).

Weak Axis:

$$L_e = KL = 0.7 \,(34') = 23.8'$$

$$\frac{KL}{r_y} = \frac{23.8' \times 12^{\text{in.}}/_{\text{ft.}}}{2.04''} = 140$$

Strong Axis:

$$L_e = L; \quad K = 1.0; \quad KL = 37'$$

$$\frac{KL}{r_x} = \frac{\left(37' \times 12^{\text{in.}}/_{\text{ft.}}\right)}{3.53'} = 125.8$$

The weak axis for this column is critical since

$$\frac{KL}{r_y} > \frac{KL}{r_x}$$

$$P_{\text{cr.}} = \frac{\pi^2 E I_y}{(KL)^2} = \frac{(3.14)^2 (29 \times 10^3 \text{ ksi})(49.1 \text{ in.}^4)}{\left(23.8' \times 12^{\text{in.}}/_{\text{ft.}}\right)^2}$$

$$= 172.1 \text{ k}$$

$$f_{\text{critical}} = \frac{P_{\text{crit.}}}{A} = \frac{172.1 \text{ k}}{11.7 \text{ in.}^2} = 14.7 \text{ ksi}$$

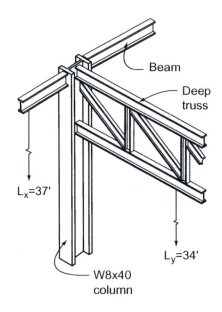

Figure 10.28   Truss/column framing.

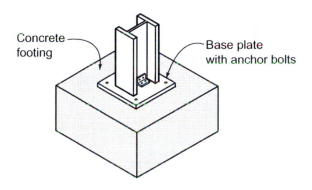

Column base connection

Figure 10.29   Column base connection.

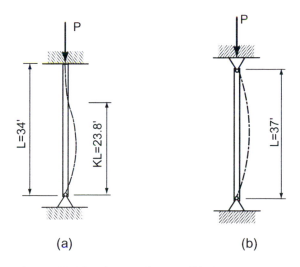

Figure 10.30   (a) Weak axis. (b) Strong axis.

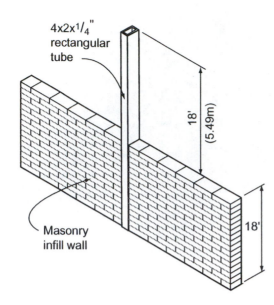

4x2x¼"
rectangular
tube

Masonry
infill wall

18'
(5.49m)

18'

Figure 10.31 *Tubular column braced by masonry infill.*

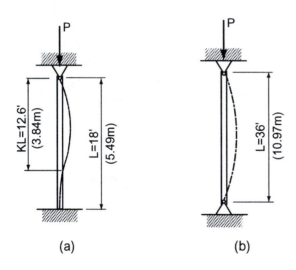

P

KL=12.6'
(3.84m)

L=18'
(5.49m)

(a)

P

L=36'
(10.97m)

(b)

Figure 10.32 *(a) Weak axis. (b) Strong axis.*

Example Problem 10.7 (Figures 10.31 and 10.32)

A rectangular steel tube is used as a 36' (10.97-m) column. It has pinned ends and its weak axis is braced at midheight by a masonry infill wall as shown. Determine the column's critical buckling load. $E = 29 \times 10^3$ ksi (200×10^3 MPa) 4"×2"×¼" rectangular tube:

$A = 2.59$ in.^{2}; $I_x = 4.69$ in.^{4}; $r_x = 1.35$"; $I_y = 1.54$ in.^{4}; $r_y = 0.77$"
($A = 100 \times 50 \times 6.35$ mm; $I_x = 1.95 \times 10^6$ mm^4; $r_x = 34.3$ mm; $I_y = 0.640 \times 10^6$ mm^4; $r_y = 19.6$ mm)

Solution:

Again, the first step in the solution must involve the determination of the critical slenderness ratio.

Weak Axis:

$$K = 0.7;$$
$$KL_y = 0.7 \times 18' = 12.6' = 151.2''$$
$$KL_y = 0.7 \times 5.49 = 3.84 \text{ m}$$
$$\frac{KL}{r_y} = \frac{151.2''}{0.77''} = 196; \qquad \frac{KL}{r_y} = \frac{3.840 \text{ mm}}{19.7 \text{ mm}} = 195$$

Strong Axis:

$$K = 1.0$$
$$KL_x = 1.0 \times 36' = 36' = 432'';$$
$$KL_x = 1.0 \times 10.97 \text{ m} = 10.97 \text{ m}$$
$$\frac{KL_x}{r_x} = \frac{(432'')}{1.35''} = 320;$$
$$\frac{KL_x}{r_x} = \frac{10,970 \text{ mm}}{34.3 \text{ mm}} = 320 \leftarrow \text{governs}$$

The strong axis governs since $\dfrac{KL_x}{r_x} > \dfrac{KL_y}{r_y}$.

$$P_{\text{crit.}} = \frac{\pi^2 E I_x}{(KL_x)^2} = \frac{(\pi^2)(29 \times 10^3 \text{ ksi})(4.69 \text{ in.}^4)}{(432 \text{ in.})^2}$$
$$= 7.19 \text{ k}$$

$$P_{\text{crit.}} = \frac{\pi^2 E I_x}{(KL_x)^2}$$
$$= \frac{(\pi^2)\left(200 \times 10^6 \text{ kN}/\text{m}^2\right)(1.95 \times 10^{-6} \text{ m}^4)}{(10.97 \text{ m})^2}$$
$$= 32 \text{ kN}$$

$$f_{\text{critical}} = \frac{P_{\text{crit.}}}{A} = \frac{7.19 \text{ k}}{2.59 \text{ in.}^2} = 2.78 \text{ ksi}$$

$$f_{\text{critical}} = \frac{P_{\text{crit.}}}{A} = \frac{32 \text{ kN}}{1.67 \times 10^{-3} \text{ m}^2} = 19.2 \times 10^3 \text{ kN}/\text{m}^2$$
$$= 19.2 \text{ MPa}$$

Example Problem 10.8 (Figures 10.33 and 10.34a, b)

Determine the buckling load capacity of a 2×4 stud 12 feet high if blocking is provided at midheight. Assume $E = 1.2 \times 10^6$ psi.

Solution:

2×4 S4S ($A = 5.25$ in.^{2}, $I_x = 5.36$ in.^{4}, $I_y = 0.984$ in.^{4})

$$r_x = \sqrt{\frac{I_x}{A}} = \sqrt{\frac{5.36 \text{ in.}^4}{5.25 \text{ in.}^2}} = 1.01"$$

$$r_y = \sqrt{\frac{I_y}{A}} = \sqrt{\frac{0.984 \text{ in.}^4}{5.25 \text{ in.}^2}} = 0.433"$$

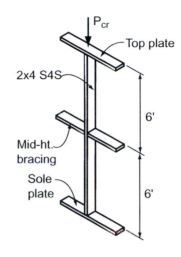

Figure 10.33   2×4 stud braced at mid-height.

Weak Axis:

$$L = 12'$$
$$K = 0.5$$
$$KL = 0.5 \times 12' = 6' = 72"$$
$$\frac{KL}{r_y} = \frac{72"}{0.433"} = 166.3$$

Strong Axis:

$$L = 12'$$
$$K = 1.0$$
$$KL = 1.0 \times 12' = 12' = 144"$$
$$\frac{KL}{r_x} = \frac{144"}{1.01"} = 142.6$$

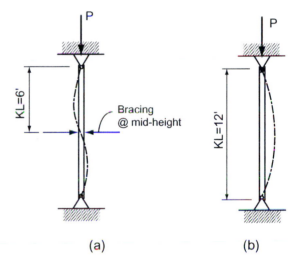

Figure 10.34   (a) Weak axis. (b) Strong axis.

The weak axis governs because $\dfrac{KL}{r_y} > \dfrac{KL}{r_x}$

$$P_{\text{crit.}} = \frac{\pi^2 E I_y}{(KL_y)^2} = \frac{\pi^2 (1.2 \times 10^6 \text{ psi})(0.984 \text{ in.}^4)}{(72")^2}$$

$$= 2250 \text{ lb.}$$

$$f_{\text{crit.}} = \frac{P_{\text{crit.}}}{A} = \frac{2250 \text{ lb.}}{5.25 \text{ in.}^2} = 429 \text{ psi}$$

Photo 10.3.1   Steel columns for a new steel framed office building.

10.3   AXIALLY LOADED STEEL COLUMNS

Much of the discussion thus far has been limited to very short columns that crush and, on the other end of the scale, long slender columns that buckle. Somewhere in-between these two extremes lies a zone where a "short" column transitions into a "long" column. Euler's buckling equation assumes that the critical buckling stress remains within the proportional limit of the material, in order that the modulus of elasticity E remains valid. Substituting the proportional limit value $F_{\text{proportional}} = 31,000$ psi for A36 steel (close to the $F_y = 36,000$ psi) into the Euler equation, the minimum slenderness ratio necessary for elastic behavior is found to be:

$$f_{\text{critical}} = \frac{P_{\text{critical}}}{A} = \frac{\pi^2 EI}{(A)(L)^2} = \frac{\pi^2 E(Ar^2)}{(A)(L)^2} = \frac{\pi^2 E}{\left(L/r\right)^2}$$

Then, solving for the slenderness ratio:

$$\frac{L}{r} = \sqrt{\frac{\pi^2 E}{P/A}} = \sqrt{\frac{\pi^2(29,000,000)}{31,000}} = 96$$

Columns (A36 steel) with slenderness ratios below $L/r \le 96$ generally exhibit characteristics of *inelastic buckling* or crushing.

The upper limit of KL/r for steel columns depends on good judgment and safe design and is usually established by the building code. Structural steel columns are limited to a slenderness ratio equal to:

$$\frac{KL}{r} \le 200$$

In reality, columns do not transition abruptly from short to long or vice versa. A transition zone exists between the two extremes; this is normally referred to as the *intermediate column range*. Intermediate columns fail by a combination of crushing (or yielding) and buckling (see Figure 10.35).

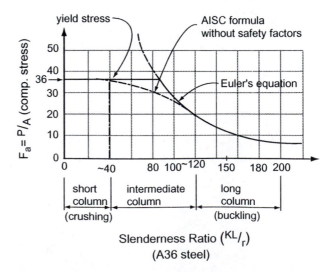

Figure 10.35   Column classification based on slenderness.

The initially flat portion of the curve (in the short-column range) indicates material yielding with no buckling taking place. On the far end of the curve ($KL/r > 120$) the compressive stresses are relatively low and buckling is the mode of failure. In the intermediate-column range ($40 < KL/r < 120$), failure has aspects of both yielding and buckling.

The load-carrying ability of intermediate-length columns is influenced by both the strength and elastic properties of the column material. Empirical design formulas, based on extensive testing and research, have been developed to cover the design of columns within the limits of each column category.

Since 1961, the American Institute of Steel Construction (AISC) has adopted a set of *column design formulas* that incorporate the use of a variable factor of safety, depending on slenderness, for determining allowable compressive stress. AISC formulas recognize only two slenderness categories: short/intermediate and long (Figure 10.36).

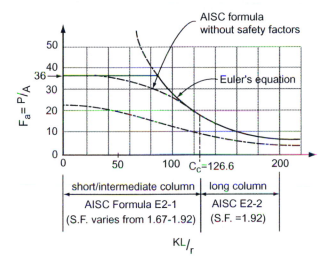

Figure 10.36   Allowable compressive stress based on AISC formulas.

Slender columns are defined as those having a KL/r exceeding a value called C_C, in which:

$$C_C = \sqrt{\frac{2\pi^2 E}{F_y}}$$

where:

   E = modulus of elasticity

   F_y = yield stress of the steel

Mild steel (A36) with an $F_y = 36$ ksi has a $C_C = 126.1$; high-strength steel with $F_y = 50$ ksi has a $C_C = 107.0$.

The C_C value represents the theoretical demarcation line between inelastic and elastic behavior (Figure 10.36).

The AISC allowable stress (F_a) formula for short/intermediate columns ($KL/r < C_C$) is expressed as:

$$F_a = \frac{\left[1 - \frac{\left(KL/r\right)^2}{2C_C^2}\right](F_y)}{\frac{5}{3} + \frac{3\left(KL/r\right)}{8C_C} - \frac{\left(KL/r\right)^3}{8C_C^3}}$$

(AISC Eq. E2-1)

where:

KL/r = the largest effective slenderness ratio of any unbraced length of column

F_a = allowable compressive stress (psi or ksi)

When axially loaded compression members have a $KL/r > C_C$, the allowable stress is computed as:

$$F_a = \frac{12\pi^2 E}{23\left(KL/r\right)^2}$$

(AISC Eq. E2-2)

Note that the two preceding equations represent actual design equations that can be used to size compression elements. These equations appear rather daunting, especially equation E2–1. Fortunately, the *AISC Manual of Steel Construction* has developed a design table for Kl/r from 1 to 200 with the respective allowable stress F_a. No computations using E2–1 and E2–2 are necessary since the equations have been used in generating these tables (see Tables 10.1 and 10.2).

In structural work, pinned ends are often assumed even if the ends of steel columns are typically restrained to some degree at the bottom by being welded to a base plate, which in turn is anchor-bolted to a concrete footing. Steel pipe columns generally have plates welded at each end, and then bolted to other parts of a structure. Such restraints, however, vary greatly and are difficult to evaluate. Thus, designers rarely take advantage of the restraint to increase the allowable stress, which therefore adds to the safety factor of the design.

Table 10.1   Allowable stress for compression members (F_y = 36 ksi and F_y = 250 MPa).

$\frac{KL}{r}$	F_a (ksi)	F_a (MPa)	$\frac{KL}{r}$	F_a (ksi)	F_a (MPa)	$\frac{KL}{r}$	F_a (ksi)	F_a (MPa)
1	21.56	148.7	41	19.11	131.8	81	15.24	105.1
2	21.52	148.4	42	19.03	131.2	82	15.13	104.3
3	21.48	148.1	43	18.95	130.7	83	15.02	103.6
4	21.44	147.8	44	18.86	130.0	84	14.90	102.7
5	21.39	147.5	45	18.78	129.5	85	14.79	102.0
6	21.35	147.2	46	18.70	128.9	86	14.67	101.1
7	21.30	146.9	47	18.61	128.3	87	14.56	100.4
8	21.25	146.5	48	18.53	127.8	88	14.44	99.6
9	21.21	146.2	49	18.44	127.1	89	14.32	98.7
10	21.16	145.9	50	18.35	126.5	90	14.20	97.9
11	21.10	145.5	51	18.26	125.9	91	14.09	97.2
12	21.05	145.1	52	18.17	125.3	92	13.97	96.3
13	21.00	144.8	53	18.08	124.7	93	13.84	95.4
14	20.95	144.5	54	17.99	124.0	94	13.72	94.6
15	20.89	144.0	55	17.90	123.4	95	13.60	93.8
16	20.83	143.6	56	17.81	122.8	96	13.48	92.9
17	20.78	143.3	57	17.71	122.1	97	13.35	92.0
18	20.72	142.9	58	17.62	121.5	98	13.23	91.2
19	20.66	142.5	59	17.53	120.9	99	13.10	90.3
20	20.60	142.0	60	17.43	120.2	100	12.98	89.5
21	20.54	141.6	61	17.33	119.5	101	12.85	88.6
22	20.48	141.2	62	17.24	118.9	102	12.72	87.7
23	20.41	140.7	63	17.14	118.2	103	12.59	86.8
24	20.35	140.3	64	17.04	117.5	104	12.47	86.0
25	20.28	139.8	65	16.94	116.8	105	12.33	85.0
26	20.22	139.4	66	16.84	116.1	106	12.20	84.1
27	20.15	138.9	67	16.74	115.4	107	12.07	83.2
28	20.08	138.5	68	16.64	114.7	108	11.94	82.3
29	20.01	138.0	69	16.53	114.0	109	11.81	81.4
30	19.94	137.5	70	16.43	113.3	110	11.67	80.5
31	19.87	137.0	71	16.33	112.6	111	11.54	79.6
32	19.80	136.5	72	16.22	111.8	112	11.40	78.6
33	19.73	136.0	73	16.12	111.1	113	11.26	77.6
34	19.65	135.5	74	16.01	110.4	114	11.13	76.7
35	19.58	135.0	75	15.90	109.6	115	10.99	75.8
36	19.50	134.5	76	15.79	108.9	116	10.85	74.8
37	19.42	133.9	77	15.69	108.2	117	10.71	73.8
38	19.35	133.4	78	15.58	107.4	118	10.57	72.9
39	19.27	132.9	79	15.47	106.7	119	10.43	71.9
40	19.19	132.3	80	15.36	105.9	120	10.28	70.9

(Continued)

Table 10.1   Allowable stress for compression members ($F_y = 36$ ksi and $F_y = 250$ MPa).   (Continued)

$\dfrac{KL}{r}$	F_a (ksi)	F_a (MPa)	$\dfrac{KL}{r}$	F_a (ksi)	F_a (MPa)
121	10.41	71.78	161	5.76	39.7
122	9.99	68.88	162	5.69	39.2
123	9.85	67.92	163	5.62	38.7
124	9.70	66.88	164	5.55	38.3
125	9.55	65.85	165	5.49	37.9
126	9.41	64.88	166	5.42	37.4
127	9.26	63.85	167	5.35	36.9
128	9.11	62.81	168	5.29	36.5
129	8.97	61.85	169	5.23	36.1
130	8.84	60.95	170	5.17	35.6
131	8.70	59.99	171	5.11	35.2
132	8.57	59.09	172	5.05	34.8
133	8.44	58.19	173	4.99	34.4
134	8.32	57.37	174	4.93	34.0
135	8.19	56.47	175	4.88	33.6
136	8.07	55.64	176	4.82	33.2
137	7.96	54.88	177	4.77	32.9
138	7.84	54.06	178	4.71	32.5
139	7.73	53.30	179	4.66	32.1
140	7.62	52.54	180	4.61	31.8
141	7.51	51.78	181	4.56	31.4
142	7.41	51.09	182	4.51	31.1
143	7.30	50.33	183	4.46	30.8
144	7.20	49.64	184	4.41	30.4
145	7.10	48.95	185	4.36	30.1
146	7.01	48.33	186	4.32	29.8
147	6.91	47.64	187	4.27	29.4
148	6.82	47.02	188	4.23	29.2
149	6.73	46.40	189	4.18	28.8
150	6.64	45.78	190	4.14	28.5
151	6.55	45.16	191	4.09	28.2
152	6.46	44.54	192	4.05	27.9
153	6.38	43.99	193	4.01	27.6
154	6.30	43.44	194	3.97	27.4
155	6.22	42.89	195	3.93	27.1
156	6.14	42.34	196	3.89	26.8
157	6.06	41.78	197	3.85	26.5
158	5.98	41.23	198	3.81	26.3
159	5.91	40.75	199	3.77	26.0
160	5.83	40.20	200	3.73	25.7

Table 10.2   *Allowable stress for compression members ($F_y = 50$ ksi and $F_y = 345$ MPa).*

$\dfrac{KL}{r}$	F_a (ksi)	F_a (MPa)	$\dfrac{KL}{r}$	F_a (ksi)	F_a (MPa)	$\dfrac{KL}{r}$	F_a (ksi)	F_a (MPa)
1	29.94	206.4	41	25.69	177.1	81	18.81	129.7
2	29.87	206.0	42	25.55	176.2	82	18.61	128.3
3	29.80	205.5	43	25.40	175.1	83	18.41	126.9
4	29.73	205.0	44	25.26	174.2	84	18.20	125.5
5	29.66	204.5	45	25.11	173.1	85	17.99	124.0
6	29.58	204.0	46	24.96	172.1	86	17.79	122.7
7	29.50	203.4	47	24.81	171.1	87	17.58	121.2
8	29.42	202.9	48	24.66	170.0	88	17.37	119.8
9	29.34	202.3	49	24.51	169.0	89	17.15	118.2
10	29.26	201.7	50	24.35	167.9	90	16.94	116.8
11	29.17	201.1	51	24.19	166.8	91	16.72	115.3
12	29.08	200.5	52	24.04	165.8	92	16.50	113.8
13	28.99	199.9	53	23.88	164.7	93	16.29	112.3
14	28.90	199.3	54	23.72	163.5	94	16.06	110.7
15	28.80	198.6	55	23.55	162.4	95	15.84	109.2
16	28.71	198.0	56	23.39	161.3	96	15.62	107.7
17	28.61	197.3	57	23.22	160.1	97	15.39	106.1
18	28.51	196.6	58	23.06	159.0	98	15.17	104.6
19	28.40	195.8	59	22.89	157.8	99	14.94	103.0
20	28.30	195.1	60	22.72	156.7	100	14.71	101.4
21	28.19	194.4	61	22.55	155.5	101	14.47	99.8
22	28.08	193.6	62	22.37	154.2	102	14.24	98.2
23	27.97	192.9	63	22.20	153.1	103	14.00	96.5
24	27.86	192.1	64	22.02	151.8	104	13.77	94.9
25	27.75	191.3	65	21.85	150.7	105	13.53	93.3
26	27.63	190.5	66	21.67	149.4	106	13.29	91.6
27	27.52	189.8	67	21.49	148.2	107	13.04	89.9
28	27.40	188.9	68	21.31	146.9	108	12.80	88.3
29	27.28	188.1	69	21.12	145.6	109	12.57	86.7
30	27.15	187.2	70	20.94	144.4	110	12.34	85.1
31	27.03	186.4	71	20.75	143.1	111	12.12	83.6
32	26.90	185.5	72	20.56	141.8	112	11.90	82.1
33	26.77	184.6	73	20.38	140.5	113	11.69	80.6
34	26.64	183.7	74	20.10	138.6	114	11.49	79.2
35	26.51	182.8	75	19.99	137.8	115	11.29	77.8
36	26.38	181.9	76	19.80	136.5	116	11.10	76.5
37	26.25	181.0	77	19.61	135.2	117	10.91	75.2
38	26.11	180.0	78	19.41	133.8	118	10.72	73.9
39	25.97	179.1	79	19.21	132.5	119	10.55	72.7
40	25.83	178.1	80	19.01	131.1	120	10.37	71.5

(Continued)

Table 10.2   Allowable stress for compression members ($F_y = 50$ ksi and $F_y = 345$ MPa).   (Continued)

$\dfrac{KL}{r}$	F_a (ksi)	F_a (MPa)	$\dfrac{KL}{r}$	F_a (ksi)	F_a (MPa)
121	10.20	70.33	161	5.76	39.7
122	10.03	69.16	162	5.69	39.2
123	9.87	68.05	163	5.62	38.7
124	9.71	66.95	164	5.55	38.3
125	9.56	65.92	165	5.49	37.9
126	9.41	64.88	166	5.42	37.4
127	9.26	63.85	167	5.35	36.9
128	9.11	62.81	168	5.29	36.5
129	8.97	61.85	169	5.23	36.1
130	8.84	60.95	170	5.17	35.6
131	8.70	59.99	171	5.11	35.2
132	8.57	59.09	172	5.05	34.8
133	8.44	58.19	173	4.99	34.4
134	8.32	57.37	174	4.93	34.0
135	8.19	56.47	175	4.88	33.6
136	8.07	55.64	176	4.82	33.2
137	7.96	54.88	177	4.77	32.9
138	7.84	54.06	178	4.71	32.5
139	7.73	53.30	179	4.66	32.1
140	7.62	52.54	180	4.61	31.8
141	7.51	51.78	181	4.56	31.4
142	7.41	51.09	182	4.51	31.1
143	7.30	50.33	183	4.46	30.8
144	7.20	49.64	184	4.41	30.4
145	7.10	48.95	185	4.36	30.1
146	7.01	48.33	186	4.32	29.8
147	6.91	47.64	187	4.27	29.4
148	6.82	47.02	188	4.23	29.2
149	6.73	46.40	189	4.18	28.8
150	6.64	45.78	190	4.14	28.5
151	6.55	45.16	191	4.09	28.2
152	6.46	44.54	192	4.05	27.9
153	6.38	43.99	193	4.01	27.6
154	6.30	43.44	194	3.97	27.4
155	6.22	42.89	195	3.93	27.1
156	6.14	42.34	196	3.89	26.8
157	6.06	41.78	197	3.85	26.5
158	5.98	41.23	198	3.81	26.3
159	5.91	40.75	199	3.77	26.0
160	5.83	40.20	200	3.73	25.7

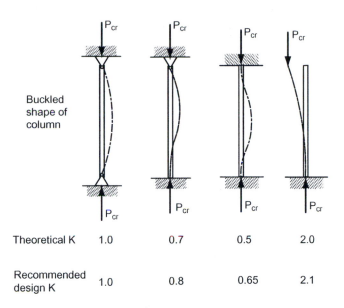

Figure 10.37   *AISC-recommended design K values.*

However, tests have indicated that in the case of fixed end conditions, the "theoretical" $K = 0.5$ values are somewhat *nonconservative* when designing steel columns. Since true joint fixity is rarely possible, the AISC recommends the use of recommended K-values (see Figure 10.37).

All examples discussed in this chapter assume that the columns are part of a braced building system. Sidesway is minimized through the use of a separate bracing system (braced frame or shearwalls), and the K-values for the braced columns need not exceed 1.0. In unbraced buildings, such as those utilizing rigid frames, sidesway can result in effective column lengths greater than the actual column length ($K > 1.0$). A much more involved analysis is required for columns with sidesway and thus will not be discussed in this text.

Analysis of Steel Columns

Column analysis implies the determination of the allowable load capacity P (see Figure 10.38) for a column. A simple analysis procedure is outlined below.

Given:

Column length, support conditions, grade of steel (F_y), applied load, and column size.

Required:

Check of the adequacy of the column. In other words, is

$$P_{actual} < P_{allowable}$$

Procedure:

a. Calculate KL/r_{min}; the largest KL/r governs.
b. Enter the appropriate AISC table (Table 10.1 or 10.2).

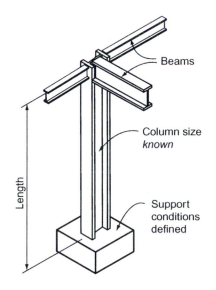

Figure 10.38   *Analysis of a steel column.*

c. Pick out the respective F_a.

d. Compute: $P_{\text{allow.}} = F_a \times A$

where:

A = cross-sectional area of the column (in.^{2})

F_a = allowable compressive stress (ksi)

e. Check the column adequacy

If: $P_{\text{allowable}} > P_{\text{actual}}$; then OK

If: $P_{\text{allowable}} < P_{\text{actual}}$; then overstressed.

Example Problem 10.9: Axially Loaded Steel Columns (Figures 10.39 and 10.40)

A W12×53 (F_y = 36 ksi) column is to be used as a primary support in a building. If the unbraced height of the column is 16 ft. with both ends assumed as pin connected, compute the allowable load on the column.

Solution:

Enter the steel section properties table (Appendix Table 6) and extract the data for the W12×53 column.

W12×53:

A = 15.6 in.^{2};

r_x = 5.23 in., r_y = 2.48 in.

Since the column is assumed to be pin connected at both ends and for both directions (axes) of buckling, the least radius of gyration (r_y) will yield the more critical (larger) slenderness ratio.

The critical slenderness ratio is then computed as:

$$\frac{KL}{r_y} = \frac{(1.0)(16' \times 12 \text{ in./ft.})}{2.48''} = 77.4$$

To determine the allowable compressive stress F_a, enter Table 10.1 with the critical slenderness ratio.

	KL/r	F_a
	77	15.69 ksi
KL/r = 77.4	Interpolating	15.65 ksi
	78	15.58 ksi

The allowable capacity of the W12×53 is computed as:

$$P_{\text{allowable}} = F_a \times A = (15.65 \text{ k/in.}^2) \times (15.6 \text{ in.}^2) = 244 \text{ k}$$

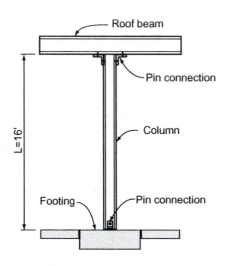

Figure 10.39   *Pinned steel column.*

Roof beam

Pin connection

Column

Footing

Pin connection

L=16'

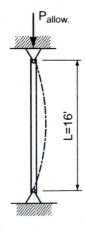

Figure 10.40   *Pins top and bottom, K = 1.0.*

$P_{\text{allow.}}$

L=16'

Example Problem 10.10 (Figure 10.41)

A 24-ft.-tall, A572 grade 50, steel column (W14×82) with an $F_y = 50$ ksi has pins at both ends. Its weak axis is braced at midheight, but the column is free to buckle the full 24 ft. in the strong direction. Determine the safe load capacity for this column.

Solution:

Properties of the W14×82:

$$A = 24.1 \text{ in.}^2, r_x = 6.05", r_y = 2.48"$$

Compute the slenderness ratio about both axes to determine the critical direction for buckling.

$$\frac{KL}{r_x} = \frac{(24' \times 12 \text{ in./ft.})}{6.05"} = 47.6$$

$$\frac{KL}{r_y} = \frac{(12' \times 12 \text{ in./ft.})}{2.48"} = 58.1$$

The larger slenderness ratio governs; therefore, the weak axis (y) buckling is used in determining F_a.

From Table 10.2:

Interpolating for $KL/r = 58.1$; $F_a = 23.04$ ksi

$$\therefore P_a = F_a \times A = 23.04 \text{ k/in.}^2 \times 24.1 \text{ in.}^2 = 555 \text{ k}$$

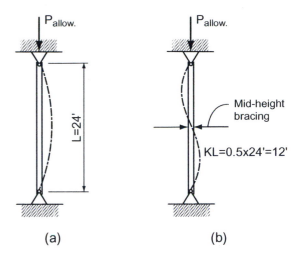

(a)   (b)

Figure 10.41   (a) Strong axis buckling. (b) Weak axis buckling.

Example Problem 10.11 (Figures 10.42 and 10.43)

A 4" ϕ standard weight steel pipe ($F_y = 50$ ksi) supports a roof framing system as shown. The timber beam-to-column connection is considered a pin, while the base of the column is rigidly embedded into the concrete. If the load from the roof is 35 kips, is the column adequate?

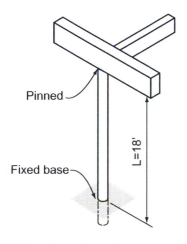

Figure 10.42   Steel pipe column.

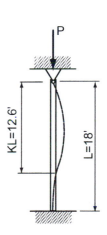

Figure 10.43   Buckled shape of column.

Solution:

4"-diameter standard weight pipe:

$$A = 3.17 \text{ in.}^2, r = 1.51 \text{ in.}$$

Although the theoretical K is 0.7 for the support condition shown, the AISC-recommended value for use in design is: $K = 0.80$.

$$\frac{KL}{r} = \frac{0.80 \times (18' \times 12 \text{ in./ft.})}{1.51''} = 114.4$$

Using Table 10.2:

$$F_a = 11.4 \text{ ksi}$$
$$P_a = F_a \times A = (11.4 \text{ k/in.}^2) \times (3.17 \text{ in.}^2)$$
$$= 36.1 \text{ k} > 35 \text{ k}$$

The column is adequate.

Example Problem 10.12 (Figures 10.44 and 10.45)

Determine the capacity of a W250×67 column, braced at midheight about the weak axis. Assume that the top and bottom are pin connections. The column is made of A572-50 steel with an $F_y = 50$ MPa.

Solution:

Check both directions of buckling to determine the critical slenderness ratio.

$$\text{W250×67:} \qquad A = 8580 \text{ mm}^2; r_x = 110 \text{ mm};$$
$$r_y = 51.1 \text{ mm}$$

$$\frac{KL}{r_x} = \frac{(1.0)(6000 \text{ mm})}{110 \text{ mm}} = 54.5$$

$$\frac{KL}{r_y} = \frac{(0.5)(6000 \text{ mm})}{51.1 \text{ mm}} = 58.7 \text{ (governs)}$$

From Table 10.2:

$$F_a = 158 \text{ MPa}$$
$$\therefore P_a = F_a \times A = (158 \text{ MN/m}^2) \times (8.58 \times 10^{-3} \text{ m}^2)$$
$$= 1.36 \text{ MN}$$

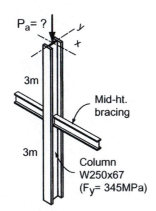

$P_a = ?$

y

x

3m

Mid-ht. bracing

3m

Column W250x67 (F_y= 345MPa)

Figure 10.44   Steel column braced at mid-height.

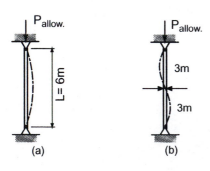

$P_{allow.}$

$P_{allow.}$

$L = 6m$

3m

3m

(a)

(b)

Figure 10.45   (a) Strong axis. (b) Weak axis.

Example Problem 10.13 (Figure 10.46)

An A572 ($F_y = 345$ MPa) W310×143 column supports a load of $P = 2.6$ MN. The top of the column is pin connected and the base is assumed fixed. Determine the adequacy of the W310×143 column.

Solution:

W310×143: $A = 18.2 \times 10^3$ mm^2 = 18.2×10^{-3} m^2

$r_x = 138$ mm; $r_y = 78.5$ mm.

Buckling will result about the weaker y axis.

Therefore;

$$\frac{KL}{r_y} = \frac{0.8(6000 \text{ mm})}{78.5 \text{ mm}} = 61.1$$

From Table 10.2:

$$F_a = 155.1 \text{ MPa} = 155.1 \text{ MN/m}^2$$

$$P_a = F_a \times A = (155.1 \text{ MN/m}^2) \times (18.2 \times 10^{-3} \text{ m}^2)$$

$$= 2.83 \text{ MN}$$

$$P_a = 2.83 \text{ MN} > P_{\text{actual}} = 2.6 \text{ MN}$$

The W310×143 column is adequate to support the load.

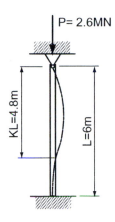

Figure 10.46   *Buckled shape of column.*

Design of Steel Columns

The design of axially loaded steel columns (in other words, the selection of an appropriate column size) is usually accomplished by using specialized column tables such as those contained in the *American Institute of Steel Construction's Manual of Steel Construction—Allowable Stress Design, 9th edition*. Structural design varies from analysis in that there are several possible answers to a problem. The selection of a column size is justly dependent on strength and safety requirements, but there are other issues (architectural as well as constructional) that may influence the final selection.

Since the *AISC Column Design Tables* are assumed not to be available (it would require the purchase of the AISC manual), steel column design will involve an iterative trial-and-error process. This methodology appears to be long and tedious, but in fact very few cycles are usually necessary to zoom in on a solution.

An earlier discussion of efficient column cross-sections for axial loads (Figure 10.10) suggested the use of circular or "boxier" wide-flange members. Along with spatial and constructional concerns, relative maximum or minimum sizes may already be specified by the architect, thus limiting the array of choices that would otherwise be available. This in no way limits the design possibilities, but in fact helps guide the structural design and choices made by the engineer. Smaller-scale steel structures may use 8" (200 mm) and 10" (250 mm) nominal size wide-flange columns, while larger buildings with heavier loads will often use 12" (310 mm) and 14" (360 mm) nominal sizes. These sections are the "boxier" or square sizes, with the depth and flange width of the cross-section being approximately equal.

One trial-and-error procedure may be outlined as follows (see Figure 10.47):

Given:

Column length, support conditions, grade of steel (F_y), applied load (P_{actual}).

Required:

Column size to safely support the load.

Procedure:

a. Guess at a size. But where does one begin? If it is a smaller-scale building, maybe try a square W8 or W10 in the middle of the weight grouping. A similar estimate using larger sections is appropriate for heavier loads.

b. Once the *trial size* has been selected, cross-sectional properties are known. Compute the critical slenderness ratio, taking into account the end conditions and intermediate bracing.

c. Using the larger KL/r value, enter Table 10.1 or 10.2. Obtain the respective F_a.

d. Calculate the $P_{allowable} = F_a \times A$ of the trial section.

e. Check to see if $P_{allowable} > P_{actual}$.

If $P_{allowable} < P_{actual}$, then the column is overstressed and a larger section should be selected next. If the trial section is much too strong, cycle again with a smaller size. One way to check the relative efficiency of the cross-section is to examine its percent of stress level.

$$\text{Percent of stress} = \frac{P_{actual}}{P_{allowable}} \times 100\%$$

A percent of stress in the 90 to 100% level is very efficient.

f. Repeat this process until an adequate but efficient section is obtained.

Note: Steps (b) through (e) are essentially the procedure used previously in the analysis of steel columns.

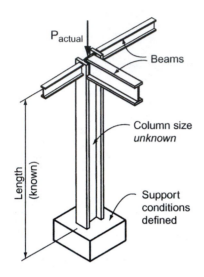

Figure 10.47   Design of a steel column.

Example Problem 10.14: Design of Steel Columns (Figure 10.48)

Select the most economical W12 × column 18' in height to support an axial load of 600 kips using A572 grade 50 steel. Assume that the column is hinged at the top but fixed at the base.

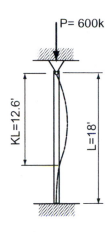

Solution:

As a first guess in this trial-and-error process, try a W12×96 (about the middle of the available "boxier" sections).

W12×96   (A = 28.2 in.^{2}, r_x = 5.44", r_y = 3.09")

Compute (KL/r) critical:

Figure 10.48   W12 steel column.

$$\frac{KL}{r_x} = \frac{(0.80)(18' \times 12 \text{ in./ft.})}{5.44"} = 31.8$$

$$\frac{KL}{r_y} = \frac{(0.80)(18' \times 12 \text{ in./ft.})}{3.09"} = 55.9$$

The larger slenderness is critical; therefore, use $KL/r = 55.9$.

Enter Table 10.2 and obtain the respective F_a:

F_a = 23.41 ksi $P_a = F_a \times A$

$$P_{\text{allowable}} = (23.41 \text{ k/in.}^2) \times (28.2 \text{ in.}^2)$$

$$= 660 \text{ k} > 600 \text{ k}$$

$$\% \text{ stress} = \frac{P_{\text{actual}}}{P_{\text{allowable}}} \times 100\%$$

$$= \frac{600 \text{ k}}{660 \text{ k}} \times 100\% = 91\%$$

This selection is quite efficient and still a bit understressed. Therefore, use the W12×96.

Example Problem 10.15 (Figures 10.49 to 10.50)

Select the most economical W8 shape column, 16' long, with P = 180 k. Assume lateral bracing is provided at mid-height in the weak axis of buckling and the top and bottom are pin connected. F_y = 36 ksi

Solution:

Again, we need to begin by guessing at a size, and then check the adequacy of the selection.

Try W8×35.   (A = 10.3 in.^{2}, r_x = 3.51", r_y = 2.03")

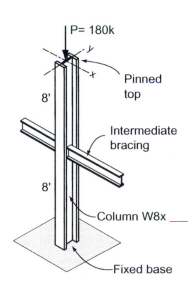

Figure 10.49   W8 column with intermediate bracing.

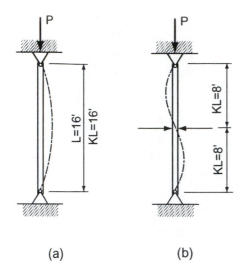

(a)   (b)

Figure 10.50   (a) Strong axis. (b) Weak axis.

Determine the critical slenderness ratio:

$$\frac{KL}{r_y} = \frac{(0.5)(16' \times 12 \text{ in.}/\text{ft.})}{2.03"} = 47.3$$

$$\frac{KL}{r_x} = \frac{(1.0)(16' \times 12 \text{ in.}/\text{ft.})}{3.51"} = 54.7$$

Buckling about the strong axis is more critical in this example because of the lateral bracing provided for the weak axis.

Therefore, the F_a value is obtained from Table 10.1 based on $KL/r = 54.7$.

$$F_a = 17.93 \text{ ksi}$$

$$P_{\text{allowble}} = F_a \times A = (17.93 \text{ k}/\text{in.}^2) \times (10.3 \text{ in.}^2)$$

$$= 184.7 \text{ k}$$

$$P_{\text{allowable}} = 184.7 \text{ k} > P_{\text{actual}} = 180 \text{ k}$$

$$\% \text{ stress} = \frac{180 \text{ k}}{184.7 \text{ k}} \times 100\% = 97.5\%$$

This is a very efficient section.

Therefore, use W8×35.

Example Problem 10.16 (Figures 10.51 to 10.53)

A four-story building has a structural steel beam–girder–column framing system. Columns are spaced at 20 ft. on center in one direction and 30 ft. on center in the perpendicular direction. An interior column supports a typical tributary floor area of 600 sq. ft.

For a preliminary design, find an economical W10 or W12 section for an interior first-floor column. Assume

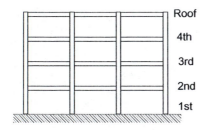

Roof

4th

3rd

2nd

1st

Figure 10.51   Building elevation.

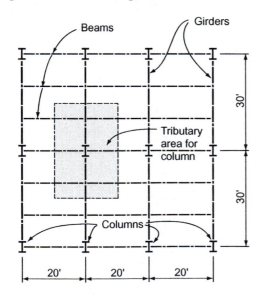

Figure 10.52   Floor framing plan.

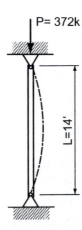

Figure 10.53   Typical interior column.

that the columns have unsupported lengths of 14' and a $K = 1.0$. $F_y = 36$ ksi

Roof loads:	DL = 80 psf	
	SL = 30 psf	
Floor loads:	DL = 100 psf	
	LL = 70 psf	

Solution:

Total roof loads: DL + SL = 80 + 30 = 110 psf
(110 psf) × (600 ft.²) = 66,000 lb. = 66 k

Total floor loads: DL + LL = 100 + 70 = 170 psf
(170 psf) × (600 ft.²) = 102,000 lb. = 102 k per floor

The load at the top of the interior first-floor column is a result of the roof plus three floor loads.

Total load on the first-floor column = P_{actual}

P_{actual} = 66 k (roof) + 3(102 k) (floors) = 372 k

Try W10×60. (A = 17.6 in.², r_y = 2.57")

The assumption being made is that the *y axis* is the critical buckling direction since no weak axis bracing is provided.

$$\frac{KL}{r_y} = \frac{(1.0)(14' \times 12 \text{ in./ft.})}{2.57"} = 65.4$$

From Table 10.1, F_a = 16.9 ksi

P_a = 16.9 ksi × 17.6 in.² = 297.4 k < 372 k

Therefore, this column section is overstressed. Select a larger section.

Try W10×77. (A = 22.6 in.², r_y = 2.60")

$$\frac{KL}{r_y} = \frac{(1.0)(14' \times 12 \text{ in./ft.})}{2.60"}$$

$$= 64.6; \quad \text{therefore } F_a = 16.98 \text{ ksi}$$

$$P_a = (16.98 \text{ k/in.}^2) \times (22.6 \text{ in.}^2)$$

$$= 383.7 \text{ k} > 372 \text{ k} \quad \therefore \text{ OK}$$

Use W10×77.

A column design using a W12 section can be carried out using an identical procedure. The resulting W12 size would be: W12×72 (P_a = 377 k).

Both of these sections are adequate for stress and efficient in material. However, the W12×72 is more economical because it is 5 pounds per foot less in weight. The final decision on selection will undoubtedly involve issues concerning dimensional coordination and construction.

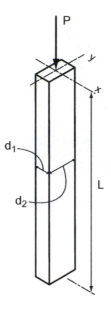

Figure 10.54 Slenderness ratio of wood columns.

10.4 AXIALLY LOADED WOOD COLUMNS

Wood columns are commonly found supporting beams and girders, which support tributary areas of roof and floor loads. Other structural members (such as bridge piers, compression cords of a truss, or the studs in a load-bearing wall) subjected to compression are also designed using the same methods utilized for building columns. As discussed in Section 10.1, long columns tend to buckle under critical load, while short columns will fail by the crushing of fibers. For wood columns, the ratio of the column length to its width is just as important as it is for steel columns. However, in wood columns, the slenderness ratio is defined as the laterally unsupported length in inches divided by the least (minimum) dimension of the column (see Figure 10.54).

$$\text{slenderness ratio} = \frac{L}{d_{\text{min.}}} = \frac{L}{d_1}$$

where:
$$d_1 < d_2$$

Wood columns are restricted to a maximum slenderness ratio:

$$\frac{L_e}{d_{\text{min.}}} \leq 50$$

which is approximately the same as the $\frac{KL}{r_{\text{min.}}} \leq 200$ used for steel columns.

A larger L/d ratio indicates a greater instability and a tendency for the column to buckle under lower axial load.

The effective length of steel columns was determined by applying a K factor (see Figure 10.37) to the unsupported length of the column to adjust for end fixity. Similar effective length factors, called K_e in wood columns, are used to adjust for the various end conditions. In fact, the recommended design K_e values are identical to those of steel columns.

Most wood construction is detailed such that translation (sidesway) is restrained but the ends of the column are free to rotate (i.e., pin connection). The K_e value is generally taken as 1.0, and the effective length is equal to the actual unsupported length. Even if some fixity may exist in the top or bottom connection, it is difficult to evaluate the degree of fixity to assume in design. Therefore, $K_e = 1.0$ is an acceptable assumption that is usually a bit conservative in some connection conditions.

Wood columns can be solid members or rectangular, round (Figure 10.55), or other shapes, or spaced columns built up from two or more individual solid members separated by blocking (Figure 10.56).

Since the majority of all wood columns in buildings are solid rectangular sections, the analysis and design methods examined in this section will be limited to these types. A more thorough treatment for the design of wood elements is usually covered in advanced structures courses. The *National Design Specification for Wood Construction (NDS–91)* approved a new standard in 1992 and incorporated a new methodology and equations for the design of wood elements and connections. Previous categorizing of wood columns into the short-, intermediate-, or long-column range resulted in three different equations for each respective slenderness range. The *NDS–91* now utilizes a single equation, providing a continuous curve over the entire range of slenderness ratios (Figure 10.57).

The allowable compressive stress for an axially loaded wood column of known size is expressed as:

$$f_c = \frac{P}{A} \le F'_C$$

where:

f_c = actual compressive stress parallel to grain
P = axial compressive force in the member
A = cross-sectional area of the column
F'_C = allowable compressive stress parallel to grain

To obtain the allowable compressive stress F'_C, many adjustments to the tabulated base stress are necessary.

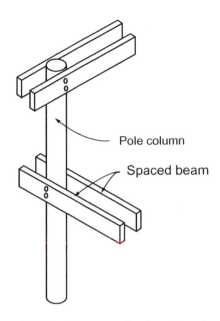

Figure 10.55   An example of a pole column.

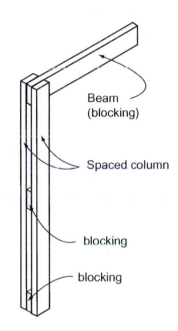

Figure 10.56   An example of a spaced column.

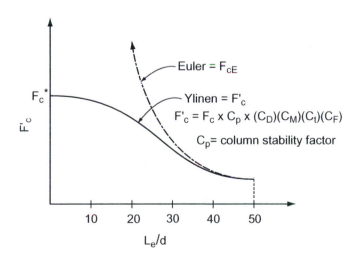

Figure 10.57   Ylinen column curve—allowable stress versus slenderness ratio.

The *NDS–91* defines the F'_C as:

$$F'_C = (F_C)(C_D)(C_M)(C_t)(C_F)(C_p)$$

where:

F'_C = allowable compressive stress parallel to grain

F_C = tabulated compressive stress parallel to grain; found in building code tables, NDS tables, and wood design handbooks

C_D = load duration factor; defined later in this section

C_M = wet service factor; accounts for moisture content in the wood;

= 1.0 for dry service conditions as in most covered structures; dry service condition is defined as:

Moisture content ≤ 19% for sawed lumber
Moisture content ≤ 16% for glu-lams

C_t = temperature factor; usually taken as 1.0 for normal temperature conditions

C_F = size factor; an adjustment based on member sizes used

C_p = column stability factor; accounts for buckling and is directly affected by the slenderness ratio

Since the objective in this book is to analyze and design structural elements in a preliminary way (rather than the full complement of equations and checks performed by a structural engineer), the preceding allowable compressive stress equation will be simplified as follows:

$$F'_C = F_c^* C_p$$

where:

$$F_c^* = F_c(C_D)(C_M)(C_t)(C_F) \cong F_c C_D$$
(for preliminary column design)

This simplification assumes C_M, C_t, and C_F are all equal to 1.0 (which is generally the case for a majority of wood columns).

Now a word about the load duration factor C_D. Wood has a unique structural property in which it can support higher stresses if the applied loads are for a short period of time. All tabulated stress values contained in building codes, NDS, or wood design manuals apply to "normal" load duration and dry service conditions. The C_D value adjusts tabulated stresses to allowable values based on the duration (time) of loading. "Normal" duration is taken as 10 years and the C_D = 1.0. Short-duration loading from wind, earthquake, snow, or impact allows C_D values higher than 1.0 but less than 2.0 (Table 10.3).

The column stability factor C_p multiplied by F_c essentially defines the column curve (equation) as shown in Figure 10.57. This equation, originally developed by Ylinen, explains the

Table 10.3   Load duration factors.

C_D—Load Duration Adjustment	
Load Duration	**Factor**
Permanent	0.9
Ten years (normal load)	1.0
Two months (snow load)	1.15
Seven days (construction load)	1.25
One day	1.33
Ten minutes (Wind or EQ)	1.6
Impact	2.0

behavior of wood columns as the interaction of two modes of failure: buckling and crushing.

$$C_p = \frac{1 + (F_{cE}/F_c^*)}{2c} - \sqrt{\left[\frac{1 + (F_{cE}/F_c^*)}{2c}\right]^2 - \frac{(F_{cE}/F_c^*)}{c}}$$

where:

F_{cE} = Euler critical buckling stress for columns

$$F_{cE} = \frac{K_{cE}E'}{(L_e/d)^2} \cong \frac{K_{cE}E}{(L_e/d)^2}$$

$F_c^* \cong F_c C_D$

E' = modulus of elasticity associated with the axis of column buckling

c = buckling and crushing interaction factor for columns

   = 0.9 for glu-lam columns

   = 0.8 for sawed lumber columns

K_{cE} = 0.30 for visually graded lumber

   = 0.418 for glu-lams

The column stability factor C_p is directly affected by the Euler buckling stress value F_{cE}, which in turn is inversely proportional to the square of a column's slenderness ratio. A table, to simplify the computations for preliminary column analysis/design, could be created by inputting slenderness ratios between 1 and 50, which results in F_{cE} values for sawed and glu-lam members. Then, if various F_{cE} values were divided by F_c^* generating ratios of (F_{cE}/F_c^*), a computer could easily calculate corresponding C_p values.

Appendix Table 14 was created for this purpose and eliminates the necessity of laborious computations for C_p.

Analysis of Wood Columns

A simple procedure can be adopted for checking the adequacy or capacity of wooden columns (Figure 10.58). This methodology is for approximate analysis and assumes the simplifications discussed in the earlier section.

Given:

Column size, column length, grade and species of lumber, and end conditions.

Required:

The allowable capacity of a column or the adequacy of a given column.

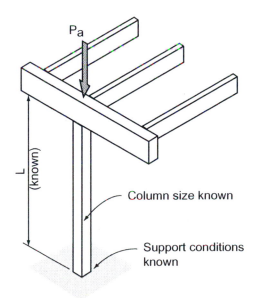

Figure 10.58   Checking the capacity of wooden columns.

Procedure:

    a. Calculate the $(L_e/d)_{min.}$
    b. Obtain F'_c (allowable compressive stress)

where:
$$F'_c = (F_c)(C_D)(C_M)(C_t)(C_F)(C_p)$$
or $F'_c = F^*_c C_p$

Compute:
$$F_{cE} = \frac{K_{cE}}{(L_e/d)^2}$$
$$K_{cE} = 0.3 \text{ (sawed lumber)}$$
$$K_{cE} = 0.418 \text{ (glu-lams)}$$
$$c = 0.8 \text{ (sawed lumber)}$$
$$c = 0.9 \text{ (glu-lams)}$$

    c. Compute $F^*_c \cong F_c C_D$

    d. Calculate the ratio: $\dfrac{F_{cE}}{F^*_c}$

    e. Enter Appendix Table 14; obtain respective C_p
    f. Calculate: $F'_c = F^*_c C_p$

$$\therefore P_{allowable} = F'_c \times A \le P_{actual}$$

where:
    A = cross-sectional area of the column

Example Problem 10.17:
Analysis of Wood Columns (Figure 10.59)

A 6×8 Douglas-fir No. 1 post supports a roof load of 20 kips. Check the adequacy of the column assuming pin support conditions at the top and bottom. From Table 6.2a, use $F_c = 1000$ psi and $E = 1.6 \times 10^6$ psi.

Solution:

6×8 S4S Douglas fir No. 1: ($A = 41.25$ in.^{2})

$$\frac{L_e}{d} = \frac{(12' \times 12 \text{ in./ft.})}{5.5''} = 26.2$$

$$F_{cE} = \frac{0.3E}{(L_e/d)^2} = \frac{0.3(1.6 \times 10^6)}{(26.2)^2} = 699 \text{ psi}$$

$$F^*_c \cong F_c C_D$$

Load duration factor for snow is $C_D = 1.15$ (15% increase in stress above "normal" condition)

$$\therefore F^*_c = (1000 \text{ lb./in.}^2)(1.15) = 1150 \text{ psi}$$

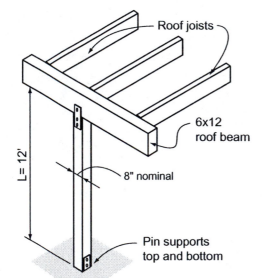

Figure 10.59   *Pin connected wood column.*

Roof joists

6x12 roof beam

L = 12'

8" nominal

Pin supports top and bottom

The column stability factor C_p can be obtained from Appendix Table 14 by entering the ratio:

$$\frac{F_{cE}}{F_c^*} = \frac{699 \text{ psi}}{1150 \text{ psi}} = 0.61$$

$$\therefore C_p = 0.506$$

$$F_c' = F_c^* C_p = (1150 \text{ lb./in.}^2) \times (0.506) = 582 \text{ psi}$$

Then:

$$P_{allowable} = P_a = F_c' \times A = (582 \text{ lb./in.}^2) \times (41.25 \text{ in.}^2)$$

$$= 24,000 \text{ lb.}$$

$$P_a = 24 \text{ k} > P_{actual} = 20 \text{ k}$$

The column is adequate.

Example Problem 10.18 (Figures 10.60 and 10.61)

An 18' tall 6×8 Southern pine column supports a roof load (dead load plus a 7-day live load) equal to 16 kips. The weak axis of buckling is braced at a point 9'6" from the bottom support. Determine the adequacy of the column.

Solution:

6×8 S4S Southern pine post: ($A = 41.25 \text{ in.}^2$, $F_c = 975$ psi, $E = 1.6 \times 10^6$ psi)

Check the slenderness ratio about the weak axis:

$$\frac{L_e}{d} = \frac{(9.5' \times 12 \text{ in./ft.})}{5.5"} = 20.7$$

The slenderness ratio about the strong axis is:

$$\frac{L_e}{d} = \frac{(18' \times 12 \text{ in./ft.})}{7.5"} = 28.8 \leftarrow \text{governs}$$

$$F_{cE} = \frac{0.3E}{(L_e/d)^2} = \frac{0.3 (1.6 \times 10^6 \text{ lb./in.}^2)}{(28.8)^2} = 579 \text{ psi}$$

$$F_c^* \cong F_c C_D = (975 \text{ lb./in.}^2)(1.25) = 1220 \text{ psi}$$

where: $C_D = 1.25$ for 7-day-duration load

From Appendix Table 14: $C_p = 0.412$

$$\therefore F_c' = F_c^* C_p = 1220 \text{ lb./in.}^2 \times 0.412 = 503 \text{ psi}$$

$$P_a = F_c' \times A = (503 \text{ lb./in.}^2) \times (41.25 \text{ in.}^2)$$

$$= 20,700 \text{ lb.}$$

$$P_a = 20.7 \text{ k} > P_{actual} = 16 \text{ k}$$

The column is adequate.

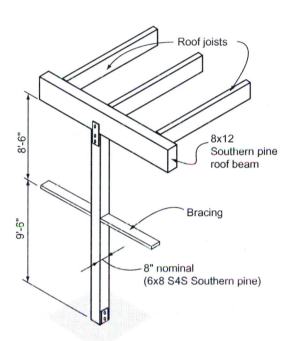

Figure 10.60   Wood column with intermediate bracing.

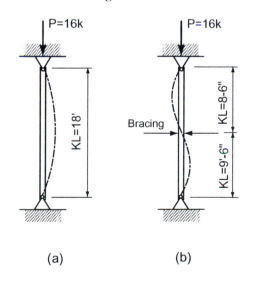

(a)   (b)

Figure 10.61   (a) Strong axis. (b) Weak axis.

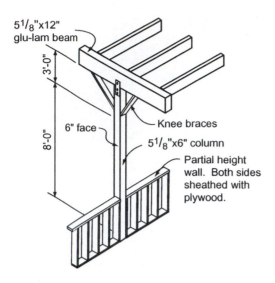

5¹⁄₈"x12" glu-lam beam

3'-0"

8'-0"

6" face

Knee braces

5¹⁄₈"x6" column

Partial height wall. Both sides sheathed with plywood.

Figure 10.62   Braced wood column.

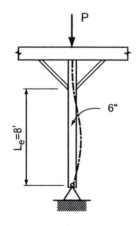

P

6"

$L_e = 8'$

Figure 10.63   Strong axis buckling.

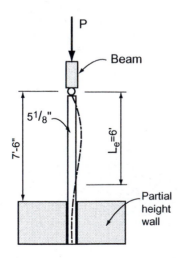

P

Beam

5¹⁄₈"

7'-6"

$L_e = 6'$

Partial height wall

Figure 10.64   Weak axis buckling.

Example Problem 10.19 (Figures 10.62 to 10.64)

An 11'-tall Douglas fir glu-lam column is used to support a roof load (DL + snow) as shown. A partial-height (3'-6")wall braces the $5\frac{1}{8}$" direction, and knee braces from the beam support the 6" face. Determine the capacity of the column.

Solution:

$5\frac{1}{8}$" × 6" glu-lam post: ($A = 30.8$ in.²; $F_c = 1650$ psi, $E = 1.8 \times 10^6$ psi)

Buckling in the plane of the 6" dimension:

$$\frac{L_e}{d} = \frac{(8' \times 12 \text{ in./ft.})}{6"} = 16$$

Buckling in the $5\frac{1}{8}$" direction:

$$\frac{L_e}{d} = \frac{(0.8 \times)(7.5' \times 12 \text{ in./ft.})}{5.125"} = 14$$

Comparing the buckling condition in both directions, the 6" direction is more critical and, therefore, governs.

$$F_{cE} = \frac{0.418E}{(L_e/d)^2} = \frac{0.418(1.8 \times 10^6 \text{ psi})}{(16)^2} = 2940 \text{ psi}$$

$$F_c^* \cong F_c C_D = (1650 \text{ psi}) \times (1.15) = 1900 \text{ psi}$$
$$\text{(snow)}$$

$$\frac{F_{cE}}{F_c} = \frac{2940 \text{ psi}}{1900 \text{ psi}} = 1.55$$

From Appendix Table 14: $C_p = 0.883$

$$\therefore F_c' = F_c^* C_p = (1900 \text{ psi}) \times (0.833) = 1580 \text{ psi}$$

$$P_a = F_c' \times A = (1580 \text{ psi}) \times (30.8 \text{ in.}^2) = 48{,}700 \text{ lb.}$$

Design of Wood Columns

Column design in wood is a *trial-and-error process*. Start by making a quick estimate on size (try out your intuition) and check out the adequacy or inadequacy by following the analysis procedure given in the previous section. Axially loaded wood columns without mid-height bracing, as shown in Figure 10.65, are generally square in cross-section, or in some cases just slightly rectangular. Fortunately, there are fewer possible wood sections to choose from compared with the wide array of sizes available in steel.

One design procedure using the trial-and-error method could be:

Given:

Column length, column load, grade and species of lumber to be used, and end conditions.

Required:

An economical column size.

Procedure:

a. Guess at a trial size; try to select a square or almost square cross-section unless the column's weak axis is braced.

b. Follow the same steps used in the analysis procedure in the previous section.

c. If $P_{allowable} \geq P_{actual}$, then OK.

d. If $P_{allowable} < P_{actual}$, pick a larger size and cycle through the analysis procedure again.

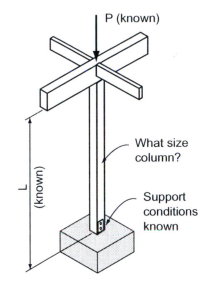

Figure 10.65   *Column design in wood.*

Example Problem 10.20:
Design of Wood Columns(Figure 10.66)

A 22'-tall glu-lam column is required to support a roof load (including snow) of 40 kips. Assuming $8\frac{3}{4}$" in one dimension (to match the beam width above), determine the minimum column size if the top and bottom are pin supported.

Select from the following sizes:

$8\frac{3}{4}$" × 9" ($A = 78.75$ in.^{2})

$8\frac{3}{4}$" × $10\frac{1}{2}$" ($A = 91.88$ in.^{2})

$8\frac{3}{4}$" × 12" ($A = 105.00$ in.^{2})

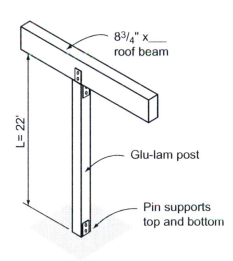

Figure 10.66   *Glu-lam column design.*

Solution:

Glu-lam column: ($F_c = 1650$ psi, $E = 1.8 \times 10^6$ psi)

Try $8\frac{3}{4}" \times 10\frac{1}{2}"$ ($A = 105.00$ in.²)

$$\frac{L_e}{d} = \frac{(22' \times 12 \text{ in./ft.})}{8.75 \text{ in.}}$$

$$= 30.2 < 50 \text{ (max. slenderness ratio)}$$

$$F_{cE} = \frac{0.418E}{(L_e/d)^2} = \frac{0.418(1.8 \times 10^6 \text{ lb./in.}^2)}{(30.2)^2} = 825 \text{ psi}$$

$$F_c^* \cong F_c C_D = (1650 \text{ psi}) \times (1.15) = 1900 \text{ psi}$$
$$\qquad\qquad\qquad\qquad\quad \underset{\text{(snow)}}{}$$

$$\frac{F_{cE}}{F_c^*} = \frac{825}{1900} = 0.43$$

From Appendix Table 14: $C_p = 0.403$

$$F_c' = F_c^* C_p = (1900 \text{ lb./in.}^2) \times (0.403) = 765 \text{ psi}$$

$$P_a = F_c' \times A = (765 \text{ lb./in.}^2) \times (91.9 \text{ in.}^2)$$

$$= 70{,}300 \text{ lb.} > 40{,}000 \text{ lb.}$$

Cycle again, trying a smaller, more economical section. Try $8\frac{3}{4}" \times 9"$ ($A = 78.8$ in.²)

Since the critical dimension is still $8\frac{3}{4}"$, the values for F_{cE}, F_c^*, and F_c' all remain the same as in trial 1. The only change that affects the capability of the column is the available cross-sectional area.

$$\therefore P_a = F_c' \times A = (765 \text{ lb./in.}^2) \times (78.8 \text{ in.}^2)$$

$$= 60{,}300 \text{ lb.}$$

$$P_a = 60.3 \text{ k} > 40 \text{ k}$$

Use $8\frac{3}{4}" \times 9"$ glu-lam section.

10.5 COLUMNS SUBJECTED TO COMBINED LOADING OR ECCENTRICITY

So far the previous sections have assumed compression members subjected to concentric loading (loads acting through the centroid of the column section). The study of axially loaded columns (Figure 10.67) was essential to the understanding of the primary issue of slenderness and its relationship to failure modes involving crushing and buckling. In practice, however, concentric loading is rarely the case. This section will introduce the idea of eccentricity

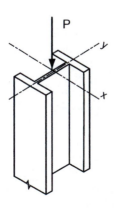

Figure 10.67   Concentrically (axially) loaded column.

(Figure 10.68) and/or side loading (Figure 10.69) and their effect on column behavior.

Many columns are subjected to bending in combination with axial compression loads. Non-uniform bearing, misalignment of the framing, or even the crookedness of a member will cause a load to miss the centroid of the column cross-section. Compression members carrying bending moment due to eccentricity or side loading in addition to compression are referred to as *beam–columns* (Figures 10.70 to 10.73).

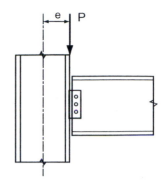

Figure 10.70   Framed beam (shear) connection. e = *eccentricity;* M = P × e.

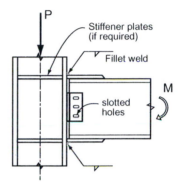

Figure 10.71   Moment connection (rigid frame). M = *Moment due to beam bending.*

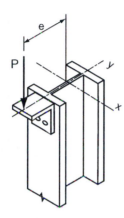

Figure 10.68   Eccentrically loaded column.

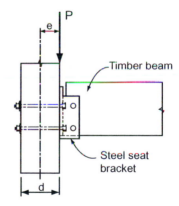

Figure 10.72   Timber beam-column connection. e = d/2 = *eccentricity;* M = P × e.

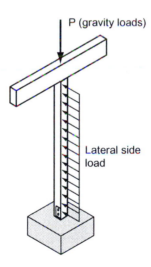

Figure 10.69   Column with compression plus side load.

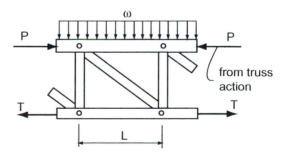

Figure 10.73   Upper cord of a truss— compression plus bending; $M = \omega L^2 / 8$.

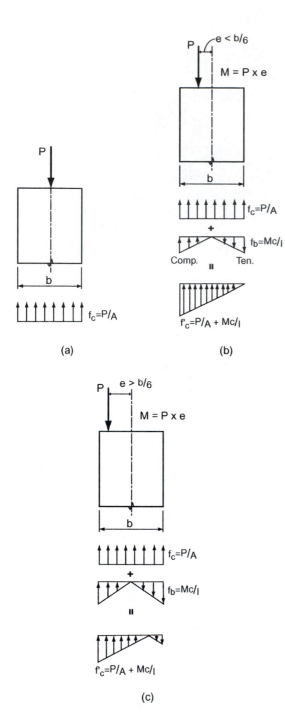

(a)   (b)

(c)

Figure 10.74   (a) Axially loaded-uniform compressive stress. (b) Small eccentricity-linearly varying stress. (c) Large eccentricity-tensile stress on part of the cross-section.

An assumption that was made for axially loaded columns was the relative uniformity of the stress distribution over the cross-sectional area, as shown in Figure 10.74a. Bending stress, which involves tension and compression stresses, must be added algebraically to the compressive stresses from gravity loading. If a beam is very flexible and the column is very rigid, the eccentricity effect will be small since most of the bending stress will be absorbed by the beam. Relatively small eccentricities alter the final stress distribution, but the cross-section will remain in compression, although non-uniform, as shown in Figure 10.74b. However, if a rigid beam is connected to a less rigid column, a considerably large eccentricity will be transmitted to the column. When large eccentricities exist, tensile stresses may develop over part of the cross-section, as shown in Figure 10.74c.

Beam–columns are evaluated using an interaction equation that incorporates the bending stress with the compressive stress (Figure 10.75). The general interaction equation is expressed as:

$$\frac{f_a}{F_a} + \frac{f_b}{F_b} \leq 1.0 \qquad \text{(interaction equation)}$$

where:

$f_a = {}^{P}\!/_{A}$; the actual compressive (axial) stress

F_a = allowable compressive stress; based on Kl/r (steel) or L_e/d (timber)

$f_b = \dfrac{Mc}{I} = \dfrac{M}{S}$ (actual bending stress)

$M = P \times e$ for eccentrically loaded members

M = bending moment due to side load or rigid frame action

F_b = allowable bending stress; values from tables

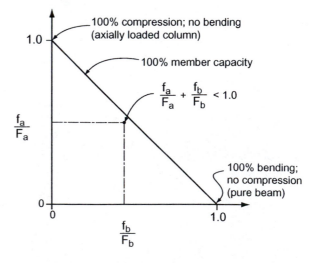

Figure 10.75   Interaction curve for compression and bending.

Analyzing and designing beam–columns using the AISC and NDS equations are more appropriately done in follow-up courses dealing specifically with steel and wood design. Oversimplification of the preceding equations does not necessarily result in appropriate approximations, even for preliminary design purposes. This text does not include problems involving the use of the interaction equation.

Supplementary Problems

Section 10.2—Euler buckling formula

10.2.1 A W8×31 steel column 20' long is pin supported at both ends. Determine the critical buckling load and stress developed in the column. $E = 29 \times 10^3$ ksi.

10.2.2 A W310×97 steel column is 6 meters tall and is pin connected at the top and bottom. Determine the critical axial load that the column supports when buckling occurs. $E = 207 \times 10^3$ MPa.

10.2.3 Determine the maximum critical length of a W10×54 (W250x80) column supporting an axial load of 250 kips (1.112.x 10³ MN). $E = 29 \times 10^3$ ksi ($E = 200 \times 10^3$ MPa).

Problem 10.2.1

10.2.4 An 8"-diameter timber pole is fixed into a large concrete footing at grade and is completely pin connected at its upper end. How high can the pole be and still just support a load of 25 kips? $E = 1.0 \times 10^6$ psi. Solve this problem assuming the diameter is 2030 mm and the load to be supported is *111 MN* ($E = 6.895 \times 10^3$ MPa).

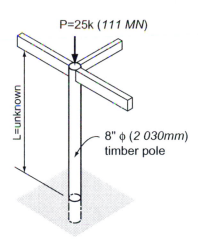

Problem 10.2.4

10.2.5 Determine the critical buckling load and stress of an 8×6×$^3/_8$" rectangular structural tube used as a column 38' long, pin connected top and bottom ($A = 9.58$ in.^{2}, $r_y = 2.36$", $I_y = 53.5$ in.^{4}).

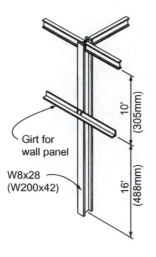

Problem 10.2.6

10.2.6 Determine the critical buckling load and stress for the W8×28 (*W200x42*) column shown. $E = 29 \times 10^3$ ksi (*E = 200 x 10^3 MPa*).

Section 10.3—Analysis of Steel Columns

10.3.1 Determine the allowable load capacity (P_a) for an $F_y = 50$ ksi (*345 MPa*) steel column, W12×65 (*W310x97*), when $L = 18'$ (*5.49 m*) and

    a. The base and top are both fixed.
    b. The base is fixed and the top is pinned.
    c. Both top and bottom are pinned.

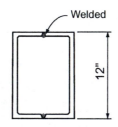

Problem 10.3.2

10.3.2 Two C12×20.7 channel sections are welded together to form a closed box section. If $L = 20'$ and the top and bottom are pinned, determine the allowable axial load capacity P_a. Assume $F_y = 50$ ksi.

10.3.3 An angle 5" $\times$ 3$\frac{1}{2}$" $\times$ $\frac{1}{2}$" (*127.0 mm x 88.9 mm x 12.7 mm*) is used as a compression member in a truss. If $L = 7'$ (*2.13 m*), determine the allowable axial load for an $F_y = 36$ ksi (*250 MPa*). $A = 4.00$ in.^{2} (*2580 mm^2*), $r_x = 1.58$ in. (*40.1 mm*), $r_y = 1.56$ in. (*39.6 mm*), $r_z = .755$ in. (*19.2 mm*).

10.3.4 Determine the maximum allowable height of an A36 column (5" ϕ standard steel pipe) if the applied load is 60 kips. Assume the top to be pin connected, and the base is fixed.

10.3.5 A two-story, continuous W12×106 column supports a roof load of 200 kips and an intermediate (second floor) load of 300 kips. Assume the top and bottom have pin connections. Is the column section shown adequate?

Note: Assume the second-floor load to be applied at the top of the column—this will result in a somewhat conservative answer. The concept of intermediate loads is much more complicated and will not be discussed further in this text.

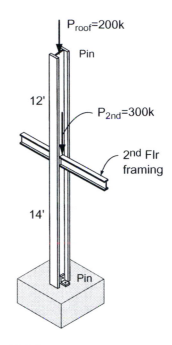

Problem 10.3.5

10.3.6 Determine the theoretical Euler buckling load and the allowable compressive load for a W8×31 column with an $F_y = 36$ ksi and a $K = 1.0$. The unsupported height of the column is 24 feet.

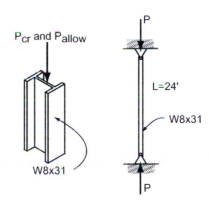

Problem 10.3.6

10.3.7 Determine the capacity of the W10×54 (*W250x80*) first-floor column. What is the tributary area (per level) that this column is capable of supporting? Assume that the steel is A572 ($F_y - 50$ ksi/345 MPa). The first floor column is 20 ft. (*6.1 m*) tall and has a pin connection at the top and rigid connection at the base. Use the recommended design K value in the computation.

Roof: Snow load = 30 psf
                   Dead load = 60 psf

Floors (2nd – 4th):
                   Live load = 50 psf
                   Dead load = 70 psf

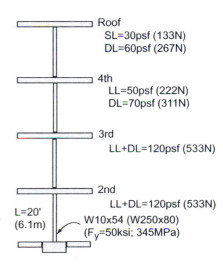

Problem 10.3.7

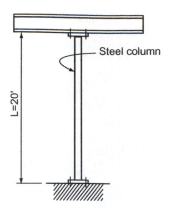

Problem 10.3.8

Section 10.3—Design of Steel Columns

10.3.8 Select the most economical steel pipe column (standard weight) to support a load of 30 k and a length of $L = 20$ ft. Assume $F_y = 36$ ksi and $K = 1.0$.

10.3.9 What is the most economical W8 (*W200*) column for Problem 10.14?

10.3.10 Select an appropriate steel column section 24 ft. long, braced at midheight about the weak axis, that supports a load of 350 kips. Use a W14 section. (See Example Problem 10.15.) Assume $F_y = 36$ ksi and $K = 1.0$.

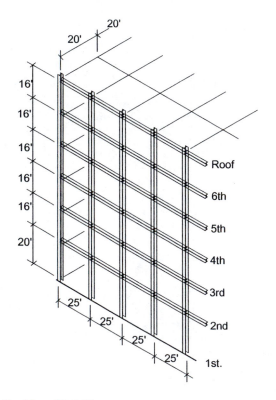

Problem 10.3.11

10.3.11 A six-story building has a structural steel beam–column frame that is appropriately fireproofed. The columns are spaced 20 ft. on centers in one direction and 25 ft. on centers in the perpendicular direction. A typical interior column supports a tributary floor area of 500 sq. ft. The governing building code specifies that the frame must be designed to withstand the dead weight of the structure, plus a roof snow load of 40 psf and a live load on each floor of 125 psf. The dead weight of the roof is estimated to be 80 psf, and each floor is 100 psf. The unsupported length of the ground-floor column is 20 ft., and the columns at the other floor levels are 16 ft. Design a typical interior third-floor column and the first-floor column using the most economical W12 section at each level.

10.3.12 Design an economical W10× section that can adequately support an axial load of $P = 400$ k with an $L = 20$ ft. and $K = 1.0$. ($F_y = 50$ ksi)

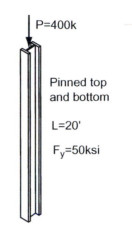

Problem 10.3.12

10.3.13 Design an adequate W10 (*W250*) to support an axial load of $P = 100$ k ($P = 445$ kN) as shown. The bottom connection is assumed as a pin. Deep trusses frame into the web of the column, which serves to fix the weak axis at the top. Small bracing beams are attached to the flanges of the column and provide a pinned condition in the strong axis of the column.

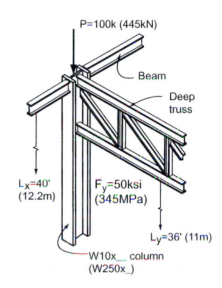

Problem 10.3.13

Section 10.4—Analysis of Wood Columns

10.4.1 A 6×6 S4S Southern pine (Dense No. 1) column is used to support headers that carry loads from roof joists. Determine the capacity of the column assuming pin connections at the top and bottom. Assume a 7-day-duration roof live load.

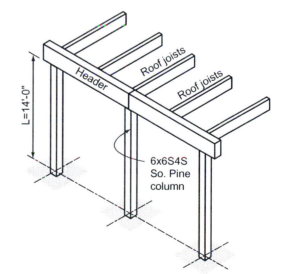

Problem 10.4.1

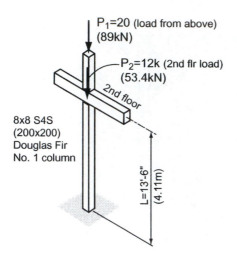

P$_1$=20 (load from above)
(89kN)

P$_2$=12k (2nd flr load)
(53.4kN)

2nd floor

8x8 S4S
(200x200)
Douglas Fir
No. 1 column

L=13'-6"
(4.11m)

Problem 10.4.2

10.4.2 An 8×8 S4S first-floor column supports a load of P_1 = 20 kips from roof and floors above and an additional second-floor load P_2 = 12 kips. Determine the adequacy of the column assuming a "normal" load duration.

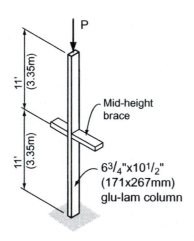

P

11' (3.35m)

11' (3.35m)

Mid-height brace

6^3/$_4$"x10^1/$_2$"
(171x267mm)
glu-lam column

Problem 10.4.3

10.4.3 Determine the axial load capacity of a $6\frac{3}{4}" \times 10\frac{1}{2}"$ glu-lam column with an area $A = 70.88$ in.^{2}, assuming lateral bracing about the weak axis at the midheight level. Assume pin connections top and bottom in both directions of buckling. ($F_c = 1650$ psi; $E = 1.8 \times 10^6$ psi)

10.4.4 An interior bearing wall in the basement of a residence utilizes 2×4 S4S studs spaced at 16" on centers to support the floor load above. Sheathing is provided on both sides of the wall and serves to prevent buckling about the weak axis of the member. Determine the permissible load ω

(in pounds per linear foot) assuming Hem-fir (joist/planks). Then, using the ω value computed, determine the bearing

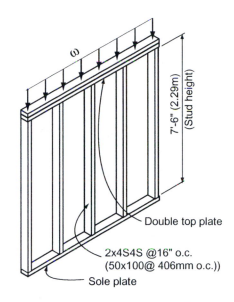

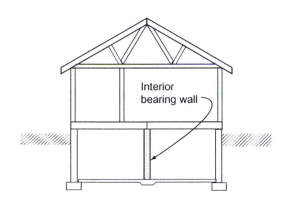

Problem 10.4.4a

Problem 10.4.4b

stress that develops between the stud and sole plate.

10.4.5 A 4×8 S4S Douglas fir column supports a roof beam as shown. The beam–column connection is to be considered a pin. The lower end is pinned for buckling about the strong axis but is fixed in the weak axis by the partial wall that measures 2 ft. tall. Determine the tributary area that

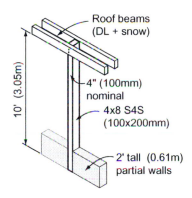

Problem 10.4.5

can be supported by this column if dead load = 20 psf and snow load = 30 psf.

10.4.6 Determine the minimum size column (Southen pine dense No. 1) required to support an axial load of $P = 25$ kips assuming an effective column length $L_e = 16$ ft.

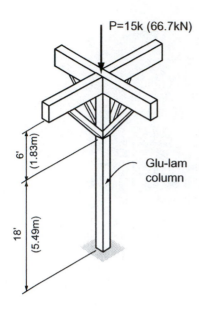

P=15k (66.7kN)

6'
(1.83m)

18'
(5.49m)

Glu-lam
column

Problem 10.4.7

10.4.7 An interior glu-lam column supports a roof load of $P = 15$ k. The total column height is 24', but knee-bracing from the beams reduces the unsupported height to 18'. Determine the minimum $6\frac{3}{4}" \times \_$ size required.

Use $F_c = 1650$ psi; $E = 1.8 \times 10^6$ psi

Appendix A

Tables

*These tables are adapted from the AISC ASD Manual of Steel Construction.

Table 1   Dimensioned Sizes—Rafters, Joists, Studs.

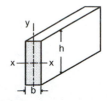

Section Properties of Dimension Western Lumber Sizes

Nominal Size (b × h) (in.)	Surfaced Dry Size (actual) (in.)	Nominal Size (b × h) (mm)	Surfaced Dry Size (actual) (mm)	Area $A = (b) \times (h)$ (in.²)	Area $A = bh$ × 10³mm²	Section Modulus $S = bh^2/6$ (in.³)	Section Modulus $S = bh^2/6$ × 10³ mm³	Moment of Inertia $I = bh^3/12$ (in.⁴)	Moment of Inertia $I = bh^3/12$ × 10⁶ mm⁴
2 ×2	1.5×1.5	50 × 50	38×38	2.25	1.44	0.56	9.12	0.42	0.17
2 ×3	1.5 × 2.5	50 × 75	38 × 64	3.75	2.43	1.56	25.9	1.95	0.83
2 ×4	1.5 × 3.5	50 × 100	38 × 89	5.25	3.38	3.06	50.2	5.36	2.23
2 ×6	1.5 × 5.5	50 × 150	38 × 140	8.25	5.32	7.58	124	20.80	8.69
2 ×8	1.5 × 7.25	50 × 200	38 × 184	10.88	6.99	13.14	214	47.63	19.7
2 ×10	1.5 × 9.25	50 × 250	38 × 235	13.88	8.93	21.39	350	98.93	41.1
2 ×12	1.5 × 11.25	50 × 300	38 × 286	16.88	10.87	31.64	518	177.98	74.1
3 ×3	2.5 × 2.5	75 × 75	64 × 64	6.25	4.10	2.60	43.7	3.26	1.40
3 ×4	2.5 × 3.5	75 × 100	64 × 89	8.75	5.70	5.10	84.5	8.93	3.76
3 ×6	2.5 × 5.5	75 × 150	64 × 140	13.75	8.96	12.60	209	34.66	14.6
3 ×8	2.5 × 7.25	75 × 200	64 × 184	18.12	11.78	21.90	361	79.39	33.2
3 ×10	2.5 × 9.25	75 × 250	64 × 235	23.12	15.04	35.65	589	164.89	69.2
3 ×12	2.5 × 11.25	75 × 300	64 × 286	28.12	18.30	52.73	872	296.63	124.7
4 ×4	3.5 × 3.5	100 × 100	89 × 89	12.25	7.92	7.15	118	12.51	5.23
4 ×6	3.5 × 5.5	100 × 150	89 × 140	19.25	12.5	17.65	292	48.53	20.4
4 ×8	3.5 × 7.25	100 × 200	89 × 184	25.38	16.4	30.66	502	111.15	46.2
4 ×10	3.5 × 9.25	100 × 250	89 × 235	32.38	20.9	49.91	819	230.84	96.3
4 ×12	3.5 × 11.25	100 × 300	89 × 286	39.38	25.4	73.83	1213	415.28	174
4 ×14	3.5 × 13.25	100 × 350	89 × 335	46.38	29.8	102.41	1664	678.48	279

Table 2   *Western Timbers—Beams and Columns.*

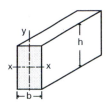

Section Properties of Western Lumber Timber Sizes

Nominal Size (b × h) (in.)	Surfaced Dry Size (actual) (in.)	Nominal Size (b × h) (mm)	Surfaced Dry Size (actual) (mm)	Area $A = (b) \times (h)$ (in.²)	Area $A = bh$ × 10³ mm²	Section Modulus $S = bh^2/6$ (in.³)	Section Modulus $S = bh^2/6$ × 10³ mm³	Moment of Inertia $I = bh^3/12$ (in.⁴)	Moment of Inertia $I = bh^3/12$ × 10⁶ mm⁴
6 × 6	5.5 × 5.5	150 × 150	140 × 140	30.25	19.6	27.7	457	76.3	32.0
6 × 8	5.5 × 7.5	150 × 200	140 × 191	41.25	26.7	51.6	851	193.4	81.3
6 × 10	5.5 × 9.5	150 × 250	140 × 241	52.25	33.7	82.7	1355	393.0	163
6 × 12	5.5 × 11.5	150 × 300	140 × 292	63.25	40.9	121.2	1989	697.1	290
6 × 14	5.5 × 13.5	150 × 350	140 × 343	74.25	48.0	167.1	2745	1127.7	471
8 × 8	7.5 × 7.5	200 × 200	191 × 191	56.25	36.5	70.3	1161	263.7	111
8 × 10	7.5 × 9.5	200 × 250	191 × 241	71.25	46.0	112.8	1849	535.9	223
8 × 12	7.5 × 11.5	200 × 300	191 × 292	86.25	55.8	165.3	2714	950.6	396
8 × 14	7.5 × 13.5	200 × 350	191 × 343	101.25	65.5	227.8	3745	1537.7	642
8 × 16	7.5 × 15.5	200 × 400	191 × 394	116.25	75.2	300.3	4942	2327.4	974
10 × 10	9.5 × 9.5	250 × 250	241 × 241	90.25	58.1	142.9	2333	678.8	281
10 × 12	9.5 × 11.5	250 × 300	241 × 292	109.25	70.4	209.4	3425	1204.0	500
10 × 14	9.5 × 13.5	250 × 350	241 × 343	128.25	82.7	288.6	4726	1947.8	810
10 × 16	9.5 × 15.5	250 × 400	241 × 394	147.25	95.0	380.4	6235	2948.1	1228
12 × 12	11.5 × 11.5	300 × 300	292 × 292	132.25	85.3	253.5	4150	1457.5	606
12 × 14	11.5 × 13.5	300 × 350	292 × 343	155.25	100	349.3	5726	2357.9	982
12 × 16	11.5 × 15.5	300 × 400	292 × 394	178.25	115	460.5	7555	3568.7	1488
12 × 18	11.5 × 17.5	300 × 450	292 × 445	201.25	130	587.0	9637	5136.1	2144

Table 3   Western Glu-Lams.

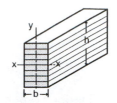

Section Properties of Western Glued Laminated Timber
Douglas fir F_b = 2400 psi, F_v = 165 psi, E = 1.8 × 10⁶psi

Finished Size (b × h) (in.)	Finished Size (b × h) (mm)	Area $A = bh$ (in.²)	Area $A = bh$ × 10³mm²	Section Modulus $S = bh^2/6$ (in.³)	Section Modulus $S = bh^2/6$ × 10³ mm³	Moment of Inertia $I = bh^3/12$ (in.⁴)	Moment of Inertia $I = bh^3/12$ × 10⁶ mm⁴
3 ⅛" × 6"	80 × 152	18.75	12.2	18.75	308	56.25	23.4
× 7.5"	× 190	23.44	15.2	29.30	481	109.9	45.7
× 9"	× 229	28.13	18.3	42.19	692	189.8	79.0
× 10.5"	× 267	32.81	21.4	57.42	942	301.5	125
× 12"	× 305	37.50	24.4	75.00	1230	450.0	187
× 13.5"	× 343	42.19	27.4	94.92	1557	640.7	267
× 15"	× 381	46.88	30.5	117.2	1922	878.9	366
× 16.5"	× 419	51.56	33.5	141.8	2326	1170	487
× 18"	× 457	56.25	36.6	168.8	2768	1519	632
5 ⅛" × 6"	130 × 152	30.75	19.8	30.75	504	92.25	38.4
× 7.5"	× 190	38.44	24.7	48.05	788	180.2	75.0
× 9"	× 229	46.13	29.8	69.19	1135	311.3	130
× 10.5"	× 267	53.81	34.7	94.17	1544	494.4	206
× 12"	× 305	61.50	39.7	123.0	2017	738.0	307
× 13.5"	× 343	69.19	44.6	155.7	2553	1051	437
× 15"	× 381	76.88	49.5	192.2	3152	1441	599
× 16.5"	× 419	84.56	54.5	232.5	3813	1919	798
× 18"	× 457	92.25	59.4	276.8	4540	2491	1036
× 19.5"	× 495	99.94	64.4	324.8	5327	3167	1317
× 21"	× 533	107.6	69.3	376.7	6178	3955	1645
× 22.5"	× 572	115.3	74.4	432.4	7091	4865	2024
× 24"	× 610	123.0	79.3	492.0	8069	5904	2456
× 25.5"	× 648	130.7	84.2	555.4	9109	7082	2946
× 27"	× 686	138.4	89.2	622.7	10212	8406	3497
6 ¾" × 7.5"	171 × 190	50.63	32.5	63.28	1038	237.3	98.7
× 9"	× 229	60.75	39.2	91.13	1495	410.1	171
× 10.5"	× 267	70.88	45.7	124.0	2034	651.2	271
× 12"	× 305	81.00	52.2	162.0	2657	972.0	404
× 13.5"	× 343	91.13	58.7	205.0	3362	1384	576
× 15"	× 381	101.3	65.2	253.1	4151	1898	790
× 16.5"	× 419	111.4	71.6	306.3	5023	2527	1051
× 18"	× 457	121.5	78.1	364.5	5979	3281	1365
× 19.5"	× 495	131.6	84.6	427.8	7016	4171	1735
× 21"	× 533	141.8	91.1	496.1	8136	5209	2167
× 22.5"	× 572	151.9	97.8	569.5	9340	6407	2665
× 24"	× 610	162.0	104	648.0	10630	7776	3235

(continued)

Table 3   Western Glu-Lams.   (Continued)

Finished Size (b × h) (in.)	Finished Size (b × h) (mm)	Area $A = bh$ (in.²)	Area $A = bh$ × 10³mm²	Section Modulus $S = bh^2/6$ (in.³)	Section Modulus $S = bh^2/6$ × 10³ mm³	Moment of Inertia $I = bh^3/12$ (in.⁴)	Moment of Inertia $I = bh^3/12$ × 10⁶ mm⁴
× 25.5"	× 648	172.1	111	731.5	12000	9327	3880
× 27"	× 686	182.3	117	820.1	13450	11072	4606
× 28.5"	× 724	192.4	124	913.8	14990	13021	5417
× 30"	× 762	202.5	130	1013	16610	15188	6318
× 31.5"	× 800	212.6	137	1116	18300	17581	7314
× 33"	× 838	222.8	143	1225	20090	20215	8409
× 34.5"	× 876	232.9	150	1339	21960	23098	9609
× 36"	× 914	243.0	156	1458	23910	26244	10920
× 37.5"	× 953	253.1	163	1582	25940	29663	12340
8 ¾" × 9"	222 × 229	78.75	50.6	118.1	1937	531.6	221
× 10.5"	× 267	91.88	59.3	160.8	2637	844.1	351
× 12"	× 305	105.0	67.7	210.0	3444	1260	524
× 13.5"	× 343	118.1	76.2	265.8	4359	1794	746
× 15"	× 381	131.3	84.7	328.1	5381	2461	1024
× 16.5"	× 419	144.4	93.1	397.0	6511	3276	1363
× 18"	× 457	157.5	102	472.5	7749	4253	1769
× 19.5"	× 495	170.6	110	554.5	9094	5407	2249
× 21"	× 533	183.8	119	643.1	10550	6753	2809
× 22.5"	× 572	196.9	127	738.3	12110	8306	3478
× 24"	× 610	210.0	135	840.0	13780	10080	4193
× 25.5"	× 648	223.1	144	948.3	15550	12091	5030
× 27"	× 686	236.3	152	1063	17430	14352	5970
× 28.5"	× 724	249.4	161	1185	19430	16880	7022
× 30"	× 762	262.5	169	1313	21530	19688	8190
× 31.5"	× 800	275.6	178	1447	23730	22791	9481
× 33"	× 838	288.8	186	1588	26040	26204	10900
× 34.5"	× 876	301.9	195	1736	28470	29942	12460
× 36"	× 914	315.0	203	1890	31000	34020	14150
× 37.5"	× 953	328.1	212	2051	33640	38452	16000
× 39"	× 990	341.3	220	2218	36380	43253	17990
× 40.5"	1029	354.4	229	2392	39230	48439	20150
× 42"	1067	367.5	237	2573	42200	54023	22470
× 43.5"	1105	380.6	245	2760	45260	60020	24970
× 45"	1143	393.8	254	2953	48430	66445	27640
10 ¾" × 10.5"	273 × 267	112.9	72.8	197.5	3239	1037	431
× 12"	× 305	129.0	83.2	258.0	4231	1548	644
× 13.5"	× 343	145.1	93.6	326.5	5355	2204	917
× 15"	× 381	161.3	104	403.1	6611	3023	1260

Table 4 Microllams.

Laminated Veneer Lumber (LVL)
$F_b = 2600$ psi, $F_v = 285$ psi, $E = 1.8 \times 10^6$ psi
$1\,^3/_4$" width:

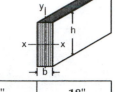

Depth (in.)	$5\,^1/_2$"	$7\,^1/_4$"	$9\,^1/_2$"	$11\,^7/_8$"	14"	16"	18"
Area (in.^{2})	9.63	12.69	16.63	20.78	24.50	28.00	31.50
S_x (in.^{3})	8.83	15.33	26.33	41.1	57.17	74.67	94.50
I_x (in.^{4})	25	55	125	245	400	595	850

Table 5 Section Properties of Parallams.

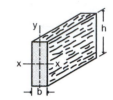

Parallel Strand Lumber (PSL)
$F_b = 2900$ psi, $F_v = 290$ psi, $F_{cll} = 2900$ psi, $E = 2.0 \times 10^6$ psi

Width	Depth	Area $A = (b) \times (h)$ (in.^{2})	Sect. Modulus $S_x = bh^2/6$ (in.^{3})	Mom. of Inertia $I_x = bh^3/12$ (in.^{4})
$1\,^3/_4$" ×	$9\,^1/_2$"	16.63	26.33	125
	$11\,^7/_8$"	20.79	41.15	245
	14"	24.50	57.17	400
	16"	28.00	74.67	595
	18"	31.50	94.50	850
$2\,^{11}/_{16}$" ×	$9\,^1/_2$"	25.53	40.42	190
	$11\,^7/_8$"	31.91	63.16	375
	14"	37.63	87.80	615
	16"	43.00	114.7	915
	18"	48.38	145.1	1305
$3\,^1/_2$" ×	$9\,^1/_2$"	33.25	52.65	250
	$11\,^7/_8$"	41.56	82.33	490
	14"	49.00	114.33	800
	16"	56.00	149.3	1195
	18"	63.00	189.0	1700
$5\,^1/_4$" ×	$9\,^1/_2$"	49.88	78.97	375
	$11\,^7/_8$"	62.34	123.5	735
	14"	73.50	171.5	1200
	16"	56.00	149.3	1790
	18"	94.50	283.5	2550
7" ×	$9\,^1/_2$"	66.50	105.3	500
	$11\,^7/_8$"	83.13	164.7	975
	14"	98.00	228.7	1600
	16"	112.0	298.7	2390
	18"	126.0	378.0	3400

Table 6   W—Wide-Flange Shapes—U.S. Customary.

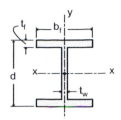

Dimensions and Properties for Preliminary Design

Section-Shape	Weight per Foot	Area	Depth of Section	Flange		Web Thickness	Axis x-x			Axis y-y		
	ω	A	d	b_f	t_f	t_w	I_x	S_x	r_x	I_y	S_y	r_y
	lb./ft.	in.²	in.	in.	in.	in.	in.⁴	in.³	in.	in.⁴	in.³	in.
W36 x	150	44.2	35.85	11.975	0.940	0.625	9040	504	14.3	270	45.1	2.47
	135	39.7	35.55	11.950	0.790	0.600	7800	439	14.0	225	37.7	2.38
W33 x	141	41.6	33.30	11.535	0.960	0.605	7450	448	13.4	246	42.7	2.43
	130	38.3	33.09	11.510	0.855	0.580	6710	406	13.2	218	37.9	2.39
	118	34.7	32.86	11.480	0.740	0.550	5900	359	13.0	187	32.6	2.32
W30 x	132	38.9	30.31	10.545	1.000	0.615	5770	380	12.2	196	37.2	2.25
	124	36.5	30.17	10.515	0.930	0.585	5360	355	12.1	181	34.4	2.23
	116	34.2	30.01	10.495	0.850	0.565	4930	329	12.0	164	31.3	2.19
	108	31.7	29.83	10.475	0.760	0.545	4470	299	11.9	146	27.9	2.15
	99	29.1	29.65	10.450	0.670	0.520	3990	269	11.7	128	24.5	2.10
W27 x	146	42.9	27.38	13.965	0.975	0.605	5630	411	11.4	443	63.5	3.21
	114	33.5	27.29	10.070	0.930	0.570	4090	299	11.0	159	31.5	2.18
	102	30.0	27.09	10.015	0.830	0.515	3620	267	11.0	139	27.8	2.15
	94	27.7	26.92	9.990	0.745	0.490	3270	243	10.9	124	24.8	2.12
	84	24.8	26.71	9.960	0.640	0.460	2850	213	10.7	106	21.2	2.07
W24 x	146	43.0	24.74	12.900	1.090	0.650	4580	371	10.3	391	60.5	3.01
	131	38.5	24.48	12.855	0.960	0.605	4020	329	10.2	340	53.0	2.97
	117	34.4	24.26	12.800	0.850	0.550	3540	291	10.1	297	46.5	2.94
	94	27.7	24.31	9.065	0.875	0.515	2700	222	9.87	109	24.0	1.98
	84	24.7	24.10	9.020	0.770	0.470	2370	196	9.79	94.4	20.9	1.95
	76	22.4	23.92	8.990	0.680	0.440	2100	176	9.69	82.5	18.4	1.92
	68	20.1	23.73	8.965	0.585	0.415	1830	154	9.55	70.4	15.7	1.87
	62	18.2	23.74	7.040	0.590	0.430	1550	131	9.23	34.5	9.80	1.38
	55	16.2	23.57	7.005	0.505	0.395	1350	114	9.11	29.1	8.30	1.34
W21 x	147	43.2	22.06	12.510	1.150	0.720	3630	329	9.17	376	60.1	2.95
	132	38.8	21.83	12.440	1.035	0.650	3220	295	9.12	333	53.5	2.93
	122	35.9	21.68	12.390	0.960	0.600	2960	273	9.09	305	49.2	2.92
	111	32.7	21.51	12.340	0.875	0.550	2670	249	9.05	274	44.5	2.90
	101	29.8	21.36	12.290	0.800	0.500	2420	227	9.02	248	40.3	2.89
	93	27.3	21.62	8.420	0.930	0.580	2070	192	8.70	92.9	22.1	1.84
	83	24.3	21.43	8.355	0.835	0.515	1830	171	8.67	81.4	19.5	1.83
	73	21.5	21.24	8.295	0.740	0.455	1600	151	8.64	70.6	17.0	1.81
	68	20.0	21.13	8.270	0.685	0.430	1480	140	8.60	64.7	15.7	1.80
	62	18.3	20.99	8.240	0.615	0.400	1330	127	8.54	57.5	13.9	1.77
	57	16.7	21.06	6.555	0.650	0.405	1170	111	8.36	30.6	9.35	1.35

(continued)

Table 6   *W—Wide-Flange Shapes—U.S. Customary.*   *(Continued)*

Section-Shape	Weight per Foot	Area	Depth of Section	Flange		Web Thickness	Axis x-x			Axis y-y		
	ω	A	d	b_f	t_f	t_w	I_x	S_x	r_x	I_y	S_y	r_y
	lb./ft.	in.²	in.	in.	in.	in.	in.⁴	in.³	in.	in.⁴	in.³	in.
	50	14.7	20.83	6.530	0.535	0.380	984	94.5	8.18	24.9	7.64	1.30
	44	13.0	20.66	6.500	0.450	0.350	843	81.6	8.06	20.7	6.36	1.26
W18 x	119	35.1	18.97	11.265	1.060	0.655	2190	231	7.90	253	44.9	2.69
	106	31.1	18.73	11.200	0.940	0.590	1910	204	7.84	220	39.4	2.66
	97	28.5	18.59	11.145	0.870	0.535	1750	188	7.82	201	36.1	2.65
	86	25.3	18.39	11.090	0.770	0.480	1530	166	7.77	175	31.6	2.63
	76	22.3	18.21	11.035	0.680	0.425	1330	146	7.73	152	27.6	2.61
W18 x	71	20.8	18.47	7.635	0.810	0.495	1170	127	7.50	60.3	15.8	1.70
	65	19.1	18.35	7.590	0.750	0.450	1070	117	7.49	54.8	14.4	1.69
	60	17.6	18.24	7.555	0.695	0.415	984	108	7.47	50.1	13.3	1.69
	55	16.2	18.11	7.530	0.630	0.390	890	98.3	7.41	44.9	11.9	1.67
	50	14.7	17.99	7.495	0.570	0.355	800	88.9	7.38	40.1	10.7	1.65
	46	13.5	18.06	6.060	0.605	0.360	712	78.8	7.25	22.5	7.43	1.29
	40	11.8	17.90	6.015	0.525	0.315	612	68.4	7.21	19.1	6.35	1.27
	35	10.3	17.70	6.000	0.425	0.300	510	57.6	7.04	15.3	5.12	1.22
W16 x	100	29.4	16.97	10.425	0.985	0.585	1490	175	7.10	186	35.7	2.52
	89	26.2	16.75	10.365	0.875	0.525	1300	155	7.05	163	31.4	2.49
	77	22.6	16.52	10.295	0.760	0.455	1110	134	7.00	138	26.9	2.47
	67	19.7	16.33	10.235	0.665	0.395	954	117	6.96	119	23.2	2.46
	57	16.8	16.43	7.120	0.715	0.430	758	92.2	6.72	43.1	12.1	1.60
	50	14.7	16.26	7.070	0.630	0.380	659	81.0	6.68	37.2	10.5	1.59
	45	13.3	16.13	7.035	0.565	0.345	586	72.7	6.65	32.8	9.34	1.57
	40	11.8	16.01	6.995	0.505	0.305	518	64.7	6.63	28.9	8.25	1.57
	36	10.6	15.86	6.985	0.430	0.295	448	56.5	6.51	24.5	7.00	1.52
	31	9.12	15.88	5.525	0.440	0.275	375	47.2	6.41	12.4	4.49	1.17
	26	7.68	15.69	5.500	0.345	0.250	301	38.4	6.26	9.59	3.49	1.12
W14 x	132	38.8	14.66	14.725	1.030	0.645	1530	209	6.28	548	74.5	3.76
	120	35.3	14.48	14.670	0.940	0.590	1380	190	6.24	495	67.5	3.74
	109	32.0	14.32	14.605	0.860	0.525	1240	173	6.22	447	61.2	3.73
	99	29.1	14.16	14.565	0.780	0.485	1110	157	6.17	402	55.2	3.71
	90	26.5	14.02	14.520	0.710	0.440	999	143	6.14	362	49.9	3.70
	82	24.1	14.31	10.130	0.855	0.510	882	123	6.05	148	29.3	2.48
	74	21.8	14.17	10.070	0.785	0.450	796	112	6.04	134	26.6	2.48
	68	20.0	14.04	10.035	0.720	0.415	723	103	6.01	121	24.2	2.46
	61	17.9	13.89	9.995	0.645	0.375	640	92.2	5.98	107	21.5	2.45
	53	15.6	13.92	8.060	0.660	0.370	541	77.8	5.89	57.7	14.3	1.92
	48	14.1	13.79	8.030	0.595	0.340	485	70.3	5.85	51.4	12.8	1.91
	43	12.6	13.66	7.995	0.530	0.305	428	62.7	5.82	45.2	11.3	1.89
	38	11.2	14.10	6.770	0.515	0.310	385	54.6	5.88	26.7	7.88	1.55
	34	10.0	13.98	6.745	0.455	0.285	340	48.6	5.83	23.3	6.91	1.53
	30	8.85	13.84	6.730	0.385	0.270	291	42.0	5.73	19.6	5.82	1.49

(continued)

Table 6 W—Wide-Flange Shapes—U.S. Customary.   *(Continued)*

Section-Shape	Weight per Foot	Area	Depth of Section	Flange		Web Thickness	Axis x-x			Axis y-y		
	ω	A	d	b_f	t_f	t_w	I_x	S_x	r_x	I_y	S_y	r_y
	lb./ft.	in.^{2}	in.	in.	in.	in.	in.^{4}	in.^{3}	in.	in.^{4}	in.^{3}	in.
W12 x	96	28.2	12.71	12.160	0.900	0.550	833	131	5.44	270	44.4	3.09
	87	25.6	12.53	12.125	0.810	0.515	740	118	5.38	241	39.7	3.07
	79	23.2	12.38	12.080	0.735	0.470	662	107	5.34	216	35.8	3.05
	72	21.1	12.25	12.040	0.670	0.430	597	97.4	5.31	195	32.4	3.04
	65	19.1	12.12	12.000	0.605	0.390	533	87.9	5.28	174	29.1	3.02
	58	17.0	12.19	10.010	0.640	0.360	475	78.0	5.28	107	21.4	2.51
	53	15.6	12.06	9.995	0.575	0.345	425	70.6	5.23	95.8	19.2	2.48
	50	14.7	12.19	8.080	0.640	0.370	394	64.7	5.18	56.3	13.9	1.96
	45	13.2	12.06	8.045	0.575	0.335	350	58.1	5.15	50.0	12.4	1.94
	40	11.8	11.94	8.005	0.515	0.295	310	51.9	5.13	44.1	11.0	1.93
	35	10.3	12.50	6.560	0.520	0.300	285	45.6	5.25	24.5	7.47	1.54
	30	8.79	12.34	6.520	0.440	0.260	238	38.6	5.21	20.3	6.24	1.52
	26	7.65	12.22	6.490	0.380	0.230	204	33.4	5.17	17.3	5.34	1.51
	16	4.71	11.99	3.990	0.265	0.220	103	17.1	4.67	2.8	1.41	0.77
W10 x	112	32.9	11.36	10.415	1.250	0.755	716	126	4.66	236	45.3	2.68
	100	29.4	11.10	10.340	1.120	0.680	623	112	4.60	207	40.0	2.65
	88	25.9	10.84	10.265	0.990	0.605	534	98.5	4.54	179	34.8	2.63
	77	22.6	10.60	10.190	0.870	0.530	455	85.9	4.49	154	30.1	2.60
	68	20.0	10.40	10.130	0.770	0.470	394	75.7	4.44	134	26.4	2.59
	60	17.6	10.22	10.080	0.680	0.420	341	66.7	4.39	116	23.0	2.57
	54	15.8	10.09	10.030	0.615	0.370	303	60.0	4.37	103	20.6	2.56
	49	14.4	9.98	10.000	0.560	0.340	272	54.6	4.35	93.4	18.7	2.54
	45	13.3	10.10	8.020	0.620	0.350	248	49.1	4.33	53.4	13.3	2.01
	39	11.5	9.92	7.985	0.530	0.315	209	42.1	4.27	45.0	11.3	1.98
	33	9.71	9.73	7.960	0.435	0.290	170	35.0	4.19	36.6	9.20	1.94
	30	8.84	10.47	5.810	0.510	0.300	170	32.4	4.38	16.7	5.75	1.37
	26	7.61	10.33	5.770	0.440	0.260	144	27.9	4.35	14.1	4.89	1.36
	22	6.49	10.17	5.750	0.360	0.240	118	23.2	4.27	11.4	3.97	1.33
	19	5.62	10.24	4.020	0.395	0.250	96.3	18.8	4.14	4.29	2.14	0.874
	17	4.99	10.11	4.010	0.330	0.240	81.9	16.2	4.05	3.56	1.78	0.845
	15	4.41	9.99	4.000	0.270	0.230	68.9	13.8	3.95	2.89	1.45	0.810
W8 x	67	19.7	9.00	8.280	0.935	0.570	272	60.4	3.72	88.6	21.4	2.12
	58	17.1	8.75	8.220	0.810	0.510	228	52.0	3.65	75.1	18.3	2.10
	48	14.1	8.50	8.110	0.685	0.400	184	43.3	3.61	60.9	15.0	2.08
	40	11.7	8.25	8.070	0.560	0.360	146	35.5	3.53	49.1	12.2	2.04
	35	10.3	8.12	8.020	0.495	0.310	127	31.2	3.51	42.6	10.6	2.03
	31	9.13	8.00	7.995	0.435	0.285	110	27.5	3.47	37.1	9.27	2.02
	28	8.25	8.06	6.535	0.465	0.285	98.0	24.3	3.45	21.7	6.63	1.62
	24	7.08	7.93	6.495	0.400	0.245	82.8	20.9	3.42	18.3	5.63	1.61
	21	6.16	8.28	5.270	0.400	0.250	75.3	18.2	3.49	9.77	3.71	1.26
	18	5.26	8.14	5.250	0.330	0.230	61.9	15.2	3.43	7.97	3.04	1.23
	15	4.44	8.11	4.015	0.315	0.245	48.0	11.8	3.29	3.41	1.70	0.876
	13	3.84	7.99	4.000	0.255	0.230	39.6	9.91	3.21	2.73	1.37	0.843
	10	2.96	7.89	3.940	0.205	0.170	30.8	7.81	3.22	2.09	1.06	0.841

Table 7   C—American Standard Channels—U.S. Customary.

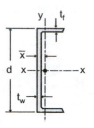

Dimensions and Properties for Preliminary Design

Section-Shape	Weight per Foot	Area	Depth of Section	Flange		Web Thickness	Axis x-x			Axis y-y			
	ω	A	d	b_f	t_f	t_w	I_x	S_x	r_x	I_y	S_y	r_y	x
	lb./ft.	in.^{2}	in.	in.	in.	in.	in.^{4}	in.^{3}	in.	in.^{4}	in.^{3}	in.	in.
C15 x	50.0	14.7	15.00	3.716	0.650	0.716	404	53.8	5.24	11.0	3.78	0.867	0.799
	40.0	11.8	15.00	3.520	0.650	0.520	349	46.5	5.44	9.23	3.36	0.886	0.778
	33.9	9.96	15.00	3.400	0.650	0.400	315	42.0	5.62	8.13	3.11	0.904	0.787
C12 x	30.0	8.82	12.00	3.170	0.501	0.510	162	27.0	4.29	5.14	2.06	0.763	0.674
	25.0	7.35	12.00	3.047	0.501	0.387	144	24.1	4.43	4.47	1.88	0.780	0.674
	20.7	6.09	12.00	2.942	0.501	0.282	129	21.5	4.61	3.88	1.73	0.799	0.698
C10 x	30.0	8.82	10.00	3.033	0.436	0.673	103	20.7	3.42	3.94	1.65	0.669	0.649
	25.0	7.35	10.00	2.886	0.436	0.526	91.2	18.2	3.52	3.36	1.48	0.676	0.617
	20.0	5.88	10.00	2.739	0.436	0.379	78.9	15.8	3.66	2.81	1.32	0.691	0.606
	15.3	4.49	10.00	2.600	0.436	0.240	67.4	13.5	3.87	2.28	1.16	0.713	0.634
C9 x	15.0	4.41	9.00	2.485	0.413	0.285	51.0	11.3	3.40	1.93	1.01	0.661	0.586
	13.4	3.94	9.00	2.433	0.413	0.233	47.9	10.6	3.48	1.76	0.962	0.668	0.601
C8 x	18.75	5.51	8.00	2.527	0.390	0.487	44.0	11.0	2.82	1.98	1.01	0.599	0.565
	13.75	4.04	8.00	2.343	0.390	0.303	36.1	9.03	2.99	1.53	0.853	0.615	0.553
	11.5	3.38	8.00	2.260	0.390	0.220	32.6	8.14	3.11	1.32	0.781	0.625	0.571
C7 x	12.25	3.60	7.00	2.194	0.366	0.314	24.2	6.93	2.60	1.17	0.702	0.571	0.525
	9.8	2.87	7.00	2.090	0.366	0.210	21.3	6.08	2.72	0.968	0.625	0.581	0.541

Table 8   Structural Pipe—U.S. Customary.

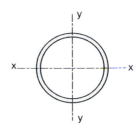

Dimensions and Properties for Preliminary Design

Nominal Diameter	Outside Diameter	Inside Diameter	Wall Thickness	Weight per Foot	Cross-Sectional Area	Moment of Inertia	Section Modulus	Radius of Gyration
ϕ	d_o	d_i	t	ω	A	I	S	r
in.	in.	in.	in.	lb./ft.	in.^{2}.	in.^{4}	in.^{3}	in.
Standard Weight								
2	2.375	2.067	0.154	3.65	1.07	.666	.561	.787
2.5	2.875	2.469	0.203	5.79	1.70	1.53	1.06	.947
3	3.500	3.068	0.216	7.58	2.23	3.02	1.72	1.16
3.5	4.000	3.548	0.226	9.11	2.68	4.79	2.39	1.34
4	4.500	4.026	0.237	10.79	3.17	7.23	3.21	1.51
5	5.563	5.047	0.258	14.62	4.30	15.2	5.45	1.88
6	6.625	6.065	0.280	18.97	5.58	28.1	8.50	2.25
8	8.625	7.981	0.322	28.55	8.40	72.5	16.8	2.94
10	10.750	10.020	0.365	40.48	11.9	161	29.9	3.67
Extra Strong								
2	2.375	1.939	0.218	5.02	1.48	.868	.731	.766
2.5	2.875	2.323	0.276	7.66	2.25	1.92	1.34	.924
3	3.500	2.90	0.300	10.25	3.02	3.89	2.23	1.14
3.5	4.000	3.364	0.318	12.50	3.68	6.28	3.14	1.31
4	4.500	3.826	0.337	14.98	4.41	9.61	4.27	1.48
5	5.563	4.813	0.375	20.78	6.11	20.7	7.43	1.84
6	6.625	5.761	0.432	28.57	8.40	40.5	12.2	2.19
8	8.625	7.625	0.500	43.39	12.8	108	24.5	2.88
10	10.750	9.750	0.500	54.74	16.1	212	39.4	3.63
Double-Extra Strong								
2	2.375	1.503	0.436	9.03	2.66	1.31	1.10	.703
2.5	2.875	1.771	0.552	13.69	4.03	2.87	2.00	.844
3	3.500	2.300	0.600	18.58	5.47	5.99	3.42	1.05
4	4.500	3.152	0.674	27.54	8.10	15.3	6.79	1.37
5	5.563	4.063	0.750	38.55	11.3	33.6	12.1	1.72
6	6.625	4.897	0.864	53.16	15.6	66.3	20.0	2.06
8	8.625	6.875	0.875	72.42	21.3	162	37.6	2.76

Table 9   L—Angles—Equal and Unequal Legs—U.S. Customary.

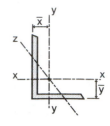

Dimensions and Properties for Preliminary Design

Size and Thickness	Weight per Foot	Area	Axis x-x				Axis y-y				Axis z-z
	ω	A	I_x	S_x	r_x	y	I_y	S_y	r_y	x	r_z
in.	lb./ft.	in.^{2}	in.^{4}	in.^{3}	in.	in.	in.^{4}	in.^{3}	in.	in.	in.
L5 × 5 × 1/2"	16.2	4.75	11.3	3.16	1.54	1.43	11.3	3.16	1.54	1.43	0.983
× 7/16"	14.3	4.18	10.0	2.79	1.55	1.41	10.0	2.79	1.55	1.41	0.986
× 3/8"	12.3	3.61	8.74	2.42	1.56	1.39	8.74	2.42	1.56	1.39	0.990
× 5/16"	10.3	34.00	7.42	2.04	1.57	1.37	7.42	2.04	1.57	1.37	0.994
L5 × 3 1/2 × 1/2"	13.6	4.00	9.99	2.99	1.58	1.66	4.05	1.56	1.01	0.906	0.755
× 7/16"	12.0	3.53	8.90	2.64	1.59	1.63	3.63	1.39	1.01	0.883	0.758
× 3/8"	10.4	3.05	7.78	2.29	1.60	1.61	3.18	1.21	1.02	0.861	0.762
× 5/16"	8.7	2.56	6.60	1.94	1.61	1.59	2.72	1.02	1.03	0.838	0.766
× 1/4"	7.0	2.06	5.39	1.57	1.62	1.56	2.23	0.830	1.04	0.814	0.770
L5 × 3 × 1/2"	12.8	3.75	9.45	2.91	1.59	1.75	2.58	1.15	0.829	0.750	0.646
× 7/16"	11.3	3.31	8.43	2.58	1.60	1.73	2.32	1.02	0.837	0.727	0.651
× 3/8"	9.8	2.86	7.37	2.24	1.61	1.70	2.04	0.888	0.845	0.704	0.654
× 5/16"	8.2	2.40	6.26	1.89	1.61	1.68	1.75	0.753	0.853	0.681	0.658
× 1/4"	6.6	1.94	5.11	1.53	1.62	1.66	1.44	0.614	0.861	0.657	0.663
L4 × 4 × 1/2"	12.8	3.75	5.56	1.97	1.22	1.18	5.56	1.97	1.22	1.18	0.782
× 7/16"	11.3	3.31	4.97	1.75	1.23	1.16	4.97	1.75	1.23	1.16	0.785
× 3/8"	9.8	2.86	4.36	1.52	1.23	1.14	4.36	1.52	1.23	1.14	0.788
× 5/16"	8.2	2.40	3.71	1.29	1.24	1.12	3.71	1.29	1.24	1.12	0.791
× 1/4"	6.6	1.94	3.04	1.05	1.25	1.09	3.04	1.05	1.25	1.09	0.795
L4 × 3 1/2 × 1/2"	11.9	3.50	5.32	1.94	1.23	1.25	3.79	1.52	1.04	1.00	0.722
× 7/16"	10.6	3.09	4.76	1.72	1.24	1.23	3.40	1.35	1.05	0.978	0.724
× 3/8"	9.1	2.67	4.18	1.49	1.25	1.21	2.95	1.17	1.06	0.955	0.727
× 5/16"	7.7	2.25	3.56	1.26	1.26	1.18	2.55	0.994	1.07	0.932	0.730
× 1/4"	6.2	1.81	2.91	1.03	1.27	1.16	2.09	0.808	1.07	0.909	0.734
L4 × 3 × 7/16"	9.8	2.87	4.52	1.68	1.25	1.30	2.18	0.992	0.871	0.804	0.641
× 3/8"	8.5	2.48	3.96	1.46	1.26	1.28	1.92	0.866	0.879	0.782	0.644
× 5/16"	7.2	2.09	3.38	1.23	1.27	1.26	1.65	0.734	0.877	0.759	0.647
× 1/4"	5.8	1.69	2.77	1.00	1.28	1.24	1.36	0.599	0.896	0.736	0.651
L3 1/2 × 3 1/2 × 7/16"	9.8	2.87	3.26	1.32	1.07	1.04	3.26	1.32	1.07	1.04	0.684
× 3/8"	8.5	2.48	2.87	1.15	1.07	1.01	2.87	1.15	1.07	1.01	0.687
× 5/16"	7.2	2.09	2.45	0.976	1.08	0.990	2.45	0.976	1.08	0.990	0.690
× 1/4"	5.8	1.69	2.01	0.794	1.09	0.968	2.01	0.794	1.09	0.968	0.694
L3 1/2 × 3 × 7/16"	9.1	2.65	3.10	1.29	1.08	1.10	2.09	0.975	0.889	0.853	0.622
× 3/8"	7.9	2.30	2.72	1.13	1.09	1.08	1.85	0.851	0.897	0.830	0.625
× 5/16"	6.6	1.93	2.33	0.954	1.10	1.06	1.58	0.722	0.905	0.808	0.627
× 1/4"	5.4	1.56	1.91	0.776	1.11	1.04	1.30	0.589	0.914	0.785	0.631
L3 1/2 × 2 1/2 × 3/8"	7.2	2.11	2.56	1.09	1.10	1.16	1.09	0.592	0.719	0.660	0.537
× 5/16"	6.1	1.78	2.19	0.927	1.11	1.14	0.939	0.504	0.727	0.637	0.540
× 1/4"	4.9	1.44	1.80	0.755	1.12	1.11	0.777	0.412	0.735	0.614	0.544
L3 × 3 × 3/8"	7.2	2.11	1.76	0.833	0.913	0.888	1.76	0.833	0.913	0.888	0.587
× 5/16"	6.1	1.78	1.51	0.707	0.922	0.865	1.51	0.707	0.922	0.865	0.589
× 1/4"	4.9	1.44	1.24	0.577	0.930	0.842	1.24	0.577	0.930	0.842	0.592

404

Table 10   W—Wide-Flange Shapes (an abridged listing)—SI Metric.

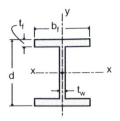

Dimensions and Properties for Preliminary Design

Section-Shape	Mass per Meter	Area	Depth of Section	Flange		Web Thickness	Axis x-x			Axis y-y		
	ω	A	d	b_f	t_f	t_w	I_x	S_x	r_x	I_y	S_y	r_y
	kg/m	mm²	mm	mm	mm	mm	10⁶ mm⁴	10³ mm³	mm	10⁶ mm⁴	10³ mm³	mm
W530 x	**182** *(122)*	23200	550	315	24	15	1230	4480	231	127	807	74.2
(W21 x)	**150** *(101)*	19200	543	312	20	13	1010	3720	229	103	661	73.4
	124 *(83)*	15700	544	212	21	13	761	2800	220	33.9	320	46.5
	101 *(68)*	12900	537	210	17	11	616	2300	218	27.0	257	45.7
	85 *(57)*	10800	535	166	17	10	487	1820	212	12.7	153	34.3
	65 *(44)*	8390	525	165	11	9	843	1340	205	8.61	104	32.0
W460 x	**177** *(119)*	22600	482	295	27	17	911	3790	201	105	736	68.3
(W18 x)	**144** *(97)*	18400	472	283	22	14	728	3080	199	83.6	592	67.3
	113 *(76)*	14400	460	280	17	11	553	2400	196	63.2	453	66.3
	97 *(65)*	12300	466	193	19	11	445	1920	190	22.8	236	42.9
	82 *(55)*	10400	460	191	16	10	370	1610	188	18.7	195	40.6
	68 *(46)*	8710	459	154	15	9	297	1290	184	9.36	122	32.8
	52 *(35)*	6640	450	152	11	8	212	945	179	6.36	84.0	31.0
W410 x	**149** *(100)*	19000	431	265	25	15	620	2870	180	77.4	585	64.0
(W16 x)	**132** *(89)*	16900	425	263	22	13	540	2540	179	67.8	515	63.2
	100 *(67)*	12700	415	260	17	10	397	1920	177	49.5	380	62.5
	74 *(50)*	9480	413	180	16	10	274	1330	170	15.5	172	40.4
	60 *(40)*	7610	407	178	13	8	215	1060	168	12.0	135	39.9
	46 *(31)*	5880	403	140	11	7	156	774	163	5.16	73.6	29.7
W360 x	**179** *(120)*	22800	368	373	24	15	574	3120	158	206	1110	95.0
(W14 x)	**162** *(109)*	20600	364	371	22	13	516	2840	158	186	1000	94.7
	134 *(90)*	17100	356	369	18	11	416	2350	156	151	818	94.0
	110 *(74)*	14100	360	256	20	11	331	1840	153	55.7	436	63.0
	91 *(61)*	11500	353	254	16	10	266	1510	152	44.5	353	62.2
	71 *(48)*	9100	350	204	15	9	202	1150	149	21.4	210	48.5
	57 *(38)*	7220	358	172	13	8	160	895	149	11.1	129	39.4
	45 *(30)*	5710	352	171	10	7	121	689	146	8.16	95.4	37.8
W310 x	**143** *(96)*	18200	323	309	23	14	347	2150	138	112	728	78.5
(W12 x)	**118** *(79)*	15000	314	307	19	12	275	1750	136	89.9	587	77.5
	97 *(65)*	12300	308	305	15	10	222	1440	134	72.4	477	76.7

(continued)

Table 10 W—Wide-Flange Shapes (an abridged listing)—SI Metric. (Continued)

Dimensions and Properties for Preliminary Design

Section-Shape	Mass per Meter	Area	Depth of Section	Flange		Web Thickness	Axis x-x			Axis y-y		
	ω	A	d	b_f	t_f	t_w	l_x	S_x	r_x	l_y	S_y	r_y
	kg/m	mm²	mm	mm	mm	mm	10⁶ mm⁴	10³ mm³	mm	10⁶ mm⁴	10³ mm³	mm
	79 *(53)*	10100	306	254	15	9	177	1160	133	39.9	315	63.0
	67 *(45)*	8510	306	204	15	9	146	953	131	20.8	203	49.3
	52 *(35)*	6640	318	166	13	8	119	748	133	10.2	123	39.1
	39 *(26)*	4930	310	165	10	6	84.9	548	131	7.20	87.6	38.4
W250 x	**149** *(100)*	19000	282	263	28	17	259	1840	117	86.1	656	67.3
(W10 x)	**131** *(88)*	16700	276	261	25	16	222	1620	115	74.5	571	66.8
	101 *(68)*	12900	264	257	20	12	164	1240	113	55.7	433	65.8
	80 *(54)*	10200	256	255	16	9	126	984	111	42.8	338	65.0
	67 *(45)*	8580	256	204	16	9	103	805	110	22.2	218	51.1
	49 *(33)*	6260	247	202	11	7	70.7	574	106	15.2	151	49.3
	39 *(26)*	4910	262	147	11	7	59.9	458	110	5.87	80.2	34.5
	28 *(19)*	3620	260	102	10	6	40.1	308	105	1.79	35.1	22.2
	22 *(15)*	2820	254	102	7	6	28.7	226	100	1.20	23.8	20.6
W200 x	**100** *(67)*	12700	229	210	24	14	113	991	94.0	36.9	351	53.8
(W8 x)	**71** *(48)*	9900	216	206	17	10	76.5	710	91.7	25.3	246	52.8
	52 *(35)*	6640	206	204	13	8	52.8	512	89.2	17.7	174	51.6
	42 *(28)*	5320	205	166	12	7	40.8	399	87.6	9.03	109	41.1
	31 *(21)*	3970	210	134	10	6	31.3	298	88.4	4.06	60.8	32.0
	22 *(15)*	2860	206	102	8	6	20.0	194	83.6	1.42	27.9	22.3
	15 *(10)*	1910	200	100	5	4	12.8	128	81.8	.869	17.4	21.4

Table 11   *Listing of W Shapes in Descending Order of* S_x *for Beam Design.*

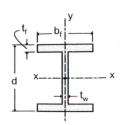

S_x			Allowable Stress Design—Selected Beam Shapes			S_x

S_x—US (in.³)	Section	S_x—SI ($10^3 \times$ mm³)	S_x—US (in.³)	Section	S_x—SI ($10^3 \times$ mm³)
448	**W33 × 141**	**7350**	188	W18 × 97	3080
439	W36 × 135	7200			
411	W27 × 146	6740	**176**	**W24 × 76**	**2890**
			175	W16 × 100	2870
406	**W33 × 130**	**6660**	173	W14 × 109	2840
380	W30 × 132	6230	171	W21 × 83	2800
371	W24 × 146	6080	166	W18 × 86	2720
			157	W14 × 99	2570
359	**W33 × 118**	**5890**	155	W16 × 89	2540
355	W30 × 124	5820			
			154	**W24 × 68**	**2530**
329	**W30 × 116**	**5400**	151	W21 × 73	2480
329	W24 × 131	5400	146	W18 × 76	2390
329	W21 × 147	5400	143	W14 × 90	2350
299	**W30 × 108**	**4900**	**140**	**W21 × 68**	**2300**
299	W27 × 114	4900	134	W16 × 77	2200
295	W21 × 132	4840			
291	W24 × 117	4770	**127**	**W21 × 62**	**2080**
273	W21 × 122	4480	127	W18 × 71	2080
			123	W14 × 82	2020
269	**W30 × 99**	**4410**	118	W12 × 87	1940
267	W27 × 102	4380	117	W18 × 65	1920
258	W24 × 104	4230	117	W16 × 67	1920
249	W21 × 111	4080			
			114	**W24 × 55**	**1870**
243	**W27 × 94**	**3990**	112	W14 × 74	1840
231	W × 119	3790	111	W21 × 57	1820
227	W21 × 101	3720	108	W18 × 60	1770
222	W24 × 94	3640	107	W12 × 79	1750
			103	W14 × 68	1690
213	**W27 × 84**	**3490**			
204	W18 × 106	3350	**98.3**	**W18 × 55**	**1610**
			97.4	W12 × 72	1600
196	**W24 × 84**	**3210**			
192	W21 × 93	3150	**94.5**	**W21 × 50**	**1550**
190	W14 × 120	3120	92.2	W16 × 57	1510

(continued)

Table 11   Listing of W Shapes in Descending Order of S$_x$ for Beam Design.   (Continued)

S_x **Allowable Stress Design—Selected Beam Shapes** S_x

S_x—US (in.³)	Section	S_x—SI (10³ × mm³)	S_x—US (in.³)	Section	S_x—SI (10³ × mm³)
92.2	W14 × 61	1510	54.6	W14 × 38	895
			54.6	W10 × 49	895
88.9	**W18 × 50**	**1460**	51.9	W12 × 40	851
87.9	W12 × 65	1440	49.1	W10 × 45	805
81.6	**W21 × 44**	**1340**	**48.6**	**W14 × 34**	**797**
81.0	W16 × 50	1330			
78.8	W18 × 46	1290	**47.2**	**W16 × 31**	**774**
78.0	W12 × 58	1280	45.6	W12 × 35	748
77.8	W14 × 53	1280	42.1	W10 × 39	690
72.7	W16 × 45	1190			
70.6	W12 × 53	1160	**42.0**	**W14 × 30**	**689**
70.3	W14 × 48	1150			
			38.6	**W12 × 30**	**633**
68.4	**W18 × 40**	**1120**			
66.7	W10 × 60	1090	**38.4**	**W16 × 26**	**630**
			35.0	W10.33	574
64.7	**W16 × 40**	**1060**			
64.7	W12 × 50	1060	**33.4**	**W12 × 26**	**548**
62.7	W14 × 43	1030	32.4	W10 × 30	531
60.0	W10 × 54	984	31.2	W8 × 35	512
58.1	W12 × 45	953			
			27.9	**W10 × 26**	**458**
57.6	**W18 × 35**	**945**	27.5	W8 × 31	451
56.5	W16 × 36	927			

Table 12   Allowable Column Loads—Selected Species/Sizes.

Eff.										4 × 4	A = 12.25	4 × 6	A = 19.25	4 × 8	A = 25.38
Col.	l/d	(l/d)sq	Fce	Fce/Fc*		Cp		F'c(psi)		Pa (k)		Pa (k)		Pa	
Len(ft)				Norm	Snow	Norm	Snow	Norm	Snow	Norm	Snow	Norm	Snow	Norm	Snow
4	13.71	188.08	2871.09	1.91	1.66	.8594	.8325	1289	1436	15.8	17.6	24.8	27.6	32.7	36.4
5	17.14	293.88	1837.50	1.23	1.07	.7573	.7138	1136	1231	13.9	15.1	21.9	23.7	28.8	31.3
6	20.57	423.18	1276.04	0.85	0.74	.6324	.5799	949	1000	11.6	12.3	18.3	19.3	24.1	25.4
7	24.00	576.00	937.50	0.63	0.54	.5184	.4611	778	795	9.5	9.7	15.0	15.3	19.7	20.2
8	27.43	752.33	717.77	0.48	0.42	.4194	.3750	629	647	7.7	7.9	12.1	12.5	16.0	16.4
9	30.86	952.16	567.13	0.38	0.33	.3439	.3035	516	524	6.3	6.4	9.9	10.1	13.1	13.3
10	34.29	1175.51	459.38	0.31	0.27	.2869	.2529	430	436	5.3	5.3	8.3	8.4	10.9	11.1
11	37.71	1422.37	379.65	0.25	0.22	.2355	.2090	353	361	4.3	4.4	6.8	6.9	9.0	9.2
12	41.14	1692.73	319.01	0.21	0.18	.2000	.1728	300	298	3.7	3.7	5.8	5.7	7.6	7.6
13	44.57	1986.61	271.82	0.18	0.16	.1728	.1544	259	266	3.2	3.3	5.0	5.1	6.6	6.8
14	48.00	2304.00	234.38	0.16	0.14	.1544	.1357	232	234	2.8	2.9	4.5	4.5	5.9	5.9

Eff.										6 × 6	A = 30.25	6 × 8	A = 41.25	6 × 10	A = 52.25
Col.	l/d	(l/d)sq	Fce	Fce/Fc*		Cp		F'c(psi)		Pa (k)		Pa (k)		Pa	
Len(ft)				Norm	Snow	Norm	Snow	Norm	Snow	Norm	Snow	Norm	Snow	Norm	Snow
5	10.91	119.01	4033.33	4.03	3.51	0.9425	0.9325	943	1072	28.5	32.4	38.9	44.2	49.2	56.0
6	13.09	171.37	2800.93	2.80	2.44	0.9119	0.896	912	1030	27.6	31.2	37.6	42.5	47.6	53.8
7	15.27	233.26	2057.82	2.06	1.79	0.8720	0.8475	872	975	26.4	29.5	36.0	40.2	45.6	50.9
8	17.45	304.66	1575.52	1.58	1.37	0.8218	0.7873	822	905	24.9	27.4	33.9	37.3	42.9	47.3
9	19.64	385.59	1244.86	1.24	1.08	0.7597	0.7169	760	824	23.0	24.9	31.3	34.0	39.7	43.1
10	21.82	476.03	1008.33	1.01	0.88	0.6944	0.6453	694	742	21.0	22.4	28.6	30.6	36.3	38.8
11	24.00	576.00	833.33	0.83	0.72	0.6235	0.5694	624	655	18.9	19.8	25.7	27.0	32.6	34.2
12	26.18	685.49	700.23	0.70	0.61	0.5586	0.5062	559	582	16.9	17.6	23.0	24.0	29.2	30.4
13	28.36	804.50	596.65	0.60	0.52	0.5000	0.4475	500	515	15.1	15.6	20.6	21.2	26.1	26.9
14	30.55	933.02	514.46	0.51	0.45	0.4406	0.3975	441	457	13.3	13.8	18.2	18.9	23.0	23.9
15	32.73	1071.07	448.15	0.45	0.39	0.3975	0.3518	398	405	12.0	12.2	16.4	16.7	20.8	21.1
16	34.91	1218.64	393.88	0.39	0.34	0.3518	0.3118	352	359	10.6	10.8	14.5	14.8	18.4	18.7
17	37.09	1375.74	348.90	0.35	0.30	0.3199	0.2785	320	320	9.7	9.7	13.2	13.2	16.7	16.7
18	39.27	1542.35	311.21	0.31	0.27	0.2869	0.2529	287	291	8.7	8.8	11.8	12.0	15.0	15.2
19	41.45	1718.48	279.32	0.28	0.24	0.2615	0.2267	262	261	7.9	7.9	10.8	10.8	13.7	13.6
20	43.64	1904.13	252.08	0.25	0.22	0.2355	0.2090	236	240	7.1	7.3	9.7	9.9	12.3	12.6
21	45.82	2099.31	228.65	0.23	0.20	0.2179	0.1910	218	220	6.6	6.6	9.0	9.1	11.4	11.5
22	48.00	2304.00	208.33	0.21	0.18	0.2000	0.1728	200	199	6.1	6.0	8.3	8.2	10.5	10.4

Eff.										8 × 8	A = 56.25	8 × 10	A = 71.25	8 × 12	A = 86.25
Col.	l/d	(l/d)sq	Fce	Fce/Fc*		Cp		F'c(psi)		Pa (k)		Pa (k)		Pa	
Len(ft)				Norm	Snow	Norm	Snow	Norm	Snow	Norm	Snow	Norm	Snow	Norm	Snow
8	12.80	163.84	2929.69	2.93	2.55	.9163	.9014	916	1037	51.5	58.3	65.3	73.9	79.0	89.4
9	14.40	207.36	2314.81	2.31	2.01	.8888	.8680	889	998	50.0	56.1	63.3	71.1	76.7	86.1
10	16.00	256.00	1875.00	1.88	1.63	.8566	.8286	857	953	48.2	53.6	61.0	67.9	73.9	82.2
11	17.60	309.76	1549.59	1.55	1.35	.8175	.7834	818	901	46.0	50.7	58.2	64.2	70.5	77.7

(continued)

Table 12   Allowable Column Loads—Selected Species/Sizes.   (Continued)

Eff. Col. Len(ft)	l/d	(l/d)sq	Fce	Fce/Fc* Norm	Snow	Cp Norm	Snow	F'c(psi) Norm	Snow	8×8 A = 56.25 Pa (k) Norm	Snow	8×10 A = 71.25 Pa (k) Norm	Snow	8×12 A = 86.25 Pa Norm	Snow
12	19.2	368.64	1302.08	1.30	1.13	.7731	.7315	773	841	43.5	47.3	55.1	59.9	66.7	72.6
13	20.8	432.64	1109.47	1.11	0.96	.7258	.6767	726	778	40.8	43.8	51.7	55.4	62.6	67.1
14	22.4	501.76	956.63	0.96	0.83	.6767	.6235	677	717	38.1	40.3	48.2	51.1	58.4	61.8
15	24.00	576.00	833.33	0.83	0.72	.6235	.5694	624	655	35.1	36.8	44.4	46.7	53.8	56.5
16	25.60	655.36	732.42	0.73	0.64	.5747	.5244	575	603	32.3	33.9	40.9	43.0	49.6	52.0
17	27.20	739.84	648.79	0.65	0.56	.5303	.4744	530	546	29.8	30.7	37.8	38.9	45.7	47.1
18	28.80	829.44	578.70	0.58	0.50	.4873	.4336	487	499	27.4	28.0	34.7	35.5	42.0	43.0
19	30.40	924.16	519.39	0.52	0.45	.4475	.3975	448	457	25.2	25.7	31.9	32.6	38.6	39.4
20	32.00	1024.00	468.75	0.47	0.41	.4122	.3673	412	422	23.2	23.8	29.4	30.1	35.6	36.4
21	33.60	1128.96	425.17	0.43	0.37	.3826	.3360	383	386	21.5	21.7	27.3	27.5	33.0	33.3
22	35.20	1239.04	387.40	0.39	0.34	.3518	.3118	352	359	19.8	20.2	25.1	25.5	30.3	30.9
23	36.80	1354.24	354.44	0.35	0.31	.3199	.2869	320	330	18.0	18.6	22.8	23.5	27.6	28.5
24	38.40	1474.56	325.52	0.33	0.28	.3035	.2615	304	301	17.1	16.9	21.6	21.4	26.2	25.9
25	40.00	1600.00	300.00	0.30	0.26	.2785	.2442	279	281	15.7	15.8	19.8	20.0	24.0	24.2
26	41.60	1730.56	277.37	0.28	0.24	.2615	.2267	262	261	14.7	14.7	18.6	18.6	22.6	22.5
27	43.20	1866.24	257.20	0.26	0.22	.2442	.2090	244	240	13.7	13.5	17.4	17.1	21.1	20.7
28	44.80	2007.04	239.16	0.24	0.21	.2267	.2000	227	230	12.8	12.9	16.2	16.4	19.6	19.8
29	46.40	2152.96	222.95	0.22	0.19	.2090	.1819	209	209	11.8	11.8	14.9	14.9	18.0	18.0
30	48.00	2304.00	208.33	0.21	0.18	.2000	.1728	200	199	11.3	11.2	14.3	14.2	17.3	17.1
	DF-L No.1	(P&T)		Fc = 1000			E = 1.6								
	DF-L No.1 & Btr	Dim.Lum		Fc = 1500			E = 1.8								

Table 13   Allowable Column Loads—Glu-Lams.

Eff. Col Len (ft.)	l/d	(l/d)sq	Fce	Fce/Fc* Norm	Snow	Cp Norm	Snow	F'c Norm	Snow	5.125×6" Pa (k) Norm	Snow	5.125×7.5" Pa (k) Norm	Snow
6	14.0	197.4	3389	2.05	1.79	.9242	.9067	1525	1721	46.9	52.9	58.6	66.2
7	16.4	268.6	2490	1.51	1.31	.8780	.8459	1449	1606	44.5	49.4	55.7	61.7
8	18.7	350.9	1906	1.16	1.00	.8113	.7597	1339	1442	41.2	44.3	51.5	55.4
9	21.1	444.1	1506	0.91	0.79	.7222	.6611	1192	1255	36.6	38.6	45.8	48.2
10	23.4	548.2	1220	0.74	0.64	.6317	.5661	1042	1074	32.1	33.0	40.1	41.3
11	25.8	663.4	1008	0.61	0.53	.5448	.4845	899	920	27.6	28.3	34.6	35.3
12	28.1	789.5	847	0.51	0.45	.4687	.4197	773	797	23.8	24.5	29.7	30.6
13	30.4	926.5	722	0.44	0.38	.4113	.3598	679	683	20.9	21.0	26.1	26.3
14	32.8	1074.6	622	0.38	0.33	.3598	.3155	594	599	18.3	18.4	22.8	23.0
15	35.1	1233.6	542	0.33	0.29	.3155	.2792	521	530	16.0	16.3	20.0	20.4
16	37.5	1403.5	477	0.29	0.25	.2792	.2423	461	460	14.2	14.1	17.7	17.7
17	39.8	1584.4	422	0.26	0.22	.2515	.2142	415	407	12.8	12.5	16.0	15.6
18	42.1	1776.3	377	0.23	0.20	.2236	.1953	369	371	11.3	11.4	14.2	14.2
19	44.5	1979.2	338	0.20	0.18	.1953	.1762	322	334	9.9	10.3	12.4	12.9
20	46.8	2193.0	305	0.18	0.16	.1762	.1571	291	298	8.9	9.2	11.2	11.5
21	49.2	2417.8	277	0.17	0.15	.1667	.1474	275	280	8.5	8.6	10.6	10.8

Eff. Col Len (ft.)	l/d	(l/d)sq	Fce	Fce/Fc* Norm	Snow	Cp Norm	Snow	F'c Norm	Snow	6.75×7.5" Pa (k) Norm	Snow	6.75×9" Pa (k) Norm	Snow
8	14.2	202.3	3306	2.00	1.74	.9213	.9027	1520	1713	77.0	86.7	92.3	104.1
9	16.0	256.0	2613	1.58	1.38	.8866	.8586	1463	1630	74.1	82.5	88.9	99.0
10	17.8	316.0	2116	1.28	1.11	.8398	.7970	1386	1513	70.2	76.6	84.2	91.9
11	19.6	382.4	1749	1.06	0.92	.7812	.7267	1289	1379	65.3	69.8	78.3	83.8
12	21.3	455.1	1470	0.89	0.77	.7129	.6496	1176	1233	59.6	62.4	71.5	74.9
13	23.1	534.1	1252	0.76	0.66	.5901	.5799	974	1101	49.3	55.7	59.2	66.9
14	24.9	619.5	1080	0.65	0.57	.5303	.5152	875	978	44.3	49.5	53.2	59.4
15	26.7	711.1	941	0.57	0.50	.4809	.4607	793	874	40.2	44.3	48.2	53.1
16	28.4	809.1	827	0.50	0.44	.4607	.4113	760	781	38.5	39.5	46.2	47.4
17	30.2	913.4	732	0.44	0.39	.3901	.3685	644	699	32.6	35.4	39.1	42.5
18	32.0	1024.0	653	0.40	0.34	.3596	.3244	593	616	30.0	31.2	36.0	37.4
19	33.8	1140.9	586	0.36	0.31	.3422	.2974	565	564	28.6	28.6	34.3	34.3
20	35.6	1264.2	529	0.32	0.28	.3065	.2700	506	512	25.6	25.9	30.7	31.1
21	37.3	1393.8	480	0.29	0.25	.2792	.2423	461	460	23.3	23.3	28.0	27.9
22	39.1	1529.7	437	0.26	0.23	.2515	.2236	415	424	21.0	21.5	25.2	25.8
23	40.9	1671.9	400	0.24	0.21	.2329	.2047	384	389	19.5	19.7	23.3	23.6
24	42.7	1820.4	367	0.22	0.19	.2142	.1858	353	353	17.9	17.9	21.5	21.4
25	44.4	1975.3	339	0.21	0.18	.2047	.1762	338	334	17.1	16.9	20.5	20.3
26	46.2	2136.5	313	0.19	0.16	.1858	.1571	307	298	15.5	15.1	18.6	18.1
27	48.0	2304.0	290	0.18	0.15	.1762	.1474	291	280	14.7	14.2	17.7	17.0
28	49.8	2477.8	270	0.16	0.14	.1571	.1378	259	262	13.1	13.2	15.7	15.9

(continued)

Table 13   Allowable Column Loads-Glu-Lams.   (Continued)

Eff. Col Len (ft.)	l/d	(l/d)sq	Fce	Fce/Fc* Norm	Snow	Cp Norm	Snow	F'c Norm	Snow	8.75×9" Pa (k) Norm	Snow	8.75×10.5" Pa (k) Norm	Snow
10	13.7	188.1	3556	2.16	1.87	.9297	.9129	1534	1733	120.8	136.4	140.9	159.2
11	15.1	227.6	2939	1.78	1.55	.9061	.8831	1495	1676	117.7	132.0	137.4	154.0
12	16.5	270.8	2469	1.50	1.30	.8767	.8439	1447	1602	113.9	126.1	132.9	147.2
13	17.8	317.9	2104	1.28	1.11	.8398	.7970	1386	1513	109.1	119.1	127.3	139.0
14	19.2	368.6	1814	1.10	0.96	.7940	.7439	1310	1412	103.2	111.2	120.4	129.7
15	20.6	423.2	1580	0.96	0.83	.7439	.6829	1227	1296	96.7	102.1	112.8	119.1
16	21.9	481.5	1389	0.84	0.73	.6881	.6255	1135	1187	89.4	93.5	104.3	109.1
17	23.3	543.6	1230	0.75	0.65	.6377	.5731	1052	1088	82.9	85.7	96.7	99.9
18	24.7	609.4	1098	0.67	0.58	.5867	.5227	968	992	76.2	78.1	88.9	91.2
19	26.1	679.0	985	0.60	0.52	.5375	.4766	887	905	69.8	71.2	81.5	83.1
20	27.4	752.3	889	0.54	0.47	.4923	.4362	812	828	64.0	65.2	74.6	76.1
21	28.8	829.4	806	0.49	0.42	.4526	.3943	747	748	58.8	58.9	68.6	68.8
22	30.2	910.3	735	0.45	0.39	.4197	.3685	693	699	54.5	55.1	63.6	64.3
23	31.5	995.0	672	0.41	0.35	.3858	.3333	637	633	50.1	49.8	58.5	58.1
24	32.9	1083.4	617	0.37	0.33	.3510	.3155	579	599	45.6	47.2	53.2	55.0
25	34.3	1175.5	569	0.34	0.30	.3244	.2883	535	547	42.2	43.1	49.2	50.3
26	35.7	1271.4	526	0.32	0.28	.3065	.2700	506	512	39.8	40.4	46.5	47.1
27	37.0	1371.1	488	0.30	0.26	.2883	.2515	476	477	37.5	37.6	43.7	43.9
28	38.4	1474.6	454	0.27	0.24	.2608	.2329	430	442	33.9	34.8	39.5	40.6
29	39.8	1581.8	423	0.26	0.22	.2515	.2142	415	407	32.7	32.0	38.1	37.4
30	41.1	1692.7	395	0.24	0.21	.2329	.2047	384	389	30.3	30.6	35.3	35.7

Table 14   *Column Stability Factor* C_p.

	C_p	$F_c' = C_p \cdot F_c'$	$F_{CE} = \dfrac{.30\,E}{(l/d)^2}$ for sawed posts	$F_{CE} = \dfrac{.418\,E}{(l/d)^2}$ for glu-lam posts

$\dfrac{F_{CE}}{F_C}$	Sawed C_p	Glu-Lam C_p	$\dfrac{F_{CE}}{F_C}$	Sawed C_p	Glu-Lam C_p	$\dfrac{F_{CE}}{F_C}$	Sawed C_p	Glu-Lam C_p	$\dfrac{F_{CE}}{F_C}$	Sawed C_p	Glu-Lam C_p
0.00	0.000	0.000	0.40	0.360	0.377	0.80	0.610	0.667	1.20	0.750	0.822
0.01	0.010	0.010	0.41	0.367	0.386	0.81	0.614	0.672	1.22	0.755	0.826
0.02	0.020	0.020	0.42	0.375	0.394	0.82	0.619	0.678	1.24	0.760	0.831
0.03	0.030	0.030	0.43	0.383	0.403	0.83	0.623	0.683	1.26	0.764	0.836
0.04	0.040	0.040	0.44	0.390	0.411	0.84	0.628	0.688	1.28	0.769	0.840
0.05	0.049	0.050	0.45	0.398	0.420	0.85	0.632	0.693	1.30	0.773	0.844
0.06	0.059	0.060	0.46	0.405	0.428	0.86	0.637	0.698	1.32	0.777	0.848
0.07	0.069	0.069	0.47	0.412	0.436	0.87	0.641	0.703	1.34	0.781	0.852
0.08	0.079	0.079	0.48	0.419	0.444	0.88	0.645	0.708	1.36	0.785	0.855
0.09	0.088	0.089	0.49	0.427	0.453	0.89	0.649	0.713	1.38	0.789	0.859
0.10	0.098	0.099	0.50	0.434	0.461	0.90	0.653	0.718	1.40	0.793	0.862
0.11	0.107	0.109	0.51	0.441	0.469	0.91	0.658	0.722	1.42	0.796	0.865
0.12	0.117	0.118	0.52	0.448	0.477	0.92	0.661	0.727	1.44	0.800	0.868
0.13	0.126	0.128	0.53	0.454	0.484	0.93	0.665	0.731	1.46	0.803	0.871
0.14	0.136	0.138	0.54	0.461	0.492	0.94	0.669	0.735	1.48	0.807	0.874
0.15	0.145	0.147	0.55	0.468	0.500	0.95	0.673	0.740	1.50	0.810	0.877
0.16	0.154	0.157	0.56	0.474	0.508	0.96	0.677	0.744	1.52	0.813	0.879
0.17	0.164	0.167	0.57	0.481	0.515	0.97	0.680	0.748	1.54	0.816	0.882
0.18	0.173	0.176	0.58	0.487	0.523	0.98	0.684	0.752	1.56	0.819	0.884
0.19	0.182	0.186	0.59	0.494	0.530	0.99	0.688	0.756	1.58	0.822	0.887
0.20	0.191	0.195	0.60	0.500	0.538	1.00	0.691	0.760	1.60	0.825	0.889
0.21	0.200	0.205	0.61	0.506	0.545	1.01	0.694	0.764	1.62	0.827	0.891
0.22	0.209	0.214	0.62	0.512	0.552	1.02	0.698	0.767	1.64	0.830	0.893
0.23	0.218	0.224	0.63	0.518	0.559	1.03	0.701	0.771	1.66	0.832	0.895
0.24	0.227	0.233	0.64	0.524	0.566	1.04	0.704	0.774	1.68	0.835	0.897
0.25	0.235	0.242	0.65	0.530	0.573	1.05	0.708	0.778	1.70	0.837	0.899
0.26	0.244	0.252	0.66	0.536	0.580	1.06	0.711	0.781	1.72	0.840	0.901
0.27	0.253	0.261	0.67	0.542	0.587	1.07	0.714	0.784	1.74	0.842	0.903
0.28	0.261	0.270	0.68	0.548	0.593	1.08	0.717	0.788	1.76	0.844	0.904
0.29	0.270	0.279	0.69	0.553	0.600	1.09	0.720	0.791	1.78	0.846	0.906
0.30	0.278	0.288	0.70	0.559	0.607	1.10	0.723	0.794	1.80	0.849	0.908
0.31	0.287	0.297	0.71	0.564	0.613	1.11	0.726	0.797	1.82	0.851	0.909
0.32	0.295	0.306	0.72	0.569	0.619	1.12	0.729	0.800	1.84	0.853	0.911
0.33	0.304	0.315	0.73	0.575	0.626	1.13	0.731	0.803	1.86	0.855	0.912
0.34	0.312	0.324	0.74	0.580	0.632	1.14	0.734	0.806	1.88	0.857	0.914
0.35	0.320	0.333	0.75	0.585	0.638	1.15	0.737	0.809	1.90	0.858	0.915
0.36	0.328	0.342	0.76	0.590	0.644	1.16	0.740	0.811	1.92	0.860	0.916
0.37	0.336	0.351	0.77	0.595	0.650	1.17	0.742	0.814	1.94	0.862	0.918
0.38	0.344	0.360	0.78	0.600	0.655	1.18	0.745	0.817	1.96	0.864	0.919
0.39	0.352	0.368	0.79	0.605	0.661	1.19	0.747	0.819	1.98	0.868	0.920

(continued)

Table 14   Column Stability Factor C_p. *   (Continued)*

"C_p"	$F_c' = C_p \cdot F_c'$ $F_{CE} = \dfrac{.30\,E}{(l/d)^2}$ **for sawed posts** $F_{CE} = \dfrac{.418\,E}{(l/d)^2}$ **for glu-lam posts**

$\dfrac{F_{CE}}{F_C}$	Sawed C_p	Glu-Lam C_p	$\dfrac{F_{CE}}{F_C}$	Sawed C_p	Glu-Lam C_p	$\dfrac{F_{CE}}{F_C}$	Sawed C_p	Glu-Lam C_p	$\dfrac{F_{CE}}{F_C}$	Sawed C_p	Glu-Lam C_p
2.00	0.867	0.921	2.40	0.894	0.940	3.40	0.930	0.962	4.40	0.948	0.972
2.02	0.869	0.922	2.45	0.897	0.941	3.45	0.931	0.963	4.45	0.949	0.973
2.04	0.870	0.924	2.50	0.899	0.943	3.50	0.932	0.963	4.50	0.949	0.973
2.06	0.872	0.925	2.55	0.901	0.944	3.55	0.933	0.964	4.55	0.950	0.974
2.08	0.874	0.926	2.60	0.904	0.946	3.60	0.934	0.965	4.60	0.950	0.974
2.10	0.875	0.927	2.65	0.906	0.947	3.65	0.936	0.965	4.65	0.951	0.974
2.12	0.876	0.928	2.70	0.908	0.949	3.70	0.937	0.966	4.70	0.952	0.975
2.14	0.878	0.929	2.75	0.910	0.950	3.75	0.938	0.966	4.75	0.952	0.975
2.16	0.879	0.930	2.80	0.912	0.951	3.80	0.938	0.967	4.80	0.953	0.975
2.18	0.881	0.931	2.85	0.914	0.952	3.85	0.939	0.968	4.85	0.953	0.975
2.20	0.882	0.932	2.90	0.916	0.953	3.90	0.940	0.968	4.90	0.954	0.976
2.22	0.883	0.932	2.95	0.917	0.954	3.95	0.941	0.969	5.00	0.955	0.976
2.24	0.885	0.933	3.00	0.919	0.955	4.00	0.942	0.969	6.00	0.963	0.981
2.26	0.886	0.934	3.05	0.920	0.956	4.05	0.943	0.969	8.00	0.973	0.986
2.28	0.887	0.935	3.10	0.922	0.957	4.10	0.944	0.970	10.0	0.979	0.989
2.30	0.888	0.936	3.15	0.923	0.958	4.15	0.944	0.970	20.0	0.990	0.995
2.32	0.889	0.937	3.20	0.925	0.959	4.20	0.945	0.971	40.0	0.995	0.997
2.34	0.891	0.937	3.25	0.926	0.960	4.25	0.946	0.971	60.0	0.997	0.998
2.36	0.892	0.938	3.30	0.927	0.961	4.30	0.947	0.972	100.0	0.998	0.999
2.38	0.893	0.939	3.35	0.929	0.961	4.35	0.947	0.972	200.0	0.999	0.999

Table developed and permission for use granted by Professor Ed Lebert, Dept. of Architecture, University of Washington.

Appendix B

Math Review

- Geometry and Properties of Simple Geometric Shapes
- Basic Trigonometric Functions
- Algebra—Simultaneous and Quadratic Equations
- Powers and Roots, Symbols, SI Units, and Conversions
- Supplementary Problems

B.1 GEOMETRY AND PROPERTIES OF SIMPLE GEOMETRIC SHAPES

As a prelude to the study of statics and the strength of materials, a mathematics refresher is included in this appendix to solidify the basis for working in an analytical environment. A fundamental knowledge of algebra, geometry, and trigonometry is sufficient to solve all of the problems contained in this text.

Geometry is a branch of mathematics that deals with space relationships. When dealing with force systems and structural elements, a knowledge of and ability to apply the principles of geometry are quite useful. This section will briefly define the common terms, symbols, and theorems/postulates commonly used in the study of statics.

Definitions (See Figure B.1)

A *point* is the position of the intersection of two or more lines. A point may also be the result of viewing down along an axis. The point has no length or thickness.

A *line* is the intersection of two surfaces. Lines have length but have no width or thickness. A *straight line* is the shortest distance connecting two points.

A *horizontal line* is a straight line that is level with the horizon line.

A *vertical line* is a straight line that is perpendicular to the horizon line.

Two lines are *perpendicular* to each other when the angles at which they intersect are all equal. *Perpendicular lines* are at *right angles* (90°) to each other.

An *oblique line* is neither horizontal nor vertical.

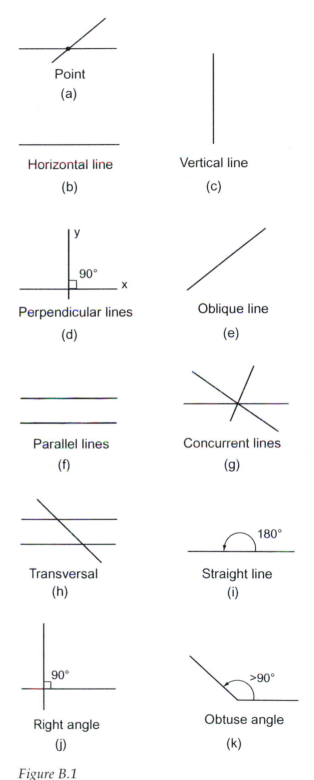

Figure B.1

Parallel lines are two or more straight lines that are equidistant from each other at all points and would never meet no matter how far they might be extended.

Concurrent lines are two or more straight lines that intersect at a common point.

A *transversal* is a line that intersects two or more lines.

An *angle* is a figure formed by two lines proceeding from a common point called the *vertex*. The lines that form an angle are called its *sides*.

A *straight angle* is one of 180°.

An *acute angle* is less than (<) 90°.

A *right angle* is one of 90°, half of a straight angle.

An *obtuse angle* is greater than (>) a right angle but less than (<) a straight angle. Obtuse angles are between 90° and 180°.

A *surface* has two dimensions, length and width. A *flat surface* is referred to as a *plane*.

A *solid* has three dimensions: length, width, and thickness.

Symbols

Common symbols used in mathematical operations in this text are listed in Table B.1.

Table B.1

$\angle$ angle	° degree
= equality	≠ not equal to
< less than	≤ less than or equal to
> greater than	≥ greater than or equal to
∴ therefore	≈ approximately equal to
⊥ perpendicular to	∝ proportional to
// parallel to	

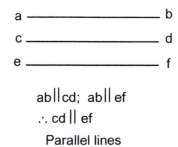

ab‖cd; ab‖ef

∴ cd ‖ ef

Parallel lines

Figure B.2   Parallel lines.

Parallel Lines

1. If two straight lines are parallel to a third straight line, they are parallel to each other (Figure B.2).

2. If two parallel lines are cut by a transversal (Figure B.3):

   a. the alternate interior angles are equal.

      $\angle a = \angle b$, and $\angle c = \angle d$

   b. the alternate exterior angles are equal.

      $\angle e = \angle f$, and $\angle g = \angle h$

   c. the corresponding angles are equal.

      $\angle d = \angle e, \angle c = \angle f, \angle b = \angle g$, and $\angle a = \angle h$

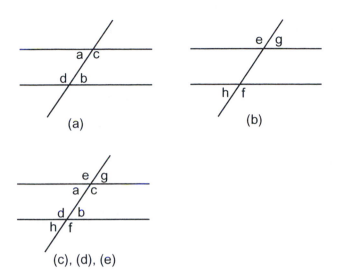

Figure B.3   *Parallel lines cut by a transversal.*

    d. the two interior angles on the same side of a transversal are supplementary. Two angles are supplementary if their sum equals 180°.

$$\angle a \text{ supplement of } \angle d, \text{ and } \angle c \text{ supp. } \angle b$$

    e. the two exterior angles on the same side of a transversal are supplementary.

$$\angle e \text{ supp. of } \angle h, \text{ and } \angle g \text{ supp. of } \angle f$$

Triangles

Triangles are three-sided figures in which the sides are composed of straight lines and are classified based on their sides. The sum of the three interior angles of a triangle is always equal to 180°. An *equilateral triangle* (Figure B.4a) has three equal sides. A *scalene triangle* (Figure B.4c) has no two sides that are equal, while an *isosceles triangle* (Figure B.4b) has two sides that are equal.

Another classification of triangles is based on their angles, such as equiangular, right, acute, and obtuse.

Equiangular triangles (Figure B.4a) contain three equal angles, each measuring 60°.

Right triangles (or right-angled triangles) (Figure B.4d) contain one right angle (often indicated by placing a small square at the 90° angle).

Acute triangles (Figure B.4b) have all three angles less than right angles but not necessarily equiangular.

Obtuse triangles (Figure B.4e) have one angle greater than a right angle.

One of the most important theorems about triangles is that no matter what the shape or size of any triangle, *the sum of the three angles of a triangle is equal to a straight angle, or 180°.*

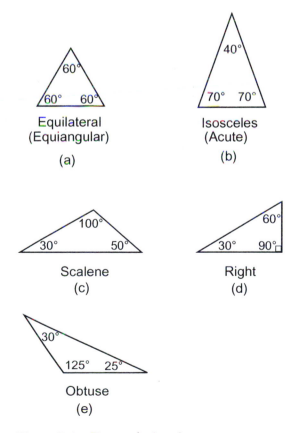

Figure B.4   *Types of triangles.*

Additional corollaries worth mentioning are:

a. *No triangle may have more than one obtuse angle or right angle.*
b. *The acute angles of a right triangle are complementary.*
c. *If two angles of one triangle are equal respective to two angles of another, the third angles are equal.*

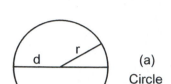

(a)
Circle

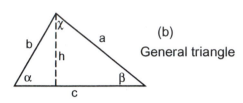

(b)
General triangle

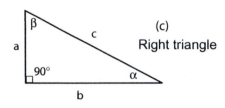

(c)
Right triangle

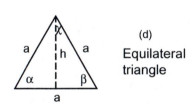

(d)
Equilateral triangle

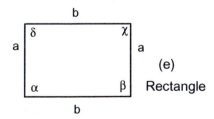

(e)
Rectangle

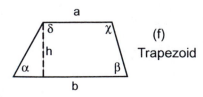

(f)
Trapezoid

Figure B.5

Table B.2   Plane Geometry Formulas for Common Geometric Shapes.

Circle (Figure B.5a):

Circumference $= 2\pi r = \pi d$

Where, r = radius; d = diameter = $2r$

$$\text{Area} = \pi r^2 = \frac{\pi d^2}{4}$$

General Triangle (Oblique Triangle) (Figure B.5b):

$\alpha + \beta + \ = 180°$

$h = a \sin \beta = b \sin \alpha$

Perimeter $= a + b + c$

Area $= \frac{1}{2}hc$

Right Triangle (Figure B.5c):

$\alpha + \beta = \ = 90°$

Perimeter $= a + b + c$

Area $= \frac{1}{2}ab = \frac{1}{2}hc$

Equilateral Triangle (Figure B.5d):

$\alpha = \beta = \chi = 60°$

Perimeter $= 3a$

Area $= \frac{1}{4}a^2\sqrt{3}$

Rectangle (Figure B.5e):

$\alpha = \beta = \chi = \delta = 90°$

Perimeter $= 2(a + b)$

Area $= ab$

Trapezoid (Figure B.5f):

$\alpha + \beta + \chi + \delta = 360°; \quad a//b$

$$\text{Perimeter} = a + b + h\left(\frac{1}{\sin \alpha} + \frac{1}{\sin \beta}\right)$$

Area $= \frac{1}{2}h(a + b)$

B.2 BASIC TRIGONOMETRIC FUNCTIONS

Trigonometry

Trigonometry is the branch of mathematics that concerns itself with the measurement of the parts of a triangle. The word trigonometry is a Greek word that means triangle measurement. Plane trigonometry is the study of triangles lying in a plane (two dimensions). A working knowledge of trigonometry is essential in surveying, navigation, and engineering mechanics (including statics).

We will restrict ourselves primarily to working with right triangles (triangles with a 90° angle) in two dimensions.

Trigonometry makes extensive use of three ideas from geometry: the Pythagorean theorem, congruence, and similarity.

Named after Pythagoras, the Greek mystic of the sixth century B.C., the *Pythagorean theorem* states that:

> *The square of the hypotenuse of a **right triangle** is equal to the sum of the squares of the other two sides (Figure B.6).*

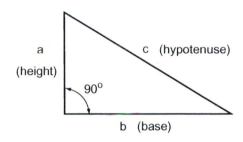

Figure B.6   Right triangle.

In other words, where:

c = hypotenuse, a = height, and b = base;

$$c^2 = a^2 + b^2$$

Taking the square root of both sides of the equation,

$$c = \sqrt{a^2 + b^2}$$

or,

$$a = \sqrt{c^2 - b^2}; \quad \text{and} \quad b = \sqrt{c^2 - a^2}$$

The hypotenuse of a right triangle is the side opposite the right angle.

Example Problem B.1 (Figure B.7)

Determine the length of the hypotenuse of a right triangle that has a base length of 4 feet and a height of 3 feet.

Solution:

$$c^2 = a^2 + b^2$$

Height:

$$a = 3 \text{ ft.}; a^2 = (3 \text{ ft.})^2 = 9 \text{ ft.}^2$$

Base:

$$b = 4 \text{ ft.}; b^2 = (4 \text{ ft.})^2 = 16 \text{ ft.}^2$$

$$\therefore c^2 = 9 \text{ ft.}^2 + 16 \text{ ft.}^2 = 25 \text{ ft.}^2$$

$$c = \sqrt{25 \text{ ft.}^2} = 5 \text{ ft.}$$

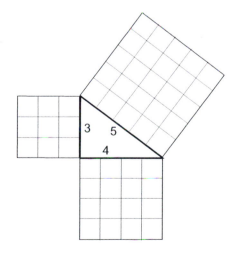

Figure B.7   Pythagorean theorem.

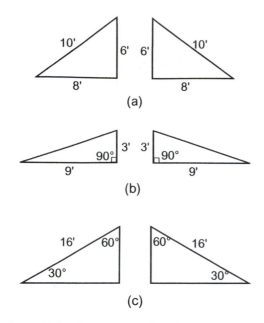

Figure B.8   *Congruent triangle sets.*

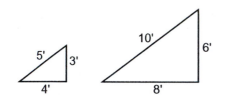

Figure B.9   *Similar triangles.*

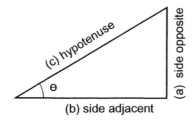

Figure B.10   *Right triangle with an acute angle.*

Objects that have both the same shape and the same size are said to be *congruent*. In geometry it is shown that triangles are congruent if they agree with respect to any of the following (Figure B.8):

   a. All three sides are equal.

   b. Two sides and the angle between them are equal.

   c. Two angles and the side between them are equal.

Triangles that are completely identical are congruent.

Objects that have the same shape are said to be *similar*, regardless of their sizes (Figure B.9). When two triangles are similar, the ratio of any two sides of one triangle equals the ratio of corresponding sides of the second triangle.

For example, if a model building is built on a scale of 1:50, the distance between any two points on the model is $\frac{1}{50}$ of the distance between corresponding points on the actual building, and the model is similar to the real building.

Trigonometry of Right Triangles

If an acute angle of one right triangle is equal to an acute angle of another right triangle, then all of the angles are equal and the triangles are similar. In similar triangles, the ratio of the lengths of any two sides in one triangle equals the corresponding ratio for the other triangle.

There are three very useful ratios that are commonly used in working with structural frameworks encountered in engineering mechanics. Trigonometric functions relate the angle θ to the ratio of the corresponding sides of a triangle. Recall that the side opposite the right angle is called the *hypotenuse*, while the other sides are called the *legs*.

The three relationships or ratios are related to the right triangle shown in Figure B.10. The angle θ is the acute angle that will be used in determining the ratios of the triangle's sides.

$$\text{sine } \theta = \sin \theta = \frac{\text{Length of opposite side}}{\text{Hypotenuse}} = \frac{a}{c}$$

$$\text{cosine } \theta = \cos \theta = \frac{\text{Length of adjacent side}}{\text{Hypotenuse}} = \frac{b}{c}$$

$$\text{tangent } \theta = \tan \theta = \frac{\text{Length of opposite side}}{\text{Length of adjacent side}} = \frac{a}{b}$$

As an angle increases from 0° to 90°, its:

sine	*increases* from 0 to 1,
cosine	*decreases* from 1 to 0,
tangent	*increases* from 0 to ∞

Example Problem B.2

A ladder 20-ft. long must not be inclined by less than 60° with the horizontal or it will begin sliding on the floor. How many feet away from the wall can you set the foot of the ladder? (See Figure B.11.)

Solution:

Since θ and *c* are known, the unknowns are *a* and *b*.

The solution for length *b* is desired here so the trigonometric function that involves θ, *c*, and *b* is the cosine.

$$\cos \theta = \frac{b}{c}; \quad \therefore b = c \times \cos \theta$$

where: θ = 60°

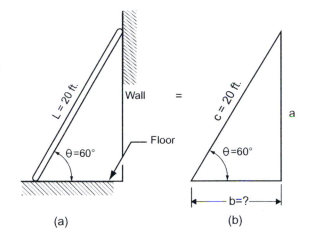

(a)   (b)

Figure B.11

Using a table of trigonometric functions or your calculator, the value of cos 60° = 0.5 is obtained.

Substituting: $b = 20 \text{ ft.} \times (0.5) = 10 \text{ ft.}$

How high up the wall does the ladder reach?

$$\sin \theta = \frac{a}{c};$$

$$\therefore a = c \times \sin 60° = (20 \text{ ft.}) \times (0.866)$$

$$= 17.32 \text{ ft.} = 17 \text{ ft.} - 4 \text{ in.}$$

or

$$\tan \theta = \frac{a}{b};$$

$$\therefore a = b \times \tan 60° = (10 \text{ ft.}) \times (1.732) = 17.32 \text{ ft.}$$

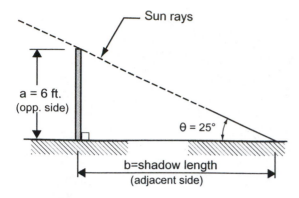

Figure B.12

Example Problem B.3 (Figure B.12)

If the sun is at a 25° angle off the horizon at 5:30 P.M., how far from your 6-ft. tall fence should you plant your garden in order to take full advantage of the sun?

Solution:

The height of the fence (6 ft.) and the angle ($\theta = 25°$) of a right triangle are known. To determine the horizontal leg dimension, the tangent trigonometric function will be utilized.

$$\tan \theta = \frac{\text{side opp.}}{\text{side adj.}} = \frac{a}{b}$$

$$\tan 25° = \frac{6 \text{ ft.}}{b}; \quad \therefore b = \frac{6 \text{ ft.}}{\tan 25°} = \frac{6 \text{ ft.}}{0.466} = 12.9 \text{ ft.}$$

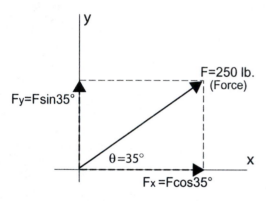

Figure B.13   Components of a force.

Example Problem B.4

Forces are *vectors*, which may be resolved into *components* along the rectangular axes x and y (the axes are at right angles to each other). Using the trigonometric functions, determine the force components (F_x and F_y) along the x and y axes for the force $F = 250 \text{ lb}$ (see Figure B.13).

Solution:

Create an imaginary rectangle in which the end points of the force F represent the diagonal. Component force F_x lies along the x axis and component F_y is projected along the y axis. Force component Fy represents the side opposite the $\theta = 35°$ angle and Fx is the side adjacent to the included angle. Therefore:

$$F_x = F \cos \theta = F \cos 35° = (250 \text{ lb.}) \times (0.819)$$
$$= 204.8 \text{ lb.}$$
$$F_y = F \sin \theta = F \sin 35° = (250 \text{ lb.}) \times (0.574)$$
$$= 143.4 \text{ lb.}$$

Law of Cosines

In cases where a triangle is not a right triangle (an *oblique triangle*), the *law of cosines* is quite helpful in finding sides and angles of any triangle. Knowing three sides, or two sides and the included angle, or two angles and a side, the unknown angle(s) and sides of an oblique triangle can be computed.

Given an oblique triangle (Figure B.14) with the angles α, β, and χ and the sides opposite, *a*, *b*, and *c*, respectively, draw a perpendicular from χ to c and label the dimension *d*. Examine the triangle *bde*, and note the following observations:

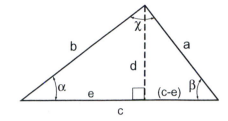

Figure B.14   Oblique triangles—law of cosines.

    a. $b^2 = e^2 + d^2$ (Pythagorean theorem)
    b. $d^2 = b^2 - e^2$ (rearranging the equation above)
    c. $a^2 = d^2 + (c - e)^2$ (Pythagorean theorem)
    d. $a^2 = (b^2 - e^2) + (c - e)^2$ (substitution)
    e. $a^2 = b^2 - e^2 + c^2 - 2ce + e^2$
    f. $a^2 = b^2 + c^2 - 2ce$
    g. $\cos \alpha = \dfrac{e}{b}$ (definition of cosine)
    h. $e = b \cos \alpha$ (rearranging the equation above)
    i. $a^2 = b^2 + c^2 - 2bc \cos \alpha$ (substitution of equation *h* into equation *f*)

Using similar geometric and algebraic properties, the law(s) of cosines are derived as:

$$a^2 = b^2 + c^2 - 2bc \cos \alpha$$

$$b^2 = a^2 + c^2 - 2ac \cos \beta$$

$$c^2 = a^2 + b^2 - 2ab \cos \chi$$

Law of Sines

The *law of sines* is another useful relationship for finding the sides and angles of any triangle, given two sides and an angle, or two angles and a side. Assume that the triangle shown in Figure B.15 has angles α, β, and χ and sides opposite *a*, *b*, and *c*, respectively.

The law of sines is expressed as:

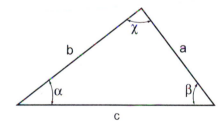

Figure B.15   Oblique triangle—law of sines.

$$\frac{\sin \alpha}{a} = \frac{\sin \beta}{b} = \frac{\sin \chi}{c} \quad \text{or}$$

$$\frac{a}{\sin \alpha} = \frac{b}{\sin \beta} = \frac{c}{\sin \chi}$$

Note that the law of cosines relates three sides and one angle. If a triangle is given with three sides, any angle can be determined using the law of cosines. If two sides and an included angle are known, the third side can be solved using the law of cosines. Then, the second angle can be determined using the law of cosines or the law of sines. The third angle is most easily obtained by using the relationship $\alpha + \beta + \chi = 180°$.

If two angles and a side are known, the third angle is obtained from $\alpha + \beta + \chi = 180°$. Then, the second and third sides are determined by applying the law of sines.

Example Problem B.5 (Figure B.16)

A sailboat, just prior to entering a fogbank, was 1200 ft. SW of a lighthouse. Compass readings indicated that the sailboat was 50° east of magnetic north. Since then the boat has sailed 2000 ft. on a course 120° off north. Determine how far the sailboat is from the lighthouse.

Solution:

The first step is to determine the angle α between the known sides b and c of the oblique triangle.

$$\alpha = 120° - 50° = 70°$$

Using the cosine law:

$$a^2 = b^2 + c^2 - 2bc \cos \alpha$$
$$a^2 = (1200 \text{ ft.})^2 + (2000 \text{ ft.})^2$$
$$- 2(1200 \text{ ft.})(2000 \text{ ft.}) \cos 70°$$
$$a^2 = (1.44 \times 10^6) + (4 \times 10^6) - (4.8 \times 10^6)(0.342)$$
$$a^2 = (3.8 \times 10^6)$$
$$\therefore a = 1.95 \times 10^3 = 1950 \text{ ft.}$$

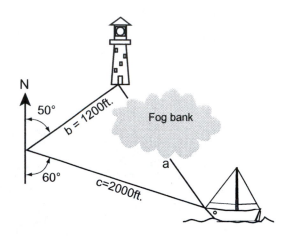

Figure B.16

Example Problem B.6 (Figure B.17)

A survey results in an oblique triangle with two angles and a side known. Determine the third angle and the distances between **AC** and **BC**.

Solution:

The third angle χ can be determined as:

$$\chi = 180° - \alpha - \beta = 180° - 30° - 50° = 100°$$

Then, using the law of sines:

$$\frac{\sin \alpha}{a} = \frac{\sin \beta}{b} = \frac{\sin \chi}{c}$$

$$\therefore a = \frac{c \sin \alpha}{\sin \chi} = \frac{(240 \text{ ft.}) \sin 30°}{\sin 100°}$$

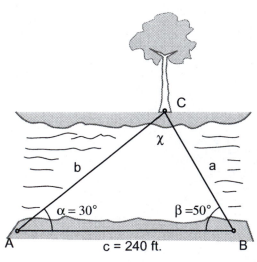

Figure B.17   Plan of river.

$$= \frac{(240 \text{ ft.})(0.50)}{(0.985)} = 122 \text{ ft.}$$

$$b = \frac{c \sin \beta}{\sin \chi} = \frac{(240 \text{ ft.}) \sin 50°}{\sin 100°}$$

$$= \frac{(240 \text{ ft.})(0.766)}{(0.985)} = 187 \text{ ft.}$$

B.3 ALGEBRA—SIMULTANEOUS AND QUADRATIC EQUATIONS

Algebra is a generalization of arithmetic involving the study of relations and properties between numbers represented by symbols and obtained by the operations of addition, subtraction, multiplication, division, raising to a power, and extracting a root. The value of algebra is its ability as a tool to solve a wide array of problems that can be expressed as equations. An *equation* is a statement that two expressions are equal. An *expression* is a collection of terms combined by addition, subtraction, or both, as in: $(4a + 3b)$, $(2a - b - 3c)$, $(3x - 5y)$.

In engineering, including the study of statics, it is common to express certain kinds of facts in *algebraic formulas*. A *formula* is a shorthand method of expressing a rule by the use of symbols or letter designations. It must be remembered that a formula is an equation.

Golden Rule of Algebra

When solving an equation, you are generally finding the value of an unknown in terms of what has been given about the other numbers in the equation. An important rule used for operating on equations must be adhered to in order to preserve equality. The following rule is sometimes referred to as the *golden rule of algebra*.

> Rule: *What is done to one side of an equation must also be done to the other side.*

Example Problem B.7

If $x - 10 = 8$, what is x?

Solution:

By adding 10 to both sides of the equation, the equality is maintained.

$$x - 10 + 10 = 8 + 10; \therefore x = 18$$

Example Problem B.8

If $x + 16 = 37$, solve for x.

Solution:

Subtract 16 from each side of the equation to maintain equality. Then,

$$x + 16 - 16 = 37 - 16; \quad \therefore x = 21$$

Example Problem B.9

If $\dfrac{x}{5} = 3$, what is x?

Solution:

Multiply both sides of the equation by 5 and solve for x.

$$\left(\frac{x}{5}\right) \times (5) = 3 \times (5); \qquad \left(\frac{x}{\cancel{5}}\right) \times (\cancel{5}) = 3 \times (5)$$

$$\therefore x = 15$$

Example Problem B.10

If $3x = 60$; solve for x.

Solution:

Divide both sides of the equation by the coefficient of x.

$$\frac{3x}{(3)} = \frac{60}{(3)}; \quad \therefore x = 20$$

Sign Laws

Learning to use signed (+ and −) numbers is essential for performing algebraic operations in solving an equation. Summarized below are the basic *sign laws* in algebra.

$$(A) = -(-A) = A$$
$$-(A) = (-A) = -A$$
$$(A)(B) = (-A)(-B) = AB$$
$$(-A)(B) = (A)(-B) = -AB$$
$$\frac{-A}{-B} = \frac{A}{B}$$
$$\frac{-A}{B} = \frac{A}{-B} = -\frac{A}{B}$$

Linear Equations in Two Unknowns

There are numerous problems in statics that require the solution of *linear equations* containing two unknowns. The general form of a linear equation with two unknowns, x and y, is one which can be written as:

$$Ax + By + C = 0$$

where it is assumed that the coefficients A and B are both non-zero.

Simultaneous equations are equations involving the same unknown quantities. For example:

$$6x - 5y = 14$$

$$3x + 2y = 16$$

are simultaneous equations since both involve the same unknowns, x and y.

Simultaneous equations involving two unknowns are solved using the general steps as follows:

    a. Eliminate one of the unknowns.

    b. Solve for the other unknown that was not eliminated.

    c. Substitute the value of the unknown found in the previous step into either of the simultaneous equations and solve for the other unknown.

Elimination of an unknown may be performed by any one of three methods:

    1. By *comparison*

    2. By *elimination* using addition or subtraction

    3. By *substitution*

Example Problem B.11 (By Comparison)

Solve for the values of x and y for the equations:

$$6x - 5y = 14 \text{ and } 3x + 2y = 16$$

Solution:

Rewrite each equation to find the value of one of the unknowns in terms of the other.

Form an equation from these equal values.

$$6x - 5y = 14, \quad \text{rewriting}; \quad x = \frac{14 + 5y}{6}.$$

$$3x + 2y = 16, \quad \text{rewriting}; \quad x = \frac{16 - 2y}{3}.$$

Equating both equations:

$$\frac{14 + 5y}{6} = \frac{16 - 2y}{3}, \text{ (both being equal to } x\text{)}$$

Multiply both sides of the equation by 6 (the common denominator):

$$(6) \times \left(\frac{14 + 5y}{6} \right) = (6) \times \left(\frac{16 - 2y}{3} \right); \text{ simplifying,}$$

$$(14 + 5y) = (2)(16 - 2y);$$

$$14 + 5y = 32 - 4y; \quad 5y + 4y = 32 - 14;$$

$$9y = 18 \quad \therefore y = \frac{18}{9} = 2$$

To solve for the other unknown x, substitute the value of $y = 2$ into any of the earlier equations.

Substituting into the equation:

$$x = \frac{14 + 5y}{6}; \quad \text{then,} \quad x = \frac{14 + 5(2)}{6} = \frac{24}{6} = 4$$

Therefore; $x = 4$ and $y = 2$.

Example Problem B.12 (By Elimination)

$$3x + y = 50, \quad \text{and} \quad x + 2y = 50$$

Solve for the values of x and y.

Solution:

Multiply one or both of the equations by such a number or numbers as will give one of the unknowns the same coefficient in both equations.

Add or subtract the equal coefficients according to the nature of their signs.

The objective in the elimination method is to eliminate one of the unknowns, either x or y, through addition or subtraction of the equations.

$$3x + y = 50 \ldots \text{Eq. (1)}$$

$$x + 2y = 50 \ldots \text{Eq. (2)}$$

In this example, the unknown x will be eliminated such that the solution for y can result. To accomplish this, multiply equation (2) throughout by the value (-3).

$$(-3) \times (x + 2y) = (-3) \times (50);$$

$$\therefore -3x - 6y = -150 \ldots \text{Eq. (2)}$$

Now that equations (1) and (2') have the same value for the x term, x can be eliminated by adding the two equations.

$$3x + y = 50 \qquad \ldots \text{Eq. (1)}$$
$$\underline{-3x - 6y = -150 \ldots \text{Eq. (2)}}$$
$$0 - 5y = -100$$

Since the remaining equation contains only the unknown y, solve for the value of y.

$$-5y = -100, \quad \text{or} \quad 5y = 100$$

$$\therefore y = \frac{100}{5} = +20$$

Substituting the calculated value of y into equation (1), (2), or (2'), solve for the unknown x.

Choosing equation (1) for this example,

$$3x + (+20) = 50 \qquad\qquad \ldots (1)$$

Simplifying,

$$3x = 50 - 20 = 30;$$

$$\therefore x = \frac{30}{3} = +10$$

In summary, $x = 10$, and $y = 20$.

Example Problem B.13 (By Substitution)

Solve for the values x and y in the two equations:

$$2x + 4y = 50 \quad \text{and} \quad 3x + 5y = 66.$$

Solution:

From one of the equations find the value of one of the unknowns in terms of the other.

Substitute the value thus found for the unknown in the other equation.

The objective in this method is to operate on one of the equations and place x in terms of the variable y or vice versa.

Hence, using equation (1) as an example,

$$2x + 4y = 50 \ldots (1); \text{ rewriting the equation,}$$

$$2x = 50 - 4y$$

$$\text{Then, } x = \frac{50 - 4y}{2} = 25 - 2y.$$

Substituting the value of x into equation (2):

$$3x + 5y = 66 \qquad\qquad\qquad \ldots (2),$$

$$3(25 - 2y) + 5y = 66; \quad \text{then, } \quad 75 - 6y + 5y = 66$$

Simplifying, $-6y + 5y = 66 - 75$;

$$\therefore -y = -9, \text{ or } y = +9.$$

Substituting 9 for y in the equation

$$x = 25 - 2y; \quad \text{then, } \quad x = 25 - 2(9) = 25 - 18 = 7$$

Summarizing, $x = 7$ and $y = 9$.

Any of the three methods just demonstrated can be used to solve a system of two linear equations in two unknowns. Select the one that appears most likely to make the solution simple and direct.

Quadratic Equations

On rare occasions, a problem in statics or strength of materials may involve an equation that takes the form of:

$$ax^2 + bx + c = 0 \text{ (assuming a} \neq 0);$$

where: a, b, and c are coefficients.

This second-degree equation, with the unknown x taken to the second power, is known as a quadratic equation and will yield two roots that may be real or complex numbers. In this text, we will be concerned with the real roots only. Quadratic equations are solved by factoring, completing the square, or by using what is known as the quadratic formula. The quadratic formula is a generalized expression that can be used in solving all quadratic-type equations.

If $ax^2 + bx + c = 0$ and $a \neq 0$, then

$$x = \frac{-b \pm \sqrt{b^2 - 4ac}}{2a}$$

The quadratic formula reduces the problem of solving a quadratic equation to an arithmetic problem.

Example Problem B.14 (Quadratic Formula)

Solve for the two roots for x in the equation:

$$x^2 - 6x + 8 = 0;$$

where: $a = +1, b = -6, c = +8$

Solution:

Substituting into the general quadratic formula, we get:

$$x = \frac{-(-6) \pm \sqrt{(-6)^2 - 4(+1)(+8)}}{2(+1)}$$

$$x = \frac{+6 \pm \sqrt{+36 - 32}}{+2} = \frac{+6 \pm \sqrt{+4}}{+2}$$

The solution will yield two roots for x, which are:

$$x = \frac{+6 + \sqrt{4}}{2} = +4$$

$$\text{and, } x = \frac{+6 - \sqrt{4}}{2} = +2$$

Therefore, the two roots for x are: $x = +4$, and $x = +2$.

B.4 POWERS AND ROOTS, SYMBOLS, SI UNITS, AND CONVERSIONS

Powers and Roots

When dealing with engineering problems that involve very large numbers (e.g., column loads) or extremely small numbers (e.g., thermal elongation in a beam), it is often convenient to express the numbers in *scientific notation*, where numbers are written as a product of some power of 10 and a number between 1 and 10. In scientific notation the *power* or *exponent* of 10 tells you where the decimal point would have to be if you were going to write the number out longhand.

A positive *exponent* (power) tells you how many places the decimal point should be shifted to the right. A negative exponent tells you how many places the decimal point should be shifted to the left. For example,

$$3858 = 3.858 \times 1000 = 3.858 \times 10^3$$

$$2{,}460{,}000 = 2.46 \times 1{,}000{,}000 = 2.46 \times 10^6$$

$$0.000687 = 6.87 \times \frac{1}{10{,}000} = 6.87 \times 10^{-4}$$

Laws of Exponents

Summarized below are the laws that govern the arithmetic operations involving exponents.

$$a^m a^n = a^{m+n} \qquad a^{1/2} = \sqrt{a}$$

$$\frac{a^m}{a^n} = a^{m-n} \qquad a^{1/3} = \sqrt[3]{a}$$

$$(a^m)^n = a^{mn} \qquad a^{1/n} = \sqrt[n]{a}$$

$$(ab)^n = a^n b^n \qquad \sqrt[n]{ab} = \sqrt[n]{a}\sqrt[n]{b}$$

$$\left(\frac{a}{b}\right)^n = \frac{a^n}{b^n} \qquad \sqrt[n]{\frac{a}{b}} = \frac{\sqrt[n]{a}}{\sqrt[n]{b}}$$

$$a^{-n} = \frac{1}{a^n} \qquad a^0 = 1$$

Example Problem B.15

Using scientific notation, simplify and solve the equation shown.

$$\frac{5 \times 1250 \times (28)^4 \times (1728)}{384 \times (30{,}000{,}000) \times (2840)} =$$

Solution:

Simplifying using scientific notation,

$$\frac{(5) \times (1.25 \times 10^3) \times (2.8 \times 10)^4 \times (1.728 \times 10^3)}{(3.84 \times 10^2)(3 \times 10^7)(2.84 \times 10^3)} =$$

Simplifying further:

$$\frac{(5) \times (1.24) \times (2.8)^4 \times (1.728) \times (10^{3+4+3})}{(3.84) \times (3) \times (2.84) \times (10^{2+7+3})} =$$

Using the law of exponents, the power of 10 becomes:

$$\frac{(5) \times (1.24) \times (2.8)^4 \times (1.728) \times (10^{10})}{(3.84) \times (3) \times (2.84) \times (10^{12})} =$$

$$\frac{(5) \times (1.24) \times (2.8)^4 \times (1.728) \times (10^{10-12})}{(3.84) \times (3) \times (2.84)} =$$

Therefore,

$$\frac{(5) \times (1.24) \times (2.8)^4 \times (1.728) \times (10^{-2})}{(3.84) \times (3) \times (2.84)}$$

$$= 20.3 \times 10^{-2} = 0.203$$

Symbols—Greek

The Greek alphabet has been the standard notation system used in equations encountered in science and engineering. Listed in Table B.3 are the common symbols that will be used in this book.

As part of an ongoing effort by the United States to convert from the U.S. customary system of units (originally the British system) to the international system of units (*SI* metric units), some examples and practice problems in this text use the SI units. Most of the world currently uses the modernized SI metric system but the architectural, manufacturing, and construction industries still rely heavily on the U.S. customary units. Therefore, both systems are included in this text and conversions between the U.S. and SI metric system are provided in Tables B.4 and B.5.

Table B.3   Greek Symbols.

Alpha	A	α
Beta	B	β
Chi	X	χ
Delta	Δ	δ
Epsilon	E	ε
Theta	Θ	θ, ϑ
Gamma	Γ	γ
Mu	M	μ
Pi	Π	π
Rho	P	ρ
Sigma	Σ	σ
Tau	T	τ
Phi	Φ	ϕ, φ
Omega	Ω	ω

Table B.4   U.S. and SI Metric Equivalents.

Measurement	U.S. Units	SI-Metric
Length	inch (in.)	millimeter (mm)
	feet (ft.)	meter (m)
Area	sq. inches (in.^{2})	sq. millimeters (mm^2)
	sq. feet (ft.^{2})	sq. meters (m^2)
Mass	pound mass (lbm)	kilogram (kg)
Force	pound (lb.)	newton (N)
	kilopound (k)	kilonewton (kN)
	(k = kip = 1000 lb.)	(kN = 1000 N)
Stress (force/area)	psi $\left(\frac{lb.}{in.^2}\right)$	pascal $\left(\frac{N}{m^2}\right)$
	ksi $\left(\frac{k}{in.^2}\right)$	
Pressure	psf $\left(\frac{lb.}{ft.^2}\right)$	kilopascal = 1000 Pa
	ksf $\left(\frac{k}{ft.^2}\right)$	
Moment (force × distance)		
	pound-ft. (lb.-ft.)	newton-meter (N-m)
	kip-ft.(k-ft.)	kilonewton-meter (kN-m)
Distributed load		
	$\omega\left(\frac{lb.}{ft.} \text{ or plf}\right)$	$\omega\left(\frac{kN}{m}\right)$
Density	$\gamma\left(\frac{lb.}{ft.^3}\right)$	$\gamma\left(\frac{kN}{m^3}\right)$

Table B.5   Conversions.

1 mm = 0.03937 in.	1 in. = 25.4 mm
1 m = 39.37 in. = 3.281 ft.	1 ft. = 0.3048 m
$1 \text{ m}^2 = 10.76 \text{ ft.}^2$	$1 \text{ ft.}^2 = 92.9 \times 10^{-3} \text{ m}^2$
$1 \text{ m}^3 = 35.31 \text{ ft.}^3$	$1 \text{ ft.}^3 = 0.02832 \text{ m}^3$
1 kg = 2.205 lb.-mass	1 lbm = 0.4536 kg
1 N = 0.2248 lb.	1 lb. = 4.448 N
1 kN = 224.8 lb.-force	1 kip = 4.448 kN
$1 \,{}^{kN}\!/_{m} = 68.53 \,{}^{lb.}\!/_{ft.}$	$1 \,{}^{lb.}\!/_{ft.} = 14.59 \,{}^{N}\!/_{m}$
$1 \text{ kPa} = 20.89 \,{}^{lb.}\!/_{ft.^2}$	$1 \,{}^{lb.}\!/_{ft.^2} = 47.87 \text{ Pa}$
$1 \,{}^{kN}\!/_{m^2} = 20.89 \,{}^{lb.}\!/_{ft.^2}$	$1 \,{}^{lb.}\!/_{ft.^2} = 0.0487 \,{}^{kN}\!/_{m^2}$
$1 \text{ MPa} = 145 \,{}^{lb.}\!/_{in.^2}$	$1 \,{}^{lb.}\!/_{in.^2} = 6.895 \text{ kPa}$
$1 \,{}^{kN}\!/_{m^3} = 6.37 \,{}^{lb.}\!/_{ft.^3}$	$1 \,{}^{lb.}\!/_{ft.^3} = 0.157 \,{}^{kN}\!/_{m^3}$
1 N-m = 0.738 lb.-ft.	1 lb.-ft = 1.356 N-m

Table B.6   Prefixes and Multiplication Factors in SI.

Symbol	Prefix	Multiplication Factor
G	giga	$10^9 = 1{,}000{,}000{,}000$
M	mega	$10^6 = 1{,}000{,}000$
k	kilo	$10^3 = 1000$
h	hecto	$10^2 = 100$
da	deca	$10^1 = 10$
d	deci	$10^{-1} = 0.1$
c	centi	$10^{-2} = 0.01$
m	milli	$10^{-3} = 0.001$
μ	micro	$10^{-6} = 0.000001$
n	nano	$10^{-9} = 0.000000001$

Example Problem B.16 (Figure B.18)

A concrete footing with the dimensions shown, in U.S. units, supports a load **P** from a column above. Convert the pedestal and footing dimensions into equivalent SI-metric units. Then, determine the volume of concrete required for the pedestal and footing base combined. Compute the volume using cubic feet (U.S. customary) and cubic meter (SI-metric) units.

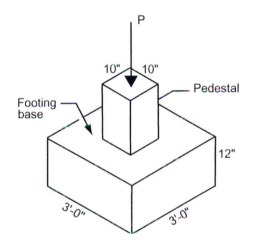

Figure B.18   Individual spread footing.

Solution:

Pedestal—

$$10 \text{ in.} = 10 \text{ in.} \times 25.4 \,{}^{mm}\!/_{in.} = 254 \text{ mm} = 0.254 \text{ m}$$

$$18 \text{ in.} = 18 \text{ in.} \times 25.4 \,{}^{mm}\!/_{in.} = 457.2 \text{ mm} = 0.457 \text{ m}$$

Footing base—

$$12 \text{ in.} = 12 \text{ in.} \times 25.4 \,{}^{mm}\!/_{in.} = 304.8 \text{ mm} = 0.305 \text{ m}$$

$$(3'-0'') = 36 \text{ in.} \times 25.4 \,{}^{mm}\!/_{in.}$$

$$= 914.4 \text{ mm} = 0.914 \text{ m}$$

Volume: U.S. customary units

$$\text{Pedestal} - \left(\tfrac{10}{12}\right)' \times \left(\tfrac{10}{12}\right)' \times \left(\tfrac{18}{12}\right)' = 1.04 \text{ ft.}^3$$

$$\text{Base} - \left(\tfrac{12}{12}\right)' \times (3)' \times (3)' = 9 \text{ ft.}^3$$

$$\text{Total volume} = 1.04 \text{ ft.}^3 + 9 \text{ ft.}^3 = 10.04 \text{ ft.}^3$$

Volume: SI-metric units

$$\text{Pedestal} - (.254 \text{ m}) \times (.254 \text{ m}) \times (.457 \text{ m}) = 0.0295 \text{ m}^3$$

$$\text{Base} - (.305 \text{ m}) \times (.914 \text{ m}) \times (.914 \text{ m}) = 0.2548 \text{ m}^3$$

$$\text{Total volume} = 0.0295 \text{ m}^3 + 0.2548 \text{ m}^3 = 0.2843 \text{ m}^3$$

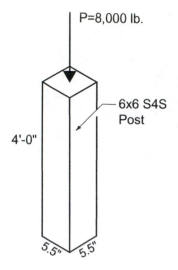

P=8,000 lb.

6x6 S4S
Post

4'-0"

5.5" 5.5"

Short Timber Post

Figure B.19

Example Problem B.17 (Figure B.19)

A short timber post (6×6 S4S) 4 ft. tall supports a compressive load equal to 8000 lb. The actual dimensions of the post are $5\frac{1}{2}" \times 5\frac{1}{2}"$.

The intensity (stress) of the force in the post is measured by dividing the applied load by the cross-sectional area of the supporting member. In this example, solve for the compressive stress in the post using the U.S. customary and SI-metric units.

Solution:

U.S. customary units:

$$\text{Stress} = \frac{\text{force}}{\text{area}} = \frac{P}{A} = \frac{8000 \text{ lb.}}{5\frac{1}{2}" \times 5\frac{1}{2}"}$$

$$= \frac{8000 \text{ lb.}}{30.25 \text{ in.}^2} = 264.5 \, {}^{\text{lb.}}/_{\text{in.}^2}$$

SI-metric:

$$P = 8000 \text{ lb.} = (8000 \text{ lb.}) \times \left(4.448 \, {}^{N}/_{\text{lb.}}\right)$$

$$= 35{,}584 \text{ N} = 35.58 \text{ kN}$$

$$5\frac{1}{2}" = (5.5 \text{ in.}) \times \left(25.4 \, {}^{mm}/_{\text{in.}}\right)$$

$$= 139.7 \text{ mm} = 0.1397 \text{ m}$$

$$\text{Area} = (0.1397 \text{ m}) \times (0.1397 \text{ m}) = 0.01952 \text{ m}^2$$

$$= 1.95 \times 10^{-2} \text{ m}^2$$

$$\text{Stress} = \frac{P}{A} = \frac{3.558 \times 10^4 \text{ N}}{1.95 \times 10^{-2} \text{ m}^2}$$

$$= 1.823 \times 10^6 \, {}^{N}/_{m^2} = 1823 \, {}^{kN}/_{m^2}$$

Another way of expressing stress in the SI-metric system is:

$$\text{Stress} = 1.823 \times 10^6 \text{ pascal} = 1.823 \times 10^6 \text{ Pa} = 1823. \text{ kPa}$$

where:

$$\text{pascal} = \text{Pa} = {}^{N}/_{m^2}$$

Also, since $1 \, {}^{\text{lb.}}/_{\text{in.}^2} = 6.895 \text{ kPa}$,

$$264.5 \, {}^{\text{lb.}}/_{\text{in.}^2} = 264.5 \times 6.895 = 1824 \text{ kPa}$$

Example Problem B.18 (Figure B.20)

A roof beam is responsible for supporting a uniformly distributed load of $\omega = 125 \, {}^{\text{lb.}}/_{\text{ft.}}$ over its entire 20-foot span. Determine the total load supported by the beam. Solve this same problem in the SI-metric unit system.

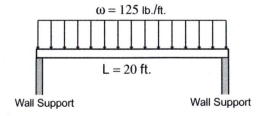

ω = 125 lb./ft.

L = 20 ft.

Wall Support   Wall Support

Figure B.20 Roof beam.

Solution:

U.S. units:

$$\text{Total load} = \omega \times L = \left(125\,^{\text{lb.}}\!/_{\text{ft.}}\right) \times (20\ \text{ft.})$$

$$= 2500\ \text{lb.}$$

SI-metric units:

$$\omega = 125\,^{\text{lb.}}\!/_{\text{ft.}} = \left(125\,^{\text{lb.}}\!/_{\text{ft.}}\right) \times \left(14.59\,^{\text{N}}\!/_{\text{m}}\right)/\left(^{\text{lb.}}\!/_{\text{ft.}}\right)$$

$$= 1824\,^{\text{N}}\!/_{\text{m}}$$

$$L = 20\ \text{ft.} = (20\ \text{ft.}) \times \left(0.3048\,^{\text{m}}\!/_{\text{ft.}}\right) = 6.096\ \text{m}$$

Total load =

$$\omega \times L = \left(1824\,^{\text{N}}\!/_{\text{m}}\right) \times (6.096\ \text{m})$$

$$= 11{,}119\ \text{N} = 11.12\ \text{kN}$$

Example Problem B.19 (Figure B.21)

A precast concrete wall panel has the dimensions as shown. Determine the total weight of the wall panel in SI units. Then, compute the weight in equivalent U.S. units. The density of concrete is: $\gamma_{\text{concrete}} = 23\,^{\text{kN}}\!/_{\text{m}^3}$.

Solution:

The volume of the concrete wall is computed as:

$$\text{Volume} = h \times L \times t = (2.5\ \text{m}) \times (6\ \text{m}) \times (0.2\ \text{m}) = 3.0\ \text{m}^3$$

The weight of the wall panel is equal to:

$$\text{Weight } W = \text{Vol.} \times \gamma = (3\ \text{m}^3) \times \left(23\,^{\text{kN}}\!/_{\text{m}^3}\right)$$

$$= 69\ \text{kN}$$

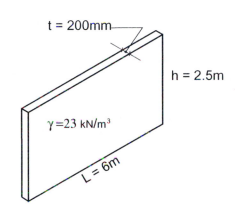

Figure B.21 *Precast concrete wall panel.*

In U.S. customary units:

$$\text{Volume} = (3\ \text{m}^3) \times \left(35.31\,^{\text{ft.}^3}\!/_{\text{m}^3}\right) = 105.93\ \text{ft}^3$$

$$\gamma = \left(23\,^{\text{kN}}\!/_{\text{m}^3}\right) \times \left(6.37\,^{\text{lb.}}\!/_{\text{ft.}^3}\right)/\left(^{\text{kN}}\!/_{\text{m}^3}\right) = 146.5\,^{\text{lb.}}\!/_{\text{ft.}^3}$$

$$W = \text{Vol.} \times \gamma = (105.93\ \text{ft.}^3) \times \left(146.5\,^{\text{lb.}}\!/_{\text{ft.}^3}\right)$$

$$= 15{,}520\ \text{lb.}$$

$$\text{or}\ \ W = (69\ \text{kN}) \times \left(224.8\,^{\text{lb.}}\!/_{\text{kN}}\right) = 15{,}511\ \text{lb.}$$

Accuracy in Computations

Computers and electronic calculators used in engineering practice are capable of producing answers that contain many decimal places, far beyond the precision necessary for the computation being performed. This degree of accuracy is misleading since the accuracy of an answer can be no more than the smallest number of significant figures in any of the factors used in the solution. Answers with more significant digits than are necessary can lead to a false sense of accuracy. In most cases, numbers rounded to three significant figures are sufficient for the problems encountered in this book. Accuracy greater than 0.5 to 1% is rarely required.

Example Problem B.20

The numbers 497, 11,800, 0.00298, and 8.00 all have three significant numbers.

Example Problem B.21

All nonzero digits in a number are significant.
The numbers 672, 9.16, 0.244, and 5.96 have three significant digits.

Example Problem B.22

Zeros are considered significant *except* when their only function is to locate the decimal point.
The numbers 0.0025, 0.093, and 0.000051 have two significant figures.

Example Problem B.23

Zeros placed after the decimal point are significant.
The numbers 49.17, 4.900, 783.0, and 0.1230 have four significant digits.

Example Problem B.24

Zeros appearing before the decimal point are not significant.
The numbers 2280, 87,700, and 143,000 have three significant numbers.

In doing arithmetic computations, the idea is to prevent loss of precision but maintain that the final answer gives a true indication of its accuracy. Retain as many significant figures in the answer as contained in the *least* precise number.

Example Problem B.25 (Addition and Subtraction)

Add: $12.6 + 87.094 = 99.694$

The least accurate number, 12.6, has three significant digits so the final sum should have three significant digits as well. Therefore, the final answer, 99.7, should have three significant figures.

Subtract: $79.03 - 4.6 = 74.43$; round to 74 (two significant digits).

Example Problem B.26 (Multiplication and Division)

Multiply: $512 \times 3.14159 = 1608.494$; rounded to 1610 (three significant figures).

Divide: $771 \div 1.01642 = 758.545$; rounded to 759 (three significant figures).

Example Problem B.27

In computations that involve an exact number, treat it as having more significant figures, equivalent to your most accurate and precise number.

If a timber post is capable of supporting 12,250 lb., what is the capacity of four (4) posts?

Multiply: 12,220 lb. $\times 4 = 48,880$ lb.

Since the 4 is an exact number, we retain as many significant figures as contained in 12,220.

In most cases, numbers rounded to three significant figures are sufficient for the problems encountered in this book. Accuracy greater than 0.5% to 1% is rarely required.

Supplementary Problems

Section B.2—Trigonometry

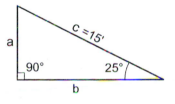

B.2.1   Determine the length of side *a* in the right triangle shown. What is the area enclosed by this triangle?

Problem B.2.1

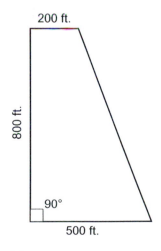

B.2.2   A parking lot has the shape of a trapezoid. Determine the number of square yards of paving material required.

Problem B.2.2

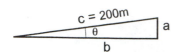

Problem B.2.3

B.2.3   A road has a constant slope of 1 in 10. How much does it rise in a distance of 2000 m?

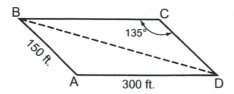

Problem B.2.4

B.2.4   A fence is to extend along the long diagonal (*BD*) of a plot of land that has the shape of a parallelogram. The sides are 150 ft. and 300 ft., respectively. Determine the required length of the fence and the area of the field. Hint: use the cosine law.

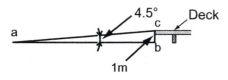

Problem B.2.5

B.2.5   A carpenter has to build a ramp that is to be used by a client in a wheelchair. The height of the deck above the ground is 1 meter and the required slope or gradient is to be $4\frac{1}{2}°$. How long must the ramp *AC* be?

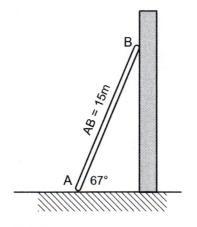

Problem B.2.6

B.2.6   The fire department has a new 15-meter ladder. The greatest angle at which it can be placed against a building with safety is at 67° with the ground. What is the maximum vertical height that the ladder can reach?

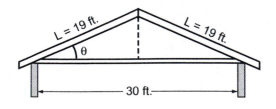

Problem B.2.7

B.2.7   A carpenter is constructing a triangular roof on a new house. The roof is to span 30 ft. across from wall to wall. If the rafters are 19 ft. long, at what angle will the rafters be laid relative to the ceiling line?

B.2.8 A 28-ft. ladder is placed against the side of a house with the foot of the ladder 5 ½ ft. away from the building. What is the angle that the ladder makes with the ground?

Problem B.2.8

B.2.9 In anticipation of constructing a bridge *BC* across a pond, a surveryor at point *B* sights along to a point *C* on the opposite bank of the pond. The transit is rotated 90° to establish a sight line *BA*. A measurement of 500 ft. is made along line *BA* to locate a station point *A*. The surveyor then moves the transit to station point *A* and sights along lines *AC* and *AB* to establish the included angle of 25°. Determine the length of the bridge from *B* to *C*.

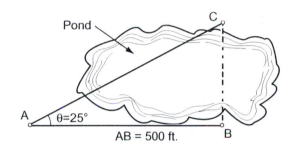

Problem B.2.9

B.2.10 What length rope, *AB*, will reach from the deck of a large ship, 18 m above water level, to the deck of a tug 1 m above the water level, if the rope makes an angle of 13° with the horizontal? Neglect any slack in the rope.

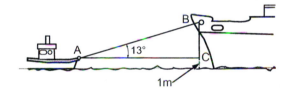

Problem B.2.10

B.2.11 A 6-ft. tall tree casts a shadow of 15 ft. late in the afternoon. If the distance from the base of the tall tree to the tip of the large tree's shadow measures 120 ft., determine the height of the tall tree.

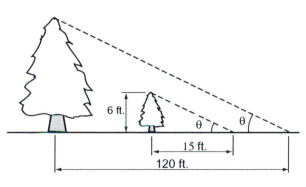

Problem B.2.11

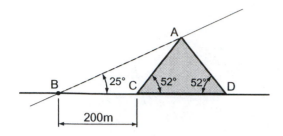

Problem B.2.12

B.2.12 An Egyptian pyramid with a square base and symmetrical sloping sides casts a shadow measuring 200 m from the base. The angle of inclination of a line *AB* from the apex of the pyramid to the tip of the shadow is 25° relative to the ground. Determine the height of the pyramid and the base dimension *DC*.

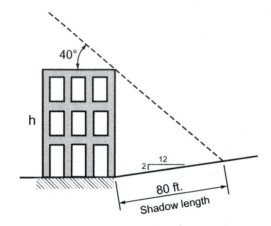

Problem B.2.13

B.2.13 A building casts a shadow of 80 ft. along a sloping ground plane that is at a 2 in 12 slope. If the sun angle measures 40° from a horizontal reference, determine the height *h* of the building.

B.2.14 A Howe truss is used as a basic structural element for supporting a roof. Assuming that the truss is constructed throughout with the same size steel member weighing 12 $^{\text{lb.}}/_{\text{ft.}}$, determine:

    a. the total length of steel required to fabricate the truss.
    b. the total weight of the truss.

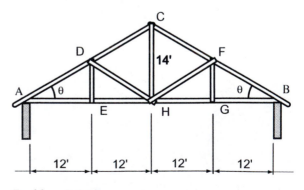

Problem B.2.14

Section B.3—Algebra—Simultaneous and Quadratic Equations
Solve Problems B.3.1 to B.3.9 using the comparison, elimination, or substitution method.

B.3.1 $3x + 4y = 11$
$2x - 3y = -4$

B.3.2 $5x + 2y = 32$
$2x - y = 2$

B.3.3 $3x + 2y = 27$
$2x - 3y = 5$

B.3.4 $x + y = 2$
$2x - 3y = 5$

B.3.5 $x + 3y = 7$
$-2x + 2y = 2$

B.3.6 $3x + 2y = 2$
$x - y = 9$

B.3.7 $3x - 5y = 5$
$7x + y = 75$

B.3.8 $3x - y = 1$
$2x + 5y = 41$

B.3.9 $2x - 3y = 17$
$x - 2y = 8$

Solve Problems B.3.10 to B.3.14 using the quadratic formula.

B.3.10 $x^2 - 8x + 15 = 0$

B.3.11 $x^2 - 0.2x - 1.2 = 0$

B.3.12 $2x^2 - 13.6x + 15.9 = 0$

B.3.13 $15y^2 - 178y + 527 = 0$

B.3.14 $2x + \dfrac{1}{x} = 3$

Section B.4—Powers and Roots, Symbols, SI Units, and Conversions

B.4.1 Convert the following quantities from U.S. customary units to the proper SI units.

a. 27 ft.-8 in. to meters
b. 135 yards to meters
c. $2\frac{3}{4}$ in. to millimeters
d. 1874.6 ft.^{2} to m^2
e. 27 ft.^{3} to m^3
f. 492.5 lb. to Newtons
g. 5747 lb. to kN.
h. 423.6 lb./ft.^{2} to Pa
i. 722 $^{lb.}/_{in.^2}$ to kPa
j. 1250 $^{lb.}/_{ft.}$ to $^{kN}/_{m}$
k. 3000 $^{lb.}/_{ft.^2}$ to $^{kN}/_{m^2}$
l. 490 $^{lb.}/_{ft.^3}$ to $^{kN}/_{m^3}$

B.4.2   Convert the following quantities from SI units to U.S. customary units.

  a. 3250 mm to inches
  b. 29.85 m to feet
  c. 6.84 m to feet and inches
  d. 18.25 m^2 to $ft.^2$
  e. 9.15 m^3 to $ft.^3$
  f. 4321 N to lb.
  g. 75.0 kN to kips
  h. 12,500 MPa to $^{lb.}/_{in.^2}$
  i. 5.2 $^{kN}/_m$ to $^{lb.}/_{ft.}$
  j. 180 $^{kN}/_{m^2}$ to $^{lb.}/_{ft.^2}$
  k. 23.5 $^{kN}/_{m^3}$ to $^{lb.}/_{ft.^3}$

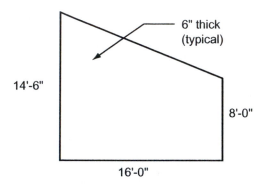

Problem B.4.3

B.4.3   A 6-in. thick precast concrete wall panel has the dimensions as shown. Determine the total volume of concrete necessary to pour this wall panel. Also, determine the total weight of the panel if concrete has a density of $\gamma_{concrete} = 150\,^{lb.}/_{ft.^3}$. Solve this same problem in SI units.

B.4.4   A portion of a continuous foundation wall supports the wall and floor loads from above. Based on the dimensions shown, determine the total volume of concrete used. If concrete has a density of $\gamma_{conc.} = 23.5\,^{kN}/_{m^3}$, what is the total weight of the foundation wall and footing? Convert the volume and weight into U.S. customary units.

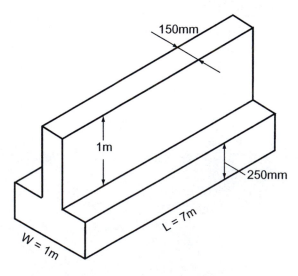

Problem B.4.4

Appendix C

Derivation of Equations for Flexure and Shear

- Derivation of the Flexural (Bending) Stress Equation
- Development of the General Shear Stress Equation
- The Elastic Curve—Radius of Curvature of a Beam

C.1 DERIVATION OF THE FLEXURAL (BENDING) STRESS EQUATION

Consider a portion of a beam that is subjected to pure bending only by couples (designated by M) at each end, as shown in Figure C.1. Since the beam is in equilibrium, the moments at each end will be numerically equal, but of opposite sense. Due to the moment couples, the beam is bent from its original straight position to the curved (deformed) shape indicated in Figure C.1.

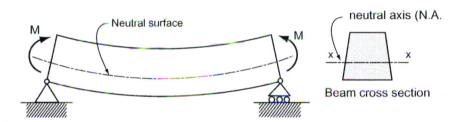

Figure C.1   Elevation of beam in bending.

Due to this bending action, we find that the lengths of the upper parts of the beam decrease, while the bottom parts of the beam undergo lengthening. This action has the effect of placing the upper portion of the beam in compression and the lower portion of the beam in tension. An equation must be obtained that will relate bending stress to the external moment and the geometric properties of the beam.

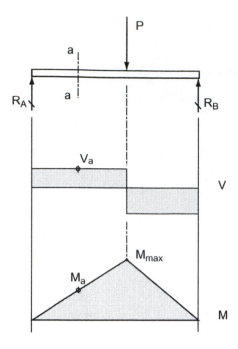

Figure C.2   Load, V, and M diagrams.

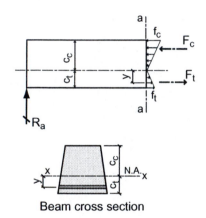

Beam cross section

Figure C.3   Bending stresses on a beam cross-section.

This can be done by examining a segment of the beam whose internal force system at any given transverse section is a moment **M**, as shown in Figures C.2 and C.3.

c_c = distance from neutral axis (N.A.) to the extreme compressive fiber

c_t = distance from N.A. to the extreme tensile fiber

y = distance from N.A. to some area ΔA

ΔA = small strip of area on the beam cross-section

If we denote the element of area at any distance y from the neutral axis (Figure C.3) as ΔA and the stress on it as f, the requirement for equilibrium of forces yields:

$$\left[\sum F_x = 0\right] \sum f_t(\Delta A) + \sum f_c(\Delta A) = 0$$

where:

f_t = tensile stress below the N.A.

f_c = compressive stress above the N.A.

$$\sum f_t(\Delta A) = F_t$$

and $\sum f_c(\Delta A) = F_c$

If each $f_y (\Delta A)$ is multiplied by its y distance above or below the neutral axis,

$$M = \sum f_y(y)(\Delta A)$$

where:

M = internal bending moment (see Figure C.3)

But remembering the relationship developed by Coulomb,

$$\frac{f_y}{y} = \frac{f_c}{c}$$

and

$$f_y = \frac{y}{c} f_c$$

Substituting the f_y relationship into the moment equation:

$$M = \sum f_y(y)(\Delta A) = \sum \frac{y}{c} f_c y(\Delta A) = \frac{f_c}{c} \sum y^2 \Delta A$$

Previously, in Chapter 7, we developed the relationship for the moment of inertia where:

$$I_x = \sum y^2(\Delta A)$$

$$\therefore M = \frac{f_c I}{c}$$

or

$$f_b = \frac{Mc}{I} \quad \text{(Flexure formula)}$$

where:

f_b = bending stress at the extreme fiber, top or bottom

c = distance from the N.A. to the extreme fiber

I = moment of inertia of the cross-section about its centroidal (or N.A.) axis

M = moment at some point along the beam length

Note: Bending stress f is directly proportional to the value c; therefore, the largest bending stress on a cross-section is obtained by selecting the largest c value for unsymmetrical cross-sections. (See Figure C.4.)

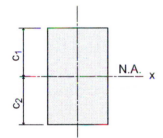
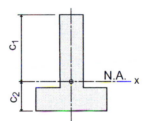

(a) Rectangular cross section

$c_1 = c_2$; $f_{top} = f_{bottom}$

(b) Unsymmetrical cross section

$c_1 > c_2$; $f_{top} > f_{bottom}$

Figure C.4   Distances to the extreme fiber for beam cross-sections.

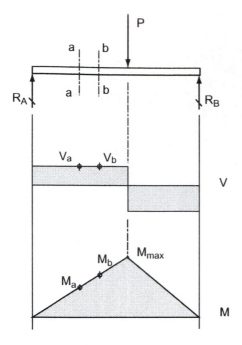

Figure C.5

C.2 DEVELOPMENT OF THE GENERAL SHEAR STRESS EQUATION

To arrive at a relationship for the shearing stress, consider the beam shown in Figure C.5.

At section a-a the moment is M_a, and at section b-b, an incremental distance to the right, the moment is M_b. (See Figures C.6a and C.7.) From the moment diagram in Figure C.5, we see that $M_b > M_a$

Therefore: $F_{ca} < F_{cb}$ and $F_{ta} < F_{tb}$

Isolating a small section of incremental beam (between sections a-a and b-b) above the neutral surface, Figure C.8 shows the distribution of tensile and compressive bending stresses. In the element cdef, the forces C_1 and C_2 are the resultants of the compressive stresses that act on the transverse planes cd and ef. Shear force V on plane df is required for horizontal equilibrium.

$$C_2 > C_1$$

$$\left[\sum F_x = 0\right] C_1 + V - C_2 = 0$$

$$\therefore V = C_2 - C_1 = (f_v)(b)(\Delta x)$$

Examine the cross-section of this isolated beam segment (Figure C.9).

$$\Delta A = \text{small increment of area}$$

$$y = \text{distance from the N.A. to the area } A$$

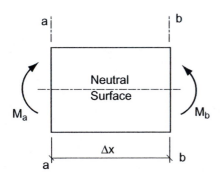

Figure C.6   Beam section between cuts a and b.

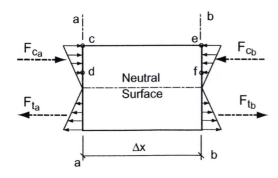

Figure C.7   Bending stresses on the beam section a-b.

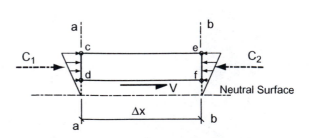

Figure C.8   FBD of beam segment cefd isolated.

From the flexure formula:

$$f_y = \frac{My}{I}$$

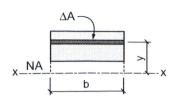

The force against the area ΔA equals

$$\Delta A(f_y) = \frac{My \Delta A}{I}$$

Figure C.9 Upper portion of the beam cross-section.

But if we sum all of the ΔAs in the shaded cross-section shown in Figure C.9:

$$\begin{bmatrix} \text{Area of the shaded} \\ \text{cross-section} \end{bmatrix} = \sum \Delta A$$

$$\begin{bmatrix} \text{Total force on the shaded} \\ \text{cross-sectional area at} \\ \text{section b-b} \end{bmatrix} = C_2$$

where:

M_b = internal bending moment at section b-b, obtained from the moment diagram (Figure C.5). See also Figure C.10.

I = moment of inertia of the *entire* beam cross-section; a constant

$\sum \Delta Ay$ = sum of all the ΔAs that compose the shaded area, times the respective y distance from the N.A.

$$\sum \Delta Ay = \Delta A \bar{y}$$

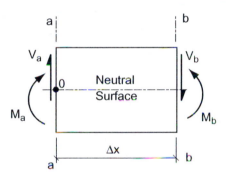

Figure C.10 Beam section between a and b.

where:

$\bar{y}$ = distance from the N.A. to the centroid of the shaded cross-section

$A\bar{y}$ is normally referred to as the statical or first moment. The symbol Q will be used to represent the value $A\bar{y}$:

$$Q = A\bar{y}$$

Next, substituting:

$$C_2 = \frac{M_b Q}{I}; \quad \text{and} \quad C_1 = \frac{M_a Q}{I}$$

But, $V = C_2 - C_1$

Therefore,

$$V = \frac{M_b Q}{I} - \frac{M_a Q}{I} = \frac{Q}{I}(M_b - M_a)$$

For beams of constant cross-section:

Q = constant, I = constant

Looking again at the section of the beam between section a-a and b-b, where $M_b > M_a$, (Figure C.10) the condition of vertical and moment equilibrium must be established:

$$\left[\sum F_y = 0\right] V_a = V_b = V_T \text{ (transverse)}$$

$$\left[\sum M_o = 0\right] + M_b - M_a - V_T(\Delta x) = 0$$

$$M_b - M_a = V_T(\Delta x)$$

Substituting back into the earlier equation:

$$V_{\text{longitudinal}} = \frac{Q}{I}(V_T \Delta x)$$

where:

V_L = shear force acting on the longitudinal beam surface; area = $b\Delta x$

f_v = shear stress (longitudinal) = $\dfrac{V}{\text{shear area}}$

$$f_v = \frac{V}{b\Delta x}$$

$$f_v = \frac{Q}{I}\frac{(V_T \Delta x)}{b\Delta x}$$

Simplifying, the resulting equation represents the general shear formula:

$$f_v = \frac{VQ}{Ib}$$

where:

f_v = unit shearing stress; transverse or longitudinal

V = shear in the beam at a given point along the beam length, usually obtained from the shear diagram

$Q = A\bar{y}$ = first moment

A = area above or below the level at which the shear stress is desired

y = distance from the beam cross-section's neutral axis (N.A.) to the centroid of the area above or below the desired plane where shear stress is being examined

$I = I_x$ = moment of inertia of the entire beam cross-section

b = width of the beam at the plane where the shear stress is being examined

C.3 THE ELASTIC CURVE—RADIUS OF CURVATURE OF A BEAM

When a beam deflects, the neutral surface of the beam assumes a curved position, which is known as the elastic curve (Figure C.11).

It is assumed from beam theory that planes a and b (Figure C.12), which were parallel before loading, will remain plane after bending so as to include a small angle dq (Figure C.11). If the curvature is small, we may assume that $R_a = R_b = R$, the radius of curvature of the neutral surface (elastic curve).

The length of the segment between sections a and b is designated as Δx; if $d\theta$ is very small and R is very large, then

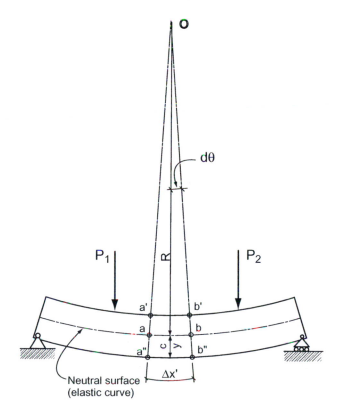

Figure C.11   Beam after loading.

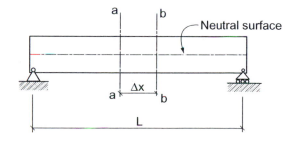

Figure C.12   Beam before loading.

$\Delta x = (R) \times d\theta$, since for very small angles $\sin \theta = \tan \theta = \theta$. From Figure C.11, $\Delta x' = (R + c)d\theta$ by the same reasoning.

Length: $ab = \Delta x$

$a'b' = \Delta x'$

The total elongation that the bottom fiber undergoes is:

$$\delta_c = \Delta x' - \Delta x$$

Substituting the values given above, we get:

$$\delta_c = (R + c)d\theta - Rd\theta = Rd\theta + cd\theta - Rd\theta$$

$$\therefore \delta_c = cd\theta$$

From $\varepsilon = \delta/L$, we get:

$$\varepsilon_c = \frac{\Delta x' - \Delta x}{\Delta x} = \frac{c(d\theta)}{R(d\theta)} = \frac{c}{R}$$

Similarly, it can be shown that the unit strain at any distance y from the neutral surface can be written as:

$$\varepsilon_y = \frac{y}{R}$$

We know that:

$$\varepsilon_y = \frac{f_y}{E} \quad \text{and} \quad f_y = \frac{My}{I};$$

so

$$\varepsilon_y = \frac{My}{EI}$$

Equating the two expressions for ε_y, we get:

$$\frac{My}{EI} = \frac{y}{R} \quad \text{or} \quad R = \frac{EI}{M} \quad \text{or} \quad \frac{1}{R} = \frac{M}{EI}$$

where:

R = radius of curvature

M = bending moment at section where R is desired

E = modulus of elasticity

I = moment of inertia of the beam cross-section

Since the flexure formula was used to obtain this relationship, it will be valid only for those members that meet the assumptions made in the derivation of the flexure formula. E and I will usually be constants for a given beam. The radius of curvature equation above is considered as a basic equation in the development of deflection formulas.

Answers to Selected Problems

Chapter 2

2.3.1 $R = 145$ N, $\theta = 46°$

2.3.3 $R = 255$ N, $\theta = 62°$

2.3.6 $F_3 = 36$ lb., $R = 230$ lb.

2.3.8 $T_2 = 240$ lb.

2.3.9 $F_x = 67.6$ lb. $(\rightarrow)$, $F_y = 145$ lb. $(\uparrow)$

2.3.11 $F_x = 800$ lb. $(\rightarrow)$, $F_y = 600$ lb. $(\uparrow)$

2.3.13 $P_x = 94.9$ lb., $P_y = 285$ lb.

2.3.15 $F_x = 1.27$ kN, $F_y = 2.15$ kN

2.3.17 $R = 714$ N, $\theta = 83.6°$

2.3.19 $R = 7.03$ k, $\theta = 35.4°$

2.3.21 $R = 10.4$ k, $\theta = 87.5°$

2.3.23 $T = 4.14$ kN, $R_y = -11.32$ kN

2.4.1 $M_A = -31,800$ lb.-ft., $M_B = -3000$ lb.-ft.

2.4.3 $M_A = -102.2$ k-ft, $M_B = +21.3$ k-ft.

2.4.5 $M_B = +82$ k-ft.

2.4.7 $M_A = -19,140$ lb.-ft., $M_B = -9738$ lb.-ft.

2.4.9 $M_A = +89.3$ N-m

2.4.11 $M_A = -7520$ lb.-ft., $M_C = +1840$ lb.-ft.

2.4.13 $R_y = -3$ kN, $M_A = -5$ kN-m, $x = 1.67$ m

2.4.15 $F_3 = 400$ lb., $R = 700$ lb.

2.4.17 $R = -37$ k, $x = 19.2$ ft.

2.4.19 $R = 4200$ lb., $M_A = 3225$ lb.-ft., $x = 0.768$ ft.

Chapter 3

3.1.1 $AB = 453$ lb., $BC = 513$ lb.

3.1.3 $CA = 20.6$ kN (T), $BC = 24.8$ kN (C)

3.1.5 $T_1 = 717$ lb., $T_2 = 163$ lb.

3.1.7 $BA = 2.2$ kN (T), $BC = 2.15$ kN (C), $DB = 2.19$ kN (T), $DE = 1.55$ kN (T)

3.1.9 $AB = 125$ lb., $BC = 150.2$ lb., $W_{max} = 83.4$ lb.

3.2.1 $A_x = 0$, $A_y = 3.8$ kN $(\uparrow)$, $B_y = 3.3$ kN $(\uparrow)$

3.2.3 $A_x = 212$ N $(\rightarrow)$, $A_y = 1177$ N $(\uparrow)$, $M_{RA} = -3428$ N-m

3.2.5 $BA = 38.6$ k (T), $C_x = 30.9$ k $(\rightarrow)$, $C_y = 6.8$ k $(\uparrow)$

3.2.7 $A_x = 2.31$ k $(\rightarrow)$, $A_y = 1.85$ k $(\downarrow)$, $E_y = 3.69$ k $(\downarrow)$

3.2.9 $A_x = 1.5$ k $(\rightarrow)$, $A_y = 2.16$ k $(\uparrow)$, $D_x = 3$ k $(\leftarrow)$, $D_y = 5.44$ k $(\uparrow)$, $M_{RD} = -52.2$ k-ft., $BC = 1.44$ k

3.3.1 $A_y = 1733$ lb. $(\uparrow)$, $B_x = 0$, $B_y = 3067$ lb. $(\uparrow)$

3.3.3 $C_x = 400$ lb. $(\rightarrow)$, $C_y = 1100$ lb. $(\uparrow)$, $M_{RC} = -9200$ lb.-ft.

3.3.5 $A_x = 0$, $A_y = 1.5$ kN $(\downarrow)$, $B_y = 10.5$ kN $(\uparrow)$

3.4.1 $W = 150$ lb.

3.4.3 $P = 200$ lb., 8 ft. of rope is needed to pull the weight up 2 ft.

3.4.5 $A_x = 0$, $A_y = 143$ lb. $(\uparrow)$, $B_y = 393$ lb. $(\uparrow)$, $C_x = 250$ lb., $C_y = 250$ lb.

Chapter 4

4.1.1 $A_y = 3$ k $(\uparrow)$, $B_x = 3$ k $(\leftarrow)$, $B_y = 5$ k $(\uparrow)$, $AD = 3.75$ k (C), $AC = 2.25$ k (T), $DE = 7.5$ k (C), $DC = 3.75$ k (T), $CE = 6.25$ k (T), $CB = 0.75$ k (T), $EB = 6.25$ k (C)

4.1.3 $A_x = 7$ kN $(\rightarrow)$, $A_y = 4.5$ kN $(\uparrow)$, $B_x = 7$ kN $(\leftarrow)$, $CE = 5.59$ kN (T), $EA = 7.83$ kN (T), $CD = 5$ kN (C), $DB = 5$ kN (C), $ED = 2$ kN (T), $EB = 2.24$ kN (C), $AB = 1$ kN (T)

4.1.5 $A_y = 1476$ lb. $(\uparrow)$, $B_x = 1073$ lb. $(\leftarrow)$, $B_y = 671$ lb. $(\uparrow)$ $AC = 2100$ lb. (C), $CD = 1200$ lb. (C), $DE = 1500$ lb. (C), $EB = 1500$ lb. (C), $AF = 1610$ lb. (T), $FB = 269$ lb. (T), $CF = 1500$ lb. (C), $FD = 671$ lb. (T), $FE = 0$

4.1.7 $A_x = 9$ k $(\leftarrow)$, $A_y = 4$ k $(\downarrow)$, $B_y = 28$ k $(\uparrow)$, $AC = 8$ k (C), $CE = 12$ k (C), $EF = 3$ k (C), $FD = 16$ k (C), $DB = 28$ k (C), $CD = 9$ k (C), $AD = 15$ k (T), $CF = 5$ k (T)

4.1.9 $A_y = 6$ kN ($\uparrow$), $B_x = 0$, $B_y = 6$ kN ($\uparrow$), $AD = BF = 12.73$ kN (C), $DE = FE = 8.48$ kN (C), $AC = CB = 9.50$ kN (T), $DC = CF = 3.16$ (C), $EC = 8$ kN (T)

4.1.11 $DE = 7.5$ k (C), $DC = 3.75$ k (T), $AC = 2.25$ k (T)

4.1.14 $AE = 7.83$ kN (T), $EB = 2.24$ kN (C), $DB = 5$ kN (C)

4.1.15 $DE = 5.69$ k (C), $EG = 2.13$ k (T), $HG = 8.25$ k (T)

4.1.17 $DC = 1.62$ k (C), $FG = 1.4$ k (T)

4.1.19 $CD = 17$ k (T), $GD = 11.3$ k (T)

4.1.21 $FC = 1.39$ k (T)

4.1.23 $A_y\ 10$ k ($\uparrow$), $C_x = 0$, $C_y = 5$ k ($\uparrow$), $FG = 1.87$ k (T), $GD = 8.33$ k (C), $AB = 4.8$ k (T)

4.1.25 $BF = 11.2$ kN (C), $FA = 10$ kN (T), $CG = 6.7$ kN (C), $GF = 3.4$ kN (C), $GH = 3$ kN (T), $HI = 3.4$ kN (C), $IA = 10$ kN (T), $DH = 6.7$ kN (C), $IE = 11.2$ kN (C)

4.1.27 $BH = 8$ k (C), $HG = 15$ k (T), $GF = 28$ k (C), $CJ = 12$ k (C), $JI = 5$ k (T), $IH = 9$ k (C), $IF = 16$ k (C), $EJ = 3$ k (C)

4.1.29 $BF = 12.7$ kN (C), $FA = 9.5$ kN (T), $GC = 8.5$ kN (C), $GF = 3.2$ kN (C), $GH = 8$ kN (T), $DH = 8.5$ kN (C), $HI = 3.2$ kN (C), $EI = 12.7$ kN (C), $IA = 9.5$ kN (T)

4.1.31 $BF = 8.8$ k (C), $FE = 14.6$ k (T), $FG = 3.6$ k (T), $GE = 14.2$ k (T), $GH = 4.5$ k (T), $CH = 17.8$ k (C), $HI = 2.7$ k (C), $ID = 22.2$ k (C), $IE = 17.8$ k (T)

4.2.1 $A_x = 900$ lb. ($\leftarrow$), $A_y = 775$ lb. ($\uparrow$), $B_x = 900$ lb. ($\rightarrow$), $B_y = 75$ lb. ($\uparrow$), $C_x = 900$ lb., $C_y = 575$ lb.

4.2.3 $A_x = 21$ kN ($\rightarrow$), $A_y = 11$ kN ($\uparrow$), $B_x = 21$ kN ($\leftarrow$), $B_y = 1$ kN ($\uparrow$), $C_x = 21$ kN, $C_y = 1$ kN

4.2.5 $A_x = 1.5$ k ($\leftarrow$), $A_y = 3.65$ k ($\uparrow$), $B_x = 5.1$ k ($\rightarrow$), $B_y = 2.75$ k ($\downarrow$), $D_x = 4.5$ k, $D_y = 3.65$ k

4.2.7 $A_x = 1.54$ kN ($\rightarrow$), $A_y = 4.5$ kN ($\uparrow$), $B_x = 1.54$ kN, $B_y = 0.90$ kN, $C_x = 1.54$ kN ($\uparrow$), $C_y = 6.3$ kN ($\uparrow$)

4.2.9 $A_x = 12$ kN ($\leftarrow$), $A_y = 21$ kN ($\downarrow$), $C_x = 12$ kN ($\leftarrow$), $C_y = 3$ kN ($\uparrow$), $B_x = 12$ kN, $B_y = 21$ kN

4.2.11 $A_y = 960$ lb. ($\uparrow$), $E_y = 960$ lb. ($\uparrow$), $B_x = 2560$ lb., $B_y = 0$, $C_x = 2560$ lb., $C_y = 0$, $D_x = 2560$ lb., $D_y = 0$

4.3.1 $A_x = 24$ k ($\leftarrow$), $A_y = 8$ k ($\uparrow$), $B_x = 24$ k ($\rightarrow$), $B_y = 8$ k ($\uparrow$), $AC = 25.3$ k, $BD = 25.3$ k, $CD = 24$ k

4.3.3 $E_x = 9.8$ kN ($\rightarrow$), $E_y = 3.5$ kN ($\uparrow$), Sag $= y_c = 2.94$ m

4.3.5 $A_x = 19$ k($\leftarrow$), $A_y = 6.27$ k($\uparrow$), $CD_x = 19$ k, $CD_y = 3.73$ k, $CB_x = 19$ k, $CB_y = 3.27$ k, Sag $= y_c = 7.4$ ft.

4.4.1 $P = 157.5$ lb, $SF = 1.43 < 1.5$, $P_{max} = 2000$ psf

4.4.3 $M_{OTM} = 3420$ lb.-ft., $M_{RM} = 8025$ lb.-ft., $SF = 2.35$ $P_{max} = 2530$ psf

4.4.5 $P = 83.2$ kN, $M_{OTM} = 152$ kN-m $M_{RM} = 457$ kN-m, $SF = 3.0$; $P_{max} = 164.4\ ^{kN}/_{m^2}$

Chapter 5

5.1.1   Beams B-1, B-2, B-3: $\omega = 250$ lb./ft. reactions = 1250 lb.

Girder G-1: reactions = 1250 lb., Girder G-2: reactions = 3750 lb.

Girder G-3: reaction @ (A-3) = 1250 lb., reaction @ (A-1) = 2500 lb.

Column loads: (A-1) 3750 lb., (A-3) = 5000 lb., (B-2) = 3750 lb., (C-2) = 3750 lb., (D-1) = 3750 lb., (D-3) = 5000 lb.

5.1.3   1. Rafters: $\omega = 66$ lb./ft.

Front wall reaction = 764 lb. per 2' = 382 lb./ft.

Roof beam reaction = 820 lb. per 2' = 410 lb./ft.

Back wall reaction = 396 lb. per 2' = 198 lb./ft.

2. Stud walls:

Top of front wall: $\omega = 382$ lb./ft.

Base of front wall: $\omega = 446$ lb./ft.

Top of back wall: $\omega = 198$ lb./ft.

Base of back wall: $\omega = 262$ lb./ft.

3. Roof beam: $\omega = 410$ lb./ft.

4. Exterior column reaction = 2460 lb.

Interior column reaction = 4920 lb.

5. Floor joist: (14' span) $\omega = 96$ lb./ft., reaction = 672 lb. per 2' (336 lb./ft.)

(12' span) $\omega = 96$ lb./ft., reaction = 576 lb. per 2' (288 lb./ft)

6. Floor beam: $\omega = 336$ lb./ft. + 288 lb./ft. = 624 lb./ft.

7. Continuous footing: Front wall: $\omega = 446$ lb./ft. + 336 lb./ft. = 782 lb./ft., Back wall: $\omega = 262$ lb./ft. + 288 lb./ft. = 550 lb./ft.

8. Critical interior footing: Floor load + the interior column load = 8664 lb.

5.1.5 Beam B-2:

$$\omega = \left(40\,{}^{\text{lb.}}/_{\text{ft.}^2} \times 6'\right) + \underset{\text{(beam wt.)}}{12\,{}^{\text{lb.}}/_{\text{ft.}}} = 252\ \text{lb.}/\text{ft.},$$

reaction = 2016 lb.

Girder G-2:

$$\omega = \left(43\,{}^{\text{lb.}}/_{\text{ft.}^2} \times 12'\right) + \underset{\text{(girder wt.)}}{40\,{}^{\text{lb.}}/_{\text{ft.}}}$$

$$= 556\ \text{lb.}/\text{ft.},$$

Four concentrated loads of 2016 lb. each; reaction = 12,372 lb.

5.1.7 1. Rafters:

$$14'\ \text{span:}\quad \underset{\text{(horiz.)}}{\omega} = \underset{\text{(snow)}}{60\ \text{lb.}/\text{ft.}} + \underset{\text{(DL adj.)}}{38\ \text{lb.}/\text{ft.}}$$

$$= 98\ \text{lb.}/\text{ft.},$$

Reaction = 686 lb. per 2' (343 lb./ft.)

$$16'\ \text{span:} - \underset{\text{(horiz.)}}{\omega} = \underset{\text{(snow)}}{60\ \text{lb.}/\text{ft.}} + \underset{\text{(DL adj.)}}{39\ \text{lb.}/\text{ft.}}$$

$$= 99\ \text{lb.}/\text{ft.},$$

Reaction = 792 lb. per 2' (396 lb./ft.).

2. Roof beam:

$$\underset{\text{(horiz.)}}{\omega} = \underset{\text{(14' rafter)}}{343\ \text{lb.}/\text{ft.}} + \underset{\text{(16' rafter)}}{396\ \text{lb.}/\text{ft.}}$$

$$+ \underset{\text{(2' brg. wall)}}{20\ \text{lb.}/\text{ft.}} = 759\ \text{lb.}/\text{ft.}$$

reaction to exterior column = (759 lb./ft.) × (5') = 3795 lb. reaction to interior column = 2 × (759 lb./ft.) × (5') = 7590 lb.

3. Exterior walls: 14' span - Top of the wall $\omega = 343$ lb./ft.,

Base of the wall

$$\omega = 343\ \text{lb.}/\text{ft.} + \underset{\text{(wall wt.)}}{80\ \text{lb.}/\text{ft.}} = 423\ \text{lb.}/\text{ft.}$$

16' span - Top of the wall $\omega = 396$ lb./ft.,

Base of the wall

$$\omega = 396\ \text{lb.}/\text{ft.} + \underset{\text{(wall wt.)}}{80\ \text{lb.}/\text{ft.}} = 476\ \text{lb.}/\text{ft.}$$

4. Interior columns: $P = 7590$ lb.

5. Floor joists: 14' span - $\omega = 66.7$ lb./ft., reaction = 467 lb. per 16" (350 lb./ft.)

16' span - $\omega = 66.7$ lb./ft., reaction = 534 lb. per 16" (400 lb./ft.)

6. Floor beam: $\omega = 350$ lb./ft. + 400 lb./ft. = 750 lb./ft. The beam also supports column loads of 3795 lb. (ext. col.) and 7590 lb. (interior col.)

7. Floor posts: Exterior post - $P = 7545$ lb.; Interior post - $P = 15,090$ lb.

8. Exterior foundation: $q = 2000$ psf,

$$q_{\text{net}} = 2000\ \text{psf} - \underset{\text{(base)}}{100\ \text{psf}} = 1900\ \text{psf}$$

$$14'\ \text{span:}\ \frac{P}{A} = \frac{1123\ \text{lb.}}{(1' \times 1.25')}$$

$$= 898\ \text{psf} < 1900\ \text{psf}\ \therefore\ \text{OK}$$

$$16'\ \text{span:}\ \frac{P}{A} = \frac{1226\ \text{lb.}}{(1' \times 1.25')}$$

$$= 980\ \text{psf} < 1900\ \text{psf}\ \therefore\ \text{OK}$$

9. Critical pier footing: $P = 15,090$ lb.,

$$q_{\text{net}} = 2000\ \text{psf} - \underset{\text{(base)}}{100\ \text{psf}} = 1900\ \text{psf}$$

$$A = x^2 = \frac{P}{q_{\text{net}}} = \frac{15,090\ \text{lb.}}{1900\ \text{lb.}/\text{ft.}^2} = 7.94\ \text{ft.}^2;$$

$$x = 2'\text{--}10"\ \text{square}$$

Chapter 6

6.1.1 $f = 32$ lb./in.^{2} = 4608 lb./ft.^{2} < $q = 5000$ psf

6.1.3 a. $f_c = 6.4$ ksi (44.1 MN/m^2 = 44.1 MPa)

b. $f_{\text{brg}} = 0.383$ ksi = 383 psi (2640 kN/m^2 = 2.64 MPa)

c. $q_{\text{net}} = q -$ ftg. wt. = 3850 psf (183.8 kN/m^2)

$A_{\text{req'd}} = 19.5$ ft.^{2} (1.82 m^2), $x = 4.4'$ square (1.35 m)

6.1.5 $f_v = 100$ lb./in.^{2} < 120 lb./in.^{2} OK; no "theoretical" need for shear reinforcement.

$f_v = 689$ kN/m2 = 689 kPa < 690 kPa

6.1.7 $L = 20"$

6.1.9 a. Member AB: $f = 1710$ kPa; Member BC: $f = 68.5$ MPa

b. $A = 2.27 \times 10^{-3}$ m^2; $f_v = 19.28$ MPa

6.1.11 $P_{\text{friction}} = 15$ k; $P_{\text{brg}} = 7.9$ k; $P_{\text{total}} = 22.9$ k

6.2.1 $\varepsilon = 0.003$ in./in.; $\varepsilon = 0.0003$ mm/mm

6.2.3 $\varepsilon = 0.0033$ in./in.

6.2.5 $\varepsilon = 1.042 \times 10^{-3}$ in./in.; $E = 24 \times 10^3$ ksi

6.2.7 $A = 10$ in.^{2} (6450 mm^2); $D = 3.57'$ (91 mm) based on deformation.

$A = 15$ in.^{2} (9675 mm^2); $D = 4.37"$ (111 mm) governs, based on tensile stress.

6.2.9   a. $D = 1.95$ in.;
        b. $\delta = 0.75$ in. (3 turns)

6.2.11 a. $153" = 12.75'$ (3.89 m)
        b. $f = 510$ psi (3.52 MPa) $< F_t = 1000$ psi
            (6.90 MPa) . . . (OK)
        c. $\delta = 6' + 12.8' = 18.3' < 25'$ . . . (OK)
            $\delta = 1.83$ m $+ 3.9$ m $= 5.73$ m < 7.63 m . . .
            (OK)

6.4.1 $\delta = 0.125$ in.

6.4.3   a. $\delta = 1.09$ in. (27.8 mm)
        b. $f = 13,190$ psi (91 MPa)
        c. $P = 263.9$ k (1.17 MN)

6.4.5   a. $f = 14,140$ psi
        b. $\delta = 0.0234$ in.

6.4.7 $\delta = 0.0468$ in., $\delta' = 0.0468" - 0.01" = 0.0368"$,
        $f = 14,820$ psi

6.4.9   a. $\delta = 0.0263"$
        b. $\Delta T = 16.9\ °F$

6.4.11 Science fiction novelist Arthur Clark
        believes that the elementary rule of
        geometry controls the world of living
        creatures. If the size of an object is doubled,
        the area increases four times but its volume,
        and thus its weight, increases by eight.
        Therefore, the stress in the bones would be
        doubled in intensity and cause distress.

Chapter 7

7.1.1 $\bar{x} = -0.19"$; $\bar{y} = +5.32"$

7.1.3 $\bar{x} = +4.2$ ft.; $\bar{y} = +3.6$ ft.

7.1.5 $\bar{y} = +18"$ (466 mm)

7.1.7 $\bar{y} = +3.45"$

7.3.1 $\bar{x} = 0$, $\bar{y} = +2.35"$ (59 mm)

        $\sum I_{xc} = 26.1$ in.^{4} $(10.94 \times 10^6$ mm$^4)$;

        $\sum Ad_y^2 = 75.4$ in.^{4} $(31.6 \times 10^6$ mm$^4)$;

        $\sum I_{yc} = 88.6$ in.^{4} $(37 \times 10^6$ mm$^4)$;

        $\sum Ad_x^2 = 0$

        $I_x = 101.5$ in.^{4} $(42.5 \times 10^6$ mm$^4)$;

        $I_x = 88.6$ in.^{4} $(37 \times 10^6$ mm$^4)$

7.3.3 $\bar{y} = +17.1"$; $I_x = 19,695$ in.^{4},
        $I_y = 2435$ in.^{4}

7.3.5 $\bar{y} = +7.53"$; $I_x = 250.7$ in.^{4}

7.3.7 $\bar{x} = -0.36"$; $\bar{y} = +7.0"$
        $I_x = 110.4$ in.^{4}; $I_y = 35.7$ in.^{4}

Chapter 8

8.2.1 $B_y = 25$ k, $D_x = 0$, $D_y = 5$ k, $V_A = -10$ k,
        $V_B = +15$ k, $V_C = -5$ k, $V_D = -5$ k'
        $M_A = 0$, $M_B = -50$ k-ft., $M_C = +25$ k-ft.,
        $M_D = 0$

8.2.3 $A_y = \dfrac{\omega L}{2}$, $B_x = 0$, $B_y = \dfrac{\omega L}{2}$,
        $V_A = \dfrac{\omega L}{2}$, $V_{(@L/_2)} = 0$, $V_B = -\dfrac{\omega L}{2}$,
        $M_A = 0$, $M_{(@L/_2)} = \dfrac{\omega L^2}{8}$, $M_B = 0$

8.4.1 $B_Y = 6$ k, $C_Y = 2$ k, $V_A = -2$ k, $V_B = +4$ k,
        $V_C = -2$ k,
        $M_A = 0$, $M_B = -10$ k, $M_{@\,4\,k\,load} = +10$ k,
        $M_{@2\,k\,load} = +10$ k-ft., $M_C = 0$

8.4.3 $A_Y = 32$ k, $C_Y = 28$ k, $V_A = +32$ k, $V_C = +16$ k
        to $+4$ k, $V_B = -28$ k,
        $M_A = 0$, $M_C = +96$ k-ft., $M_{Max.} = +98$ k-ft. (at
        5' *from A*), $M_B = 0$

8.4.5 $A_Y = 1.5$ kN, $B_Y = 6.5$ kN, $V_A = +1.5$ kN,
        $V_D = -2.5$ kN, $V_B = +4$ kN,
        $V_C = +4$ kN, $M_A = 0$, $M_D = -1$ kN-m,
        $M_B = -6$ kN-m, $M_C = 0$

8.4.7 $M_A = 10$ k-ft., $A_Y = 3$ k, $V_A = +3$ k, $V_B = +3$ k,
        $V_C = 0$ to 3 k, $V_D = 0$,
        $M_A = -10$ k-ft., $M_B = +5$ k-ft. to -15 k-ft.,
        $M_C = 27.5$ k-ft., $M_D = 0$

8.4.9 $B_Y = -5$ k, $D_y = 14$ k, $V_A = 0$, $V_B = -9$ k to
        $+16$ k, $V_C = +2$ k to -4 k,
        $V_D = -14$ k, $M_A = 0$, $M_B = -18$ k-ft.,
        $M_C = +45$ k-ft., $M_D = 0$

8.4.11 $M_A = 400$ k-ft., $A_Y = 50$ k, $D_Y = 50$ k,
        $M_D = 400$ k-ft.,
        $V_A = +50$ k, $V_B = 30$ k, $V_{mid-span} = 0$,
        $V_C = -30$ k, $V_D = -50$ k,
        $M_A = -400$ k-ft., $M_B = 0$, $M_{mid-span} = +150$ k-
        ft., $M_C = 0$, $M_D = -400$ k-ft.

Chapter 9

9.1.1 $V_{Max} = -2000$ lb., $M_{Max} = -18,000$ lb.-ft.

$f_b = 14.2$ ksi < 22 ksi ... OK

9.1.3 $V_B = -12$ k to $+24$ K (-53 kN to $+107$ kN),
$V_{at\ 6\ k\ load} = +12$ k to $+6$ k ($+53$ kN to $+27$ kN),

$V_C = -18$ k (-80 kN),
$M_B = -24$ k-ft. (-32.5 kN-m),
$M_{Max.} = +54 (+73.2$ kN-m),

$f_b = 20.8$ ksi < 22 ksi. ... OK;
($f_b = 143$ MPa < 150 MPa. ... OK)

9.1.5 $V_L = 5200$ lb., $V_R = -4400$ lb.,
$M_{Max} = 24,200$ lb.-ft.

$f_b = 19.1$ ksi < 22 ksi ... OK

9.1.7 $V_{max} = -5800$ lb. (-25.8 kN),
$M_{max} = -25,600$ lb.-ft. (-34.7 kN-m)

$f_b = 1900$ psi < 2400 psi ... OK; ($f_b = 13$ MPa
< 16.5 MPa ... OK)

9.1.9 $V_{max} = +10$ k, $M_{max} = 50$ k-ft.; $S_{req'd} = 27.3$
in.^{3}, Use: W14×22 ($S_x = 29.0$ in.^{3})

9.1.11 $V_{max} = -8.75$ k, $M_{max} = 43.75$ k-ft.; for the
cross-section: $\bar{y} = 3.07"$, $I_x = 113.2$ in.^{4},

$f_b = 27.5$ ksi < 30 ksi. ... OK;

$f_v = 1.36$ ksi (at the neutral axis); $f_v = 1.19$
ksi (at the flange)

9.1.13   For the cross-section:

$\bar{y} = 6.35"$, $I_x = 207.3$ in.^{4}

$M_{allow} = \dfrac{F_b I_x}{c} = 720$ k-in. $= 60$ k-ft.,

$M_{allow} = \dfrac{\omega L^2}{8} = 60$ k-ft., $\therefore \omega = 1.2$ k/ft.

$V_{max} = 12$ k, $f_v \underset{@\ flange}{} = 0.156$ k/in.^{2}

9.1.15 $M_{allow} = F_b \times S_x = 230$ k-in. $= 19.2$ k-ft.

$M_{max} = 10.5\ \omega; \omega = 1.83$ k/ft. (bending);

$V_{allow} = \dfrac{F_v I_x b}{Q} = 4.08$ k;

$V_{Max} = 3\ \omega, = 1.36$ k/ft. (shear); Shear
governs

9.1.17 $I_x = 434$ in.^{4}, $c = 5.5"$, $S_x = 79$ in.^{3}

$M_{max} = 1735$ k-in. $= 145$ k-ft. $= \dfrac{\omega L^2}{8}$;

$\omega = 2.9$ k/ft.

$Q = 26.3$ in.^{3}, $b = 2 \times 2.74" = 5.48"$,

$V = \dfrac{\omega L}{2} = 29$ k; $f_v = 0.32$ ksi

9.1.19 $\bar{y} = 3.45"$, $I_x = 188.8$ in.^{4};

$M_{max} = -42.25$ k-ft.,

$f_b = 19$ ksi < 22 ksi ... OK

$V_{max} = 6.5$ k, $Q = 15.5"$, $b = 1"$, $I_x = 188.8$ in.^{4},

$f_v = 0.54$ ksi < 14.5 ksi (at the web/flange
intersection) ... OK

9.1.21 $V_{max} = 10$ k, $M_{max} = 48$ k-ft., $S_{req'd} = 26.2$ in.^{3},
W8×31 ($S_x = 27.5$ in.^{3}),

$M_{beam\ wt.} = 992$ lb. ft., $S_{bm.wt.} = 0.54$ in.^{3}, S_{total}
$= 26.74$ in.$^3 < 27.5$ in.^{3} ... OK

$f_v = 4.38$ ksi < 14.5 ksi ... OK; $\Delta_{total} = 0.286"$
$+ 0.185" = 0.47"$

9.1.23 $V_{max} = 2000$ lb., $A_{req'd} = 27.3$ in.^{2}, $M_{max} =$
12,000 lb.-ft., $S_{req'd} = 92.9$ in.^{3}

4×14 S4S; $M_{bm.\ wt.} = 34$ lb.-ft., $S_{bm\ wt.} = 2.9$
in.^{3}; $S_{total} = 95.8$ in.$^3 < 102.4$ in.^{3} ... OK

$\Delta_{total} = 0.48"$ (includes the beam weight),
$\Delta_{allow} = 0.8"$, ... OK

$f_{bearing} = 109$ psi $< F_C = 410$ psi ... OK

Chapter 10

10.2.1 $P_{cr.} = 184.2$ k, $f_{cr.} = 20.2$ ksi

10.2.3 $L = 28.6'$, ($L = 8.72$ m)

10.2.5 $P_{cr.} = 73.64$ k, $f_{cr.} = 7.7$ ksi

10.3.1   a. $F_a = 24.89$ ksi (171.6 MPa); $P_a = 475$ k
(2.11 MN)

       b. $F_a = 23.2$ ksi (159.9 MPa); $P_a = 443$ k
(1.97 MN)

       c. $F_a = 20.65$ ksi (142.4 MPa); $P_a = 394$ k
(1.75 MN)

10.3.3 $\dfrac{KL}{r_z} = 111;$

$F_a = 11.54$ ksi (79.6 MPa);

$P_a = 46$ k (0.205 MN)

10.3.5 Weak axis:

$\dfrac{KL}{r_y} = 54,\ F_a = 17.99$ ksi, $P_a = 561$ k ... OK

Strong axis:

$\dfrac{KL}{r_x} = 57,\ F_a = 17.71$ ksi, $P_a = 553$ k ... OK

(Assume the 2nd floor load applied at the top of the column)

10.3.7 $\dfrac{KL}{r_y} = 75,\ F_a = 19.99$ ksi, $P_A = 316$ k

Area $= 702$ ft.^{2}, or 26.5' square

10.3.9 $L = 20',\ P_{actual} = 30$ k, W8×24; $P_a = 47.6$ k

10.3.11 3rd floor column: $P_{actual} = 397.5$ k, W12×79;
$F_a = 17.14$ ksi, $P_a = 398$ k

1st floor column: $P_{actual} = 622.5$ k,
W12×136;
$F_a = 15.79$ ksi, $P_a = 630$ k

Roof load $= 60$ k; Floor loads $= 112.5$ k/ft.

10.3.13 W10×49 (W250×73); weak axis governs

$\dfrac{KL}{r_y} = 136,\ F_a = 8.07$ ksi, $P_a = 116$ k (516 kN)

10.4.1 $\dfrac{l_e}{d} = 30.5,\quad F_c^* = 1219$ psi, $\quad F_c' = 495.5$ psi,

$P_a = 13,900$ lb.

10.4.3 Strong axis governs; $\dfrac{l_e}{d} = 25.2,$

$F_c' = F_c^* \times C_D = 1021$ psi, $\quad P_{allow} = 72.4$ k

10.4.5 Weak axis governs: $\dfrac{l_e}{d} = 16.55,$

$F_c' = 691$ psi,

$P_a = 17,540$ lb., $\quad A_{trib.} = 351$ ft.^{2}

10.4.7 Use 6$\frac{3}{4}$"×7$\frac{1}{2}$" glu-lam; ($A = 50.63$ in.)

Appendix B–Math Review

B.2.1 $a = 6.34',$ Area $= 43.1$ ft.^{2}

B.2.2 Area $= 280,000$ ft.$^2 = 31.1$ yd.^{2}

B.2.3 $a = 19.9'$

B.2.4 $BD = 420',\ A = 31,800$ ft.^{2}

B.2.5 $ac = 12.7$ m

B.2.6 $h = 13.8$ m

B.2.7 $\theta = \cos^{-1}(.789) = 37.9°$

B.2.8 $\theta = \cos^{-1}(.196) = 78.7°$

B.2.9 $BC = 233'$

B.2.10 $AB = 75.6$ m

B.2.11 $\theta = 21.8°,\ h = 48'$

B.2.12 $h = 146.8$ m, $DC = 229.4$ m

B.2.13 $h = 79.4'$

B.2.14 a. Total length $= 159.4$ ft.;
b. Total weight $= 1913$ lb.

B.3.1 $x = 1,\ y = 2$

B.3.2 $x = 4,\ y = 6$

B.3.3 $x = 7,\ y = 3$

B.3.4 $x = 2\frac{1}{5},\ y = -\frac{1}{5}$

B.3.5 $x = 1,\ y = 2$

B.3.6 $x = 4,\ y = -5$

B.3.7 $x = 10,\ y = 5$

B.3.8 $x = 2.7,\ y = 7.1$

B.3.9 $x = 10,\ y = 1$

B.3.10 $x = 5,\ x = 3$

B.3.11 $x = 1.2,\ x = -1.0$

B.3.12 $x = 5.3,\ x = 1.5$

B.3.13 $y = 6.2,\ y = 5.67$

B.3.14 $x = 1,\ x = \frac{1}{2}$

B.4.1 a. $27'-8" = 8.43$ m,
b. 135 yd. $= 123.4$ m,
c. $2\frac{3}{4}" = 69.9$ mm,
d. $1,874.6$ ft.$^2 = 174$ m^2,
e. 27 ft.$^3 = 0.765$ m^3,
f. 492.5 lb. $= 2.19$ kN,

g. 5,747 lb. = 25.6 kN,

h. 423.6 lb./ft.^{2} = 2028 Pa,

i. 722 lb./in.^{2} = 4980 kPa,

j. 1250 lb./ft. = 18.24 kN/m,

k. 3000 lb./ft.^{2} = 143.7 kN/m^2,

l. 490 lb./ft.^{3} = 76.9 kN/m^3

B.4.2 a. 3250 mm = 130 in.,

b. 29.85 m = 98 ft.,

c. 6.84 m = 22 ft. $-5\frac{1}{4}$"

d. 18.25 m^2 = 196.4 ft.^{2},

e. 9.15 m^3 = 323 ft.^{3},

f. 4321 N = 971.4 lb.,

g. 75.0 kN = 16.9 k,

h. 12,500 MPa = 1,812,500 lb./in.^{2},

i. 5.2 kN/m = 356 lb./ft.,

j. 180 kN/m^2 = 3760 lb./ft.^{2},

k. 23.5 kN/m^3 = 150 lb./ft.^{3}

B.4.3 U.S. customary: W = 13,500 lb. = 13.5 k;

SI Metric: W = 59.9 kN

B.4.4 SI Metric: W = 65.8 kN;

U.S. Customary: W = 14,850 lb. = 14.85 k

Index

459